U.S. CIVIL
AIRCRAFT SERIES
VOLUME 4

This work is dedicated to the preservation and perpetuation of a fond memory for the men and the planes that made a future for our air industry. And, to help kindle a knowledge and awareness within us of our debt of gratitude we owe to the past. . . . .

© 1993 by TAB Books
© 1967 by AERO Publishers, Inc.
First TAB printing 1993
Second TAB printing 1994
Published by TAB Books
TAB Books is a division of McGraw-Hill, Inc.

Library of Congress Catalog Card Number 62-15967

U.S. CIVIL AIRCRAFT SERIES

VOLUME 4

(ATC 301 - ATC 400)

Joseph P. Juptner

TAB *AERO*
Division of McGraw-Hill, Inc.
Blue Ridge Summit, PA 17294-0850

ACKNOWLEDGEMENTS

Any historian soon learns that in the process of digging for obscure facts and information, he must oftentimes rely on the help of numerous people, unselfish and generous people, many who were close to or actually participated in various incidents or events that make up this segment of history recorded here and have been willing to give of their time and knowledge in behalf of this work. To these wonderful people I am greatly indebted and I feel a heart-warming gratitude; it is only fitting then that I proclaim their identity in appreciation.

My thanks to Alfred V. Verville; Charles W. Meyers; H. Lloyd Child; George A. Page, Jr.; Clayton Folkerts; Will D. Parker; John H. Livingston; Melba Beard; Gene J. Poiron; Tom Towle; to F. J. Delear of Sikorsky Aircraft Div.; Gordon S. Williams of The Boeing Co.; American Airlines; Ken M. Molson of National Museum of Canada; the staff at National Air Museum, Smithsonian Institution; Pan American World Airways; The Texas Co.; to Harvey Lippincott of Pratt & Whitney Div.; Convair Div.; Fairchild-Stratos; to T.K. Rinehart of Fairchild-Hiller; Beech Aircraft Corp.; United Air Lines; Alexander Film Co.; to Gerald O. Deneau of Cessna Aircraft and the following group of dedicated aviation historians; Earl C. Reed; Roger F. Besecker; Stephen J. Hudek; Richard Sanders Allen; Henri D'Estout; James H. Harvey; Roy Oberg; Chas. E. Lebrecht; Joe Christy; Marion Havelaar; Peter M. Bowers; Wm. T. Larkins; Page Shamburger; John W. Underwood; John R. Ellis; James W. Bott; Erwin C. Eshelman; Louis M. Lowry; Ralph Nortell; Chas. F. Schultz; and Gerald H. Balzer.

FOREWORD

The "Great American Depression" was less than a year old but already was getting a good toe-hold in the hip-pockets of the aviation industry. Dazed and unbelieving, many could not realize yet just what had happened; but one by one, like felled warriors, they pulled in the remnants of what they could salvage and made plans to carry on as best they could, if they could. Over-production in 1929 left many inventories at a high level and in some cases were cleared away at bargain-basement prices; others that had no inventory problems were still forced to pinch the budget and warmed-over old designs were being offered with tongue-in-cheek as something new. Those fortunate enough not to have old inventories to move and had a healthy budget that allowed for some new development, were bringing out interesting airplanes that caused a slight revival in airplane buying.

The glider movement took the country's fancy for a while and then the inexpensive low-powered airplane made its formal debut. This gave thousands a chance to keep flying, who otherwise would have been forced to quit. The cold fact of economics led many to the light, less expensive airplane and for nearly the next decade, the lightplane was to carry on its back the destiny of aviation.

The lush days of 1929 were now gone and there was no great promise on the horizon just yet; but there were now several new avenues for channeling energies and these held some promise of keeping things going. It is thus that we came up through a period of doubt, perhaps even verging on despair, and then slowly and cautiously moved into a period of new confidence that was based on more sober planning. The forlorn outlook worn by most people of the aviation industry, earlier in the year of 1930, had been brightened somewhat by the end of the year and 1931 began to look much better. One of the most outstanding qualities of the aviation pioneer was the almost religious fervor he had for the "aviation game" and his great inner passion to further its development regardless of the privations it imposed upon him.

Joseph P. Juptner

TERMS

To make for better understanding of the various information contained herein, we should clarify a few points that might be in question. At the heading of each new chapter, the bracketed numerals under the ATC number, denote the first date of certification; any amendments made are noted in the text. Unless otherwise noted, the title photo of each chapter is an example of the model that bears that particular certificate number; any variants from this particular model, such as prototypes and special modifications, are identified. Normally accepted abbreviations and symbols are used in the listing of specifications and performance data. Unless otherwise noted, all maximum speed, cruising speed, and landing speed figures are based on sea level tests; this method of performance testing was largely the custom during this early period. Rate of climb figures are for first minute at sea level, and the altitude ceiling given is the so-called service ceiling. Cruising range in miles or hours duration is based on the engine's cruising r.p.m., but even at that, the range given must be considered as an average because of pilot's various throttle habits.

At the ending of each chapter, we show a listing of registered aircraft of a similar type; most of the listings show the complete production run of a particular type and this information we feel will be valuable to historians, photographers, and collectors in making correct identification of a certain aircraft by its registration number.

In each volume there are separate discussions on 100 certificated airplanes and we refer to these discussions as chapters, though they are not labeled as such; at the end of each chapter there is reference made to see the chapter for an ATC number pertaining to the next development of a certain type. As each volume contains discussions on 100 aircraft, it should be rather easy to pin-point the volume number for a chapter of discussion that would be numbered as A.T.C. #93, or perhaps A.T.C. #176, as an example. The use of such terms as "prop," "prop spinner," and "type," are normally used among aviation people and should present no difficulty in interpreting the meaning.

TABLE OF CONTENTS

AMERICAN EAGLE, MODEL D-430

Fig. 1. American Eagle model D-430 with 165 h.p. Wright J6-5 engine, shown here in prototype.

The Model D-430 for the year 1930 was an enlarged and slightly improved version of the earlier Wallace "Touroplane" as designed by Stan Wallace, a craft that was built by "American Eagle" as the model 330 or B-330 (refer to ATC #119 of U. S. CIVIL AIRCRAFT, Vol. 2). The 3 place "Touroplane" was primarily designed and developed for general-purpose work as usually carried out by the 3 place open cockpit biplane. It was felt as inevitable, in some circles, that the small cabin monoplane of this type would soon replace the old-fashioned open cockpit biplane, but the versatile biplane was hard to unseat from its place on top of the heap. Though equally capable to handle most flying-service chores, the 3 place light cabin monoplane, on the average, made only a token showing for this type of work. The Kinner-powered "Touroplane B" was selling only in small numbers as compared to its expected acceptance. So in an effort to salvage the usefulness of this design, and to fall in line with requirements prevailing for more seating capacity, it was enlarged to carry four. It could then be offered as a family-type airplane, a cabin version that had come into a certain amount of popularity by this time. Because of the way it was arranged, the basic design of the "Touroplane" needed very little alteration, and very little enlargement to carry the fourth passenger. The addition of 65 extra horsepower was enough to promise a fairly good performance. Powered now with 165 h.p. engines the new 4-place version was capable of many and varied uses. It

was ideal as a family-airplane for Dad, Mom, and the kids to travel or sight-see in, and its economical performance per seat-mile was well suited also for men of business. However, the few examples which were built in this new version were mostly used by the flying services for air-taxi and charter work. With a fairly good performance and a practical utility on nominal power this new series was rightfully hailed as the ideal for the needs of business and sport, but in terms of demand from the large selection that was available, we might say that the market place was glutted with this type of airplane. The American Eagle 430-series found very few buyers and it remains as one of this company's rarer types.

The American Eagle model D-430 was a high-winged cabin monoplane with comfortable seating for 4, and was quite typical of the Wallace "Touroplane" as introduced earlier except for the extra seating and a fairly handsome boost in horsepower. The powerplant for this particular model was the 5 cyl. Wright J6 (R-540) engine of 165 h.p., and the slight increase in airframe bulk and passenger seating still left a power reserve to offer an improvement in performance. Tests on this new series of cabin monoplanes were completed by Feb. of 1930 and its first formal showing was at the St. Louis Air Exposition held during that same month. We can understand that American Eagle Aircraft would be quite proud of the new series but their bold exaggeration of this craft's capabilities was way out of line. Even under the best possible

Fig. 2 Wings of D-430 folded for space-saving storage.

conditions, performance figures shown in this account are more in keeping with the actual performance as demonstrated in normal service. Flight characteristics and general behavior were most often described as pleasant. For a 4-seated airplane, offering ample room even for big people, featuring ease of entry and exit to and from any seat, plus the economy of a smaller engine than usually used for a craft of this type, we could

agree that the "430" was a good value for the money invested. The type certificate number for the American Eagle model D-430, as powered with the 165 h.p. Wright R-540 engine, was issued 3-18-30. Probably no more than one or two examples of this model were manufactured by the American Eagle Aircraft Corp. on Fairfax Airport in Kansas City, Kansas.

Listed below are specifications and performance data for the American Eagle monoplane model D-430 as powered with the Wright 5 cyl. engine of 165 h.p.; length overall 24'7"; height overall 8'1"; wing span 41'0"; wing chord 70"; total wing area 233 sq. ft.; airfoil USA-45-B (unverified); wt. empty 1864 lbs.; useful load 1133 lbs.; payload with 60 gal. fuel was 553 lbs.; gross wt. 2997 lbs.; max. speed 115; cruising speed 100; landing speed 50; climb 700 ft. first min. at sea level; ceiling 14,000 ft.; gas cap. 60 gal.; oil cap. 7 gal.; cruising range at 9.5 gal. per hour was approx. 600 miles; price at the factory field first quoted at $7395 and lowered to $6395 early in 1931.

The fuselage framework was built up of welded chrome-moly steel tubing faired to shape with wooden formers and fairing strips, then fabric covered. The cabin walls were insulated

Fig. 3. Interior of D-430 provided comfort for four big people

Fig. 4. American Eagle D-430 was scaled-up "Touroplane".

and upholstered neatly with seating arrangement to provide ample room for 4 people. The forward and side-windows were of "Triplex" safety glass and a large sky-light was provided in the cabin roof for vision overhead. The wing framework, in two halves, was built up of laminated spruce spar beams that were routed out for lightness with spruce and mahogany plywood gusseted truss-type wing ribs. The leading edges were covered with dural sheet and the completed framework was covered in fabric. A 30 gal. fuel tank was mounted in the root end of each wing half flanking the fuselage; a baggage compartment to the rear of the cabin had allowance for 58 lbs. The robust landing gear of 70 in. tread was now of the out-rigger type and was incorporated into the wing bracing struts to form a rigid truss, using oil-draulic shock absorbing struts. Wheels were 30x5 and brakes were standard equipment. The fabric covered tail-group was built up of welded steel tubing; the fin was ground adjustable and the horizontal stabilizer was adjustable in flight. The next development in the American Eagle 430 series was the model E-430 as described in the chapter for ATC #302 in this volume.

Listed below are model D-430 entries as gleaned from registration records:
NC-200N; Model D-430 (#901) Wright R-540.
X-578H was a model A-430 and the prototype of 430 series; serial #901 later converted into model E-430 with 165 h.p. Continental A-70 engine.

A.T.C. #302
(3-18-30)
AMERICAN EAGLE, MODEL E-430

Fig. 5. American Eagle model E-430 with 165 h.p. Continental A-70 engine, shown here over Kansas City.

The "American Eagle" model E-430 was a companion to the D-430 monoplane (as described just previous) except for the powerplant installation, which in this case was the 7 cyl. Continental A-70 engine of 165 h.p. Of the 2 different versions which were built in the 430-series, the model E-430 was the more popular because American Eagle Aircraft got a much better price on the A-70 engine. This amounted to a craft with the same horsepower, the same performance, and the same amount of equipment for a price that delivered to the customer some $400.00 cheaper. With all tests completed by March, the series was somewhat slow in selling, so Larry D. Ruch (American Eagle's test pilot) picked a model E-430 to fly in the 1930 National Air Tour in hopes of stirring up a little interest for the new offering. Flying hard in some fierce competition, Ruch finished into 13th spot and showed the plane off to good advantage in many parts of the country, but no flood of orders developed from this demonstration. With its capacity for ample payload and a good performance on nominal power, this model offered utility that was bound to show a profit. One example was selected by an operator for "bush flying" duties in Alberta, Canada. The design of the 430-series, as originally planned, was calculated to mount several engines in the 165 h.p. range, including the "Comet" 7-E.

High-performance versions were also planned to mount engines up to 225 h.p. Had better sales warranted the expense of new developments, it is likely that the "Four Thirty" would have evolved into some very interesting versions. By mid-1930 all of the previous "American Eagle" biplanes, except the "Model 201", had been phased out and were no longer in regular production. The "Touroplane B" light monoplane was more or less discontinued, the 430-series weren't selling too well either and things were getting worse by leaps and bounds. Several accounts reported that American Eagle Aircraft was approaching dire straits but then had the good fortune of acquiring some additional income from wells which were drilled on its property on Fairfax Field; these wells were producing natural gas. With a little more "cash in the till" providing an incentive to carry on, Ed Porterfield initiated design work on the well-remembered "Eaglet" which was up and flying near the end of 1930 and was to put a new spurt of life into the progressively failing firm.

The American Eagle model E-430 was also a high-winged cabin monoplane with seating for 4 and was basically typical to the model D-430 in all respects except for the powerplant installation and a few minor modifications necessary for this combination. Powered with the

Fig. 6. E-430 had large interior and 3 doors for entry.

7 cyl. Continental A-70 engine of 165 h.p. its performance was more or less comparable and there was very little difference between models to help one choose one or the other, except the delivered price which at this particular time would certainly have a strong bearing on the customer's decision. The type certificate number for the model E-430 as powered with the Continental A-70 engine, was issued 3-18-30. It is possible that 5 examples of this model were manufactured by the American Eagle Aircraft Corp. at Kansas City, Kansas. Ship number X-578H was the prototype for the 430-series and first flown as the model A-430 with the Wright J6-5-165 engine. After completion of tests this engine was replaced with the Continental A-70 and this same aircraft was then tested as the model C-430. Production models varied slightly from the prototype in some respects and became the models D-430 and E-430. Ed E. Porterfield was the president; Jack E. Foster was chief engineer; Stanley Wallace was project engineer on "Model 430" development; Larry D. Ruch was chief pilot in charge of test and promotion; and John Carroll Cone was sales manager. Noel Hockaday who had been draftsman for Stan Wallace during "Touroplane" manufacture in Chicago, moved over with Wallace to American Eagle Aircraft where he specialized in preliminary design. Hockaday was to figure more prominently some years later when he designed the Wyandotte "Pup" that became the basis for "Porterfield" airplanes. During the early part of 1930 one account credits Harvey J. Stoneburner as being appointed chief engineer for A. E. company but detailed research failed to substantiate this so it must have been a temporary assignment. D. H. Webber, formerly an assistant engineer, replaced Jack Foster when he left the company. Because of the depressed times, many aircraft firms operated with threats of bankruptcy hovering over them and this resulted in an unsettled household; consequently, this caused some shifting of personnel in the higher levels and also in the shops.

Listed below are specifications and performance data for the American Eagle model E-430 as powered with the 165 h.p. Continental A-70 engine: length overall 24'7"; height overall 8'1"; wing span 41'0"; wing chord 70"; total wing area 233 sq. ft.; airfoil USA-45-B (unverified); wt. empty 1875 lbs.; useful load 1133 lbs.; payload with 60 gal. fuel was 553 lbs.; gross wt. 3008 lbs.; max. speed 115; cruising speed 100; landing speed 50; climb 700 ft. first min. at sea level; ceiling 14,000 ft.; gas cap. 60 gal.; oil cap. 7 gal.; cruising range at 9.5 gal. per hour was 600 miles; price at the factory field first quoted at $6995. lowered to $5995. early in 1931.

Construction details and general arrangement of the model E-430 were typical to the model D-430 (as described in the previous chapter of this volume), including the following, which would apply to both models. Entry to the cabin was without effort because of 3 large doors which were placed 2 in front of one right rear; upholstery was rich mohair fabric. A large skylight in the forward portion of the cabin roof afforded vision upward to the front occupants. For added safety a total of 22.5 sq. ft. of "Triplex" safety-glass was used in the windshield and the side-windows; split panels of glass slid open for ventilation and an exhaust pipe "heater-muff" provided cabin heat. Similar to early "Touroplanes" the wings folded back to allow storage in a space of 13x25 feet. Main wheels were 30x5 with Bendix brakes and a swivel type tail wheel allowed good ground maneuvering. Examples in service sometime later were often equipped with low pressure "airwheels". An air-operated Heywood engine starter was optional equipment. Some examples were at times equipped with a low drag speed-ring engine cowling which added measurably to the cruising speeds. The model E-430 was also built in small number but it has been evident that most served for many useful years. The next American Eagle development was the flivver-type "Eaglet" as described in the chapter for ATC #380 in this volume.

Listed below are model E-430 entries as gleaned from registration records:
NC-200N; Model E-430 (#901) Continental A-70.
NC-284N; " (#902) "
NC-228N; " (#904) "
NC-457V; " (#905) "
Serial #901 was first as a model D-430; serial #903 may have been exported to Canadian service; registration number for serial #905 unverified; X-205N was listed as a model F-430 (ser. #900) with a Kinner engine believed to be the 160 h.p. model R5.

Fig. 8. Pilot's station of E-430, good visibility is evident.

Fig. 7. Low operating costs made E-430 ideal family-type airplane.

Fig. 9. Driggs "Skylark" 3 with 75 h.p. Michigan-Rover engine.

Reflecting the versatile design genius of Ivan H. Driggs, the "Skylark" 3 posed as one of the finest light sport-trainer biplanes of this period. Though not nationally known, as some of the other more numerous sport-trainer types, the trim "Skylark" was indeed a handsome and lovable little ship that could boast of an enthusiastic group of boosters and admirers. Designed and developed as a sport-type craft for the private owner with inherent characteristics also well suited for the small flying schools, the "Skylark" was necessarily robust of frame and character; stout hearted enough to absorb the strains of a playful mood, or the unintentional abuse sometimes dished out by fledgling pilots learning to fly. Basically of a sure-footed and playful nature, the "Skylark" nevertheless held itself well in check when under the guidance of a timid hand with cautious movements, but responded eagerly and precisely when spurred on by the determined hand of experience, matching its behavior to the tempo of the mood. Born and reared amidst rather modest surroundings, and never bally-hooed to any great extent, the Driggs "Skylark" was known mostly by word-of-mouth and its slowly accumulated reputation would be the envy of any good airplane.

The first commercial aircraft design effort by Ivan Driggs was the little DJ-1 sport monoplane of 1924 which was progressively improved for 1925 and 1926, becoming just about the finest ultra-light airplane in this country at that time. The Driggs "Coupe" was a cabin two-seater also introduced in 1926, and the unique "Dart" 2 sport-biplane was developed in 1927. Enthusiasm for the little "Dart" biplane was rather light and even skeptical, so the "Skylark" series was developed later in 1928 to take its place. The Driggs "Skylark" was an evolution of the earlier "Dart" model 2 (refer to chapter for ATC #15 of U. S. CIVIL AIRCRAFT, Vol. 1). The first experimental development in this new series was actually a "Dart" 2 airframe (X-10033) fitted with the 4 cyl. inverted in-line "Rover" engine of 55 h.p. After tests in this combination it was felt this configuration was unwieldy and fell somewhat short of the requirements for this type of craft, so it was redesigned considerably to a more normal pattern. The prototype for the "Skylark," as it later came to be known, now had tandem cockpits both behind the center-section braces. The lower wing with added area was now stretched out to a greater span, and the landing gear was redesigned to the sturdy out-rigger type with a wide tread. Meeting with fairly good acceptance in this form, the "Skylark" 3 remained in nominal production until Dec. of 1933, at which time the manufacturing rights had been sold to the Skylark Aircraft Corp. of Muskegon, Mich. Because of the severe "depression" still rampant, the design lay more or less dor-

Fig. 11. "Skylark" was easy to handle and fun to fly.

mant for a time, and was finally sold to the Western Airmotive Corp. of Van Nuys, Calif. who planned to revive it as the Western "Pirate" in 1937. Plans for manufacture did not develop beyond some experimentation, so the design was apparently sold again, this time to Phillips Aviation Co. of Van Nuys, Calif. They finally revived it, with only a little modification, as the Phillips "Skylark" CT-1 with the 4 cyl. Menasco B-4 engine of 95 h.p. A prototype version was built for tests (NX-18989) in 1939, but it is not known at this time if the new series was built in many more examples. However, it was approved on ATC #731. This prototype airplane

by Phillips remained in California for several years and was still flying around in the early "sixties".

The Driggs "Skylark" model 3 was a lean and sporty looking open cockpit biplane seating two in tandem, in a proportion quite typical to other sport-trainer types of this period. Leveled especially at the sportsman-pilot of modest means the "Skylark" was endowed with arrangement for efficient operation. A hardy nature, calculated to keep costs and repairs at a minimum, allowed many care-free hours of flying on a small budget. Several flying schools found these attributes also to their liking;

Fig. 10. Rugged of frame, hardy of nature, "Skylark" was excellent trainer.

Fig. 12. Revival of "Skylark" design shown here as Western "Pirate."

every hour from the flight line translated into fair profit. Both instructors and fledgling pilots enjoyed their association with this airplane. Powered with the 4 cyl. inverted in-line "Michigan-Rover" engine of 75 h.p. the "Skylark" offered surprisingly good performance with this nominal amount of power, pointing to intelligent use of basic design features and aerodynamic arrangement. Flight characteristics were enjoyable and precise with a slight bent towards a frisky nature; hundreds of log books have recorded hours in this craft as time well spent and well remembered. Of hardy nature, the "Skylark" wore well and at least 16 were still flying actively in 1939. The type certificate number for the Driggs "Skylark" 3, as powered with the 75 h.p. Michigan-Rover engine, was issued 3-24-30 (first approved on Group 2 certificate numbered 2-180 issued 2-12-30 but this was superseded by ATC #303) and some 17 or more examples of this model were manufacturered by the Driggs Aircraft Corp. at Lansing, Mich. Harry F. Harper was president; Ivan Howard Driggs was V. P., chief engineer and general manager; E. C. Shields was secretary; Hugo B. Lundberg was treasurer; and the test-pilot was a happy fellow named Matthews.

Listed below are specifications and performance data for the Driggs "Skylark" model 3 (as powered with the 75 h.p. Rover engine). Figures in brackets are for amended allowance: length overall 22'6"; height overall 8'3"; wing span upper 28'3"; wing span lower 27'0"; wing chord upper & lower 44"; wing area upper 97 sq. ft.; wing area lower 88 sq. ft.; total wing area 185 sq. ft.; airfoil (N.A.C.A.) M-18; wt. empty 878 [912] lbs.; useful load 501 [501] lbs.; payload 17 gal. fuel was 213 [213] lbs.;

gross wt. 1379 [1413] lbs.; max. speed 105; cruising speed 87; landing speed 40; climb 650 ft. first min. at sea level; ceiling 13,500 ft.; gas cap. 17 gal.; oil cap. 2 gal.; cruising range at 5 gal. per hour was 300 miles; price at the factory was $3485. lowered to $2985. in March of 1931. The following figures are for the 1931 models; wt. empty 951 lbs.; useful load 500 lbs.; payload with 22 gal. fuel was 182 lbs.; gross wt. 1451 lbs.; landing speed 45; climb 615 ft. first min.; ceiling 12,500 ft.; all else was more or less same.

The fuselage framework was built up of welded chrome-moly steel tubing, then lightly faired to shape and fabric covered. The turtle-back portion beyond the rear cockpit was covered with a removable panel of dural metal sheet. The cockpits were deep and well protected, bucket-type seats had wells for parachute pack and instrument panels were lined with foam-rubber crash pads. The wing framework was built up of solid spruce spar beams with stamped-out aluminum alloy wing ribs; the leading edges were covered with either dural metal or bakelite sheet and. the completed framework was covered in fabric. A fuel tank of either 17 or 22 gal. capacity was mounted in the center-section panel of the upper wing; a direct-reading fuel gauge was visible from either cockpit. Two large ailerons of the Friese balanced-hinge type were mounted in the lower panels, operated by push-pull tubes and bellcranks. All interplane struts were of streamlined aluminum alloy tubing and interplane bracing was of streamlined steel wire. A 5 deg. dihedral angle in the lower wing and 24 in. stagger between wing panels promoted both good stability and good visibility. The sturdy out-rigger type landing gear of 72 in. tread used oleo-rubber shock absorbing

Fig. 13. "Skylark" was also popular for sport; this ship fitted with sliding canopy.

struts of 10 in. travel to absorb extreme shock; wheels were 24x4 or 6.50x10 and wheel brakes were optional. The fabric covered tail-group was built up of welded steel tubing and the horizontal stabilizer was adjustable in flight. A Hartzell wooden propeller, spring-leaf tail skid, wiring for navigation lights, baggage bin, streamlined head-rest for rear cockpit and "Alemite" grease fittings at all points of wear were standard equipment. A metal propeller, engine starter, navigation lights, streamlined metal cuffs on struts of landing gear, dual joy-stick controls, tool kit, and tail wheel were optional. The next development in the "Skylark" series was the model 3-95 as powered with the 95 h.p. A.C.E. Cirrus Hi-Drive engine on a Group 2 approval numbered 2-281.

Listed below are Driggs "Skylark" 3 entries as gleaned from registration records:

X-592E; Skylark 3	(#18)	Rover 55-75.
C-665E;	"	(#19)	Rover 75.
C-664E;	"	(#20)	"
NC-894N;	"	(#3001)	"
NC-895N;	"	(#3002)	"
NC-896N;	"	(#3003)	"
NC-339V;	"	(#3005)	"
NC-537V;	"	(#3006)	"
NC-542V;	"	(#3007)	"
NC-300W;	"	(#3008)	"
NC-301W;	"	(#3009)	"
NC-302W;	"	(#3010)	"
NC-11112;	"	(#3011)	"
NC-11136;	"	(#3012)	"
NC-11190;	"	(#3013)	"
NC-11196;	"	(#3014)	"
NC-817Y;	"	(#3015)	"
NC-11301;	"	(#3016)	"
NC-11328;	"	(#3017)	"
NC-11378;	"	(#3018)	"

Serial #18 was prototype (X-592E) for "Skylark" 3 series with 55 h.p. Rover engine, later fitted with 75 h.p. Rover engine and later as model 3-95 with Cirrus Hi-Drive engine on Group 2 approval 2-281; serial #19 first with Rover 75, tested with Cirrus Hi-Drive and later had 85 h.p. D.H. Gipsy engine as model 3-95A on a Group 2 approval 2-369; serial #20 later as model 3-95A with Gipsy 85-95 also on Group 2 approval 2-369; serial #3015-3016-3017 as model 3-95 with Cirrus Hi-Drive engine (inverted) on Group 2 approval 2-281; registration number for serial #3002 unverified; registration number for serial #3004 unknown; serial #3016 later on Group 2 approval 2-567; no listing available for serial #3019; serial #3020 on Group 2 approval 2-469.

A.T.C. #304
(3-26-30)
STEARMAN "JUNIOR SPEEDMAIL"
4-C

Fig. 14. Stearman 4-C "Jr. Speedmail"; exhaust collector-ring of Wright J6-9 formed leading edge of engine cowling.

As a progressive development and further refinement of the classic "Stearman" design, the new 4-series biplane was introduced as a high performance sport craft with open cockpits. It was also readily adaptable to any number of commercial uses. Somewhat bigger and bulkier now than the standard model C3R but not nearly as big as the earlier "Speedmail" model M-2 mail-carrier, the new "Junior Speedmail" series was an in-between model that was probably the handsomest airplane that Stearman ever built. Carefully fitted with the bulbous N.A.C.A. low-drag engine cowling as an integral part of the configuration, the "Stearman 4" was the first production biplane to take advantage of this deep engine fairing with any measure of success. The resulting performance of this combination caused other heretofore skeptical manufacturers to take a new look at the advantages to be gained with this type of air-cooled engine streamlining. With a "wait and see" policy, the new series Stearman was first formally introduced to the flying public as the model C4W (4-E) (with the 420 h.p. Pratt & Whitney "Wasp" engine), specifically as a thundering and very colorful high performance airplane for the sportsman-pilot or for the business-house planning to do promotion that would catch the public eye. With the initial entry of the series soon well proven and firmly entrenched in the minds

of those in the industry, it was a small matter to modify the basic design slightly with various power units to adapt it to the specific needs of private-owner and air-line customers. As a consequence, the company fared quite well with the "Stearman 4" which was finally produced in 3 models. Each of these models was convertible to models for carrying air-mail and cargo, plus a few versions which were arranged for special purposes. The almost flawless design, the careful workmanship and built in safety that had been characteristic of the "Stearman" airplane in the past, seemed to be even more outstanding in the make-up of the 4-series.

As pictured here in various interesting views the Stearman model 4-C under discussion, was an open cockpit biplane of beautiful line and proportion that was carefully engineered around the big N.A.C.A. type engine cowling; though seemingly large for the seating of just 3 people, the large cross-section of the oval fuselage was necessary to fair out the large diameter up front. Powered with the 9 cyl. Wright J6 "Nine" (R-975) engine of 300 h.p. the model 4-C had a better than average performance range that translated into utility for the business man or happy hours for the sportsman. Among those taking advantage of the utility in this biplane was the Richfield Oil Co., the Fenestra Steel Products, the Airways Branch of Dept. of Commerce and at

Fig. 15. Stearman 4-C at factory, prepared for delivery to Canada.

least one was exported to Canada for service. Though primarily arranged as a sport-craft with seating for 3 (with allowance for plenty of baggage and extra equipment) the payload capacity of some 660 lbs. was easily adapted to the hauling of mail and cargo. With outward appearance modified considerably, the aerodynamic arrangement of the model 4-C was still recognizable to the basic design as laid down a few years back, so flight characteristics and general behavior were more or less comparable to those which had made the Stearman biplane a long-time favorite. The rugged structure and inherent reliability built into the new "Stearman 4", plus its popularity in almost any service, naturally promoted longevity. It is not at all surprising that at least 20 examples of various models in the 4-series were still in active service some 8 or 9 years later. The Stearman model 4-C was first announced in October of 1929 as the C4A and the first examples of this model received a Group 2 approval numbered 2-155 which was issued 11-22-29. The type certificate number for the model 4-C, as powered with the 300 h.p. Wright J6 engine, was issued 3-26-30. Four examples of this particular model were manufactured by the Stearman Aircraft Co. at Wichita, Kansas.

Listed below are specifications and performance data for the Stearman model 4-C "Junior Speedmail" (as powered with the 300 h.p. Wright J6 engine): length overall 26'11"; height overall 10'2"; wing span upper 38'0"; wing span lower 28'0"; wing chord upper 66"; wing chord lower 51"; wing area upper 204 sq. ft.; wing area lower 103 sq. ft.; total wing area 307 sq. ft.; airfoil Goettingen 436; wt. empty 2256 lbs.; useful load 1544 lbs.; payload with 106 gal. fuel was 663 lbs.; gross wt. 3800 lbs.; max. speed 145; cruising speed 120; landing speed 50; climb 1050 ft. first min. at sea level; climb in 10 min. was 7000 ft.; ceiling 15,000 ft.; gas cap. 106 gal.; oil cap. 10 gal.; cruising range at 16 gal. per hour was 720 miles; price at the factory field was $12,500.

The fuselage framework was built up of welded chrome-moly steel tubing, extensively faired to an oval shape in the forward portion with wooden formers and fairing strips, then covered with fabric. The cockpits were deep and well protected with a door on the left side for entry to the front cockpit. There was a small baggage locker of 5 cu. ft. capacity with allowance for 50 lbs. just in back of the rear cockpit and a larger baggage compartment forward of the front cockpit with an allowance for 273 lbs. The wing framework was built up of solid spruce spar beams which were routed out for lightness with spruce and plywood Warren truss wing ribs; the leading edges were covered with dural metal sheet and the completed framework was covered with fabric. The large N.A.C.A. engine cowling was removable in sections for easy servicing and the lower wing was neatly faired where it joined the fuselage. The large exhaust collector ring was an unusual feature in that it formed the leading edge of the circular engine cowling, providing cooling and expansion space for the hot gases; silencing of the exhaust noises was very effective. The main fuel tank of 62 gal. capacity was mounted high in the fuselage just ahead of the front cockpit and extra fuel was carried in a 44 gal. tank mounted in the center-section panel of the upper wing. The split-axle landing gear of 96 in. tread was of the out-rigger type using "Aerol" (air-oil) shock absorbing struts; wheels were 30x5 and Bendix brakes were standard equipment. The 10x3 in. swiveling tail wheel was mounted on the very end of the fuselage with a notch in the rudder to provide working clearance. The fabric covered tail-group was built up of welded steel tubing and the horizontal stabilizer was adjustable for trim in flight. A ground adjustable metal propeller, hand crank inertia-type engine starter, navigation lights, tail wheel and wheel brakes were standard equipment. Dual joy-stick controls, duplicate instrument panel in front cockpit, parachute flares, radio receiver, battery, generator, electric engine starter, landing lights, wheel streamlines, and

Fig. 16. Stearman 4-C in Canadian service.

9.50x12 low pressure semi-airwheels were optional. The next development in the "Junior Speedmail" series was the model 4-D as described in the chapter for ATC #305 in this volume.

Listed below are Stearman model 4-C entries as gleaned from registration records:

NS-2;   Model 4-C (#4001) Wright R-975.
NC-8839;   "   (#4002)   "
NC-665K;   "   (#4003)   "
NC-666K;   "   (#4006)   "
NC-667K;   "   (#4007)   "
NC-770H;   "   (#4010)   "
NC-772H;   "   (#4012)   "
CF-CCG;   "   (#4013)   "
CF-CCH;   "   (#4014)   "
NC-778H;   "   (#4018)   "

Serial #4001-4002-4003-4007 on Group 2 approval numbered 2-155; #4001 later as model 4-E; #4003 later as 4-E (NS-665K); #4006-4007 later as model 4-E; #4010 later as models 4-CM, 4-E, 4-EM; #4012 later as model 4-CM; serial number for CF-CCG unverified; this certificate (ATC) for #4010-4012-4014-4018 only; all models eligible with improved Wright R-975 of 330 h.p.

A.T.C. #305
(4-3-30)
STEARMAN "JUNIOR SPEEDMAIL",
4-D

Fig. 17. Stearman "Junior Speedmail" model 4-D with 300 h.p. "Wasp Jr." engine; shown here at factory.

Almost paralleling the development of the "Junior Speedmail" models 4-C and 4-E was the Stearman model 4-D, a craft that was basically typical but powered with the new 9 cyl. Pratt & Whitney "Wasp Junior" engine of 300 h.p. Also introduced as a high performance sport-plane, the model 4-D was also quite flexible and easily adaptable to various business and commercial service uses. This particular version was the first certificated (ATC) airplane with the new "Wasp Junior" engine, and quite naturally the first example of the model 4-D was bought by Pratt & Whitney for test and promotion of their new "baby" engine. "Bernie" Whelan of P & W flew the new "Stearman 4-D" on an extensive promotion tour throughout the country to acquaint the aircraft manufacturers and other prospective engine buyers with the new 300 h.p. engine. This same ship, with many miles already to its credit, was then used as the "official pathfinder" for the 1931 National Air Tour. Western Air Express also used the model 4-D on its sprawling western mail routes, and the Texaco Oil Co. added a sparkling 4-D to their growing fleet of aircraft as the "Texaco 14". The model 4-D with the "Wasp Jr." engine, and the model 4-C with the 300 h.p. Wright engine, were companion models in this series. They were almost identical in every respect, except for minor personality traits, so selection of either of these versions would be but a matter of engine preference. To stimulate buying amongst operators that were rather hard-pinched for money at this particular time, Stearman airplanes were now offered on the time payment plan. The plan, though creating the urge to buy, actually did not create many extra sales. The people who could venture to buy a $12,000. sport-plane were very few and far between at this particular time. Frank Fuller, Jr., noted sportsman, was one of the few who could afford an airplane like the 4-D and he flew his strictly for sport.

As pictured here in several views, the Stearman model 4-D was an open cockpit biplane of flowing lines and beautiful proportion that was carefully engineered around the big N.A.C.A. engine fairing. Though seemingly large for the seating of just 3 people, the large cross-section was necessary to fair out the large diameter up front. As a consequence, the added bulk provided ample room in the cockpits and yet provided extra space for stowing large amounts of baggage or a worthwhile paying load. Powered with the 9 cyl. "Wasp Jr." engine of 300 h.p. the model 4-D was also blessed with a better than average performance which translated into a practical utility for the busy business man, or many happy hours for the flying sportsman. Though primarily arranged as a 3 place sport-craft with allowance for plenty of

Fig. 18. High performance of 4-D was suitable for business or sport.

extra baggage load or extra equipment, the pay-load capacity of some 600 lbs. was easily adapted to profitable cartage of mail or cargo. With outward appearance altered extensively the aerodynamic arrangement of the model 4-D still held closely to the basic design as laid out a few years back. Therefore its inherited flight characteristics and general behavior were more or less comparable to those which had made the "Stearman" biplane a perennial favorite. The

Fig. 19. Stearman 4-D during 1931 National Air Tour.

Fig. 20. Stearman 4-D in service with Western Air Express; later modified to 4-DM.

rugged structure and sure-footed reliability built into the 'new "Stearman 4", plus its popularity in almost any service, naturally promoted a longevity for the breed and it was not unusual to see this series well represented in actual service many years later. The type certificate number for the Stearman model 4-D was issued 4-3-30 and some 8 or more examples of this model were manufactured by the Stearman Aircraft Co. at Wichita, Kansas.

Listed below are specifications and performance data for the Stearman model 4-D as powered with the 300 h.p. Wasp Jr. engine; length overall 26'11"; height overall 10;2"; wing span upper 38'0"; wing span lower 28'0"; wing chord upper 66"; wing chord lower 51"; wing area upper 204 sq. ft.; wing area lower 103 sq. ft.; total wing area 307 sq. ft.; airfoil Goettingen 436; wt. empty 2297 lbs.; useful load 1503 lbs.; payload with 106 gal. fuel was 622 lbs.; gross wt. 3800 lbs.; max. speed 145; cruising speed 120; landing speed 50; climb 1050 ft. first min. at sea level; climb in 10 min. was 7000 ft.; ceiling 15,000 ft.; gas cap. 106 gal.; oil cap. 10 gal.; cruising range at 16 gal. per hour was 720 miles; price at the factory field was $12,500. with standard equipment.

The construction details and general arrangement of the model 4-D were typical of the model 4-C as described in the chapter for ATC #304 in this volume, including the following: a baggage compartment of 5 cu. ft. capacity with allowance for 40 lbs. was just in back of the rear cockpit and a larger compartment forward of the front cockpit had allowance for 162 lbs. The model 4-D with the "Wasp Jr." engine was 41 lbs. heavier when empty than the previously

described 4-C. This difference was only in engine weights so the 4-D had to sacrifice this amount in its total useful load. Large Friese type (balanced-hinge) ailerons were in the upper wing only, and were actuated by a streamlined push-pull strut through connecting rods, tubes and bellcranks in the lower wing. The front passenger's cockpit could be covered over by a metal panel when not in use. This added appreciably to the top speed because it eliminated drag-producing turbulence instigated by the gaping open cockpit. Wheel streamlines were also available to add further to the normal cruising speeds. All standard equipment and all optional equipment available for the model 4-C was also available for the model 4-D. The next development in the "Stearman" biplane was the mail-carrying "Sr. Speedmail" model 4-EM, a single-seater as described in the chapter for ATC #322 in this volume.

Listed below are Stearman model 4-D entries as gleaned from various records:

NC-769H;	Model 4-D (#4009) Wasp Jr.	
NC-774H;	" (#4011)	"
NC-776H;	" (#4015)	"
NC-779H;	" (#4019)	"
NC-792H;	" (#4024)	"
NC-796H;	" (#4025)	"
NC-11724;	" (#4026)	"
NC-563Y;	" (#4027)	"

Serial #4011 later converted to model 4-DM for Western Air Express; serial #4019 later as model 4-E; serial #4024 later converted to model 4-DX on Group 2 approval numbered 2-406 (R-792H) for Chas. W. Deeds.

A.T.C. #306
(4-2-30)
MONOCOUPE 90 (90-A)

Fig. 21. The beautiful Monocoupe 90-A against a back-drop of fleecy clouds.

The friendly charm of the earlier "Monocoupe" had captured a good many hearts, and its country-wide popularity pointed directly to the fact that it was certainly an ideal combination for the average private owner-flyer. To keep pace with this reputation, under widening demands, Mono Aircraft now offered the "Model 90" for year 1930. This offering proved soon to be a lively combination of soft, round curves that inherited all of the best from the earlier "Model 113", plus several new innovations that had marked it as quite an outstanding little airplane. Basically endowed with exceptional performance and good, sharp maneuverability, the new model 90 was soon put to the acid-test in up-coming tours, air-derbies, and the air-races, where it practically pulled a run-away in the small cabin airplane field. In so-called efficiency races it was even hard on the heels of the "Bellanca", a perennial winner in these events. In the National Air Tour for 1930, handsome Bart Stevenson flew the model 90 to show it off across the country and managed to garner a 12th place in the grueling grind against the best machines the industry had to offer. Well-attended national events had a way of providing notice and giving proof of the relative merits of the new Model 90; therefore it was in fairly good production despite the general sag in aircraft buying later that year. A good solid bargain for the money, the new "Ninety" sold quite well even through 1931

despite the weakening economy which greatly affected the market. In early 1932 it was announced that the Curtiss-Wright organization would be the national distributor for the "Monocoupe", thus offering the service convenience of its many airport-stations. With no let-up in the general economic distress, year 1933 found everyone in the depths of the depression, with very little or no development taking place, so the "Monocoupe 90" for that year had not changed very much since its introduction back in 1930.

Beset with times that broke the back of many good organizations, the popularity of the "Monocoupe" still tended to hold up well and then spanned over a good many years. Enough years, actually, to compile a fair-sized opus on the subject, so it must be sufficient here to only highlight the progressive development of this outstanding series. Only minor changes garnished the frame of the Model 90 into early 1934; the main attraction had been a gay paint job, a few useful gimmicks and a new low price of $2885. The year 1934 was shaping well and showed some promise of an up-swing in aircraft buying of certain types, so Monocoupe took the time and groomed the new "90 DeLuxe", a sharp looking craft with a new tunnel-type engine cowling, several other basic improvements, and a performance that made other similar airplanes green-eyed with envy. The price was upped to $3485. but it was still a terrific value.

Fig. 22. The Monocoupe 90 with 90 h.p. Lambert engine, shown here in prototype.

By the time mid-1934 rolled around, about 1000 "Monocoupes" had been built and sold. This figure did much to prove that the little "Coupe" had practically monopolized the small cabin airplane market since 1928. Innovations, tried and proven in the "Model 90 DeLuxe", paved the way for further developments along these lines, which were then incorporated into the new Model 90-A for 1935. Although the price for this new gem was upped again, this was a combination that was particularly outstanding so it was held in favor throughout flying circles

beyond 1940. It is quite remarkable to note, and to dwell on the fact, that the "Gay 90" in all its variants survived the test of changing times for approximately 30 years, on ATC #306 that was first issued in April of 1930.

The "Monocoupe 90" was a high-winged cabin monoplane seating two side-by-side, in a frame with well rounded and delicately graceful lines that set the pattern for all variants produced in at least the next 10 years. As compared to the previous Model 113 of 1929, this 1930 "Monocoupe" was a foot or so longer, at least 4

Fig. 23. The "Ninety" shown here over cornfields in Kansas.

Fig. 24. Monocoupe 90 during 1930 National Air Tour, flown by Bart Stevenson.

in. wider for more room in the cabin and was about 8 in. taller on its new-type landing gear. The wing spanned the same with new elliptical tips that cut down some on the area, and it weighed only some 10 lbs. more when empty. The useful load was increased to 130 lbs. more and the gross weight topped off to somewhat the same amount; an increase of 18 m.p.h. in the top speed, and some 15 m.p.h. in cruising speed, was testimony to its aerodynamic cleanness in order to achieve such ample increases with only a little more horsepower. Landing speed was just a little faster, climb was greatly improved, and about 4000 feet more altitude was easily available. Economy wise, the "Ninety" matched the very best and extensive flights were possible on pennies per mile. Many hours later, after engine

parts got a bit worn and sloppy, a motor overhaul kit was available for only $75.00! At first hand, one would think that improvement on this caliber of performance was hardly possible, but it was; the "Monocoupe 90" kept getting better and better. Judicious use of careful streamlining, improvements in engine cowling and landing gear for the next 5 years, brought steadily mounting increases in speed to the point where the sprightly "Monocoupe" was now cruising at three-quarter throttle with the same speed the 1930 model had attained with everything pushed to the firewall! Of course weights were steadily increased as the years went by but this didn't seem to affect general all-round performance to any great extent. Always a pleasure to fly, the "Monocoupe" retained its inherent ability to make

Fig. 25. Monocoupe 90 on Edo floats.

Fig. 27 The 90 Deluxe had unusual tunnel-type engine cowling.

Fig. 26. Many flying schools found Monocoupe 90 excellent for training.

every flight a lark but its fast-stepping cousins of some years later (90-A etc.) were definitely out of the puddle-jumper class that characterized the models of 1928-29. Strictly a "let's-go-some-where" type of airplane, the later "Monocoupe" was not recommended for leisurely and casual flights around the home-field pattern.

The type certificate number for the "Mono-coupe 90" (90-A etc.), as powered with the 5 cyl. Lambert R-266 engine of 90 h.p., was issued 4-2-30 and continued in force with occasional amendments throughout the active life of this model, which spanned a time of nearly 30 years. Except for a period in the bottom of the "de-pression" (1933), the Model 90 remained so popular that it was selling when many others were just sitting around hoping. By mid-1934 nearly 1000 "Monocoupes" had been built and sold and there is no way to tell accurately the actual number built by the time it was finally phased out. At its introduction in 1930, the "Ninety" was manufactured by the Mono Air-craft Corp. at Moline, Ill. Don A. Luscombe was the president; J. A. Love was V.P.; Clayton Folkerts was chief engr.; and various pilots (some of whom were: Vern Roberts, Stub Quimby, and Johnnie Livingston) handled chores of test and promotion. Discontent with airport conditions at Moline, Ill. spurred plans

for a move, and by late 1931 the "Monocoupe" already had new quarters at Lambert Field in Robertson, Mo. Reorganized into the Mono-coupe Corp., Don A. Luscombe was retained as president and Frederick Knack replaced Clayton Folkerts as chief engineer. Folkerts had left the firm to develop small high-performance racing airplanes. In 1933 Don Luscombe had left for Kansas City where he busied himself for a year or more in developing the Luscombe "Phan-tom." By 1935 Wooster Lambert was president of Monocoupe Corp.; John C. Nulsen, formerly of Ryan Aircraft, was V.P. and general manager; the genial Clare W. Bunch was sales manager and Tom Towle, the noted flying-boat engineer, was chief of engineering. In 1936 Clare Bunch was promoted to president, general manager and sales manager; Al W. Mooney was the chief engineer. At this time Al Mooney designed the low-winged "Monoprep" and "Monosport" prototypes. They were never produced in quantity by the Monocoupe Corp. but were sold out-right to another firm and later turned up to fly as the "Dart." By 1939-40 Clare Bunch was still president, general manager and sales man-ager; G. L. Miller was V.P.; C. Widner was factory superintendent and M. K. Smart was the chief engineer. Shortly after, the company was dissolved, reorganized and moved to Orlando,

Fig. 28. By 1934 the Monocoupe 90 was more streamlined, a gradual process that netted extra performance.

Fig. 29. A striking view of the 90 Deluxe.

Florida as a subsidiary company of Universal Moulded Products. After World War 2 the company was purchased by another group, also in Florida, to carry on with limited production of the 90-AF and 90-AL series.

Listed below are specifications and performance data for the Monocoupe model 90 (1930-31) as powered with the 90 h.p. Lambert engine; length overall 20'10"; height overall 6'11"; wing span 32'0"; wing chord 60"; total wing area 132 sq. ft.; airfoil Clark Y; wt. empty 859 [888] lbs., useful load 631 [631] lbs., payload with 30 gal. fuel was 266 [266] lbs.; gross wt. 1490 [1519] lbs.; figures in brackets are for landplane with speed-ring cowling and engine starter; max. speed 115 [118]; cruising speed 100 [102]; landing speed 40; climb 850 ft. first min. at sea level; ceiling 15,000 ft.; gas cap. 30 gal.; oil cap. 2.5 gal.; cruising range at 5.5 gal. per hour was 500 miles; price at the factory was $3375, lowered to $2885 in 1933. The following figures are for the Model 90 as seaplane on Edo

Fig. 30. Introduced in 1934, the Model 90-A was probably the most popular version.

Fig. 31. A 1941 model 90-AF, with Franklin engine, a similar Lycoming-powered version was the 90-AL.

H. twin-float gear; wt. empty 989 lbs.; useful load 591 lbs.; payload with 30 gal. fuel was 226 lbs.; gross wt. 1580 lbs.; max. speed 108; cruising speed 92; landing speed 45; climb 800 ft. first min. at sea level; ceiling 14,000 ft.; gas cap. 30 gal.; oil cap. 2.5 gal.; cruising range 450 miles.

The following figures are for the "Model 90 DeLuxe" with tunnel-type engine cowling, wheel pants and landing flaps; length overall 20'6"; height overall 6'11"; wing span 32'0"; wing chord 60"; total wing area 132 sq. ft.; airfoil Clark Y; wt. empty 935 lbs.; useful load 650 lbs.; payload with 28 gal. fuel was 293 lbs.; gross wt. 1585 lbs.; max. speed 135; cruising speed 115; landing speed (with flaps) 40; no flaps 45; climb 800 ft. first min. at sea level; ceiling 14,000 ft.; gas cap. 28 gal.; oil cap. 2.5 gal.; cruising range at 5.7 gal. per hour was 515 miles; price at the factory was $3485 with standard equipment.

The following figures are for the 1936-39 model 90-A; length overall 20'6"; height overall 6'10"; wing span 32'0"; wing chord 60"; total wing area 132 sq. ft.; airfoil Clark Y; wt. empty 967 lbs.; useful load 643 lbs.; payload with 28 gal. fuel was 286 lbs.; gross wt. 1610 lbs.; baggage allowance 116 lbs.; max. speed 130; cruising speed 112; landing speed (with flaps) 42; no flaps 48; climb 780 ft. first min. at sea level; ceiling 14,000 ft.; gas cap. 28 gal.; oil cap. 2.5 gal.; cruising range at 5.5 gal. per hour was 525 miles; price at the factory was $3485; also available as seaplane on Edo 1525 floats.

The following figures are for the model 90-A of 1939-40; length overall 20'6"; height overall 6'10"; wing span 32'0"; wing chord 60"; total wing area 132 sq. ft.; landing flap area 11.22 sq. ft.; airfoil Clark Y; wt. empty 973 lbs.; useful load 637 lbs.; payload with 28 gal. fuel was 280 lbs.; gross wt. 1610 lbs.; baggage allowance 110 lbs.; max. speed 130; cruising speed 115; landing speed (with flaps) 42; no flaps 48; climb 780 ft. first min. at sea level; ceiling 14,000 ft.; gas cap. 28 gal.; oil cap. 2.5 gal.; cruising range at 5.7 gal. per hour was 515 miles; price at factory was $3175 with standard equipment.

The following figures are for the 1954-57 model 90-AL; length overall 21'2"; height overall 6'10"; wing span 32'0"; total wing area 151 sq. ft.; landing flap area 11.25 sq. ft.; wt. empty 1000 lbs.; useful load 610 lbs.; payload with 28 gal. fuel was 260 lbs.; gross wt. 1610 lbs.; max. speed 130; cruising speed 120; landing speed (with flaps) 45; no flaps 49; climb 1000 ft. first min.; ceiling 16,500 ft.; gas cap. 28 gal.; oil cap. 1.5 gal.; range at 6 gal. per hour was 540 miles; baggage allowance 90 lbs. in 4 cu. ft. compartment; engine was Lycoming O-235-C1 of 115 h.p. max. at take-off.

The fuselage framework of the model 90 series was built up of welded 1025 and 4130 steel tubing in a rigid truss form, heavily faired to shape with dural metal sheet formers and wooden fairing strips, then fabric covered. The cabin interior of the 90 was widened 4 in. to provide a bit more room, and the upholstery was both practical and handsome. A large skylight provided vision overhead and a baggage compartment behind the seat back had an allowance for up to 95 lbs. Flight controls were of the joy-stick type and a dual set was available. The pattern for the fuselage exterior and interior remained fairly typical throughout the series except for wider entry doors, an adjustment of seats for better posture, increase in the baggage compartment allowance to 116 lbs., and latest models had doors on each side. The wing framework of one-piece construction was built up of solid spruce spars that were routed to an I-beam section, with wing ribs of basswood webs and spruce cap-strips; the leading edges

Fig. 32. Monocoupe 90 shown was first production example of 1930, shown here some 30 years later and still flying!

were covered with dural metal sheet and the completed framework was covered in fabric. The 1934 model introduced the "landing flap" which was an air-brake to reduce landing speeds; control was manual for the 11.22 sq. ft. flap. The gravity-feed fuel tanks were mounted in the wing, one flanking each side of the fuselage. Models to early 1934 were available with the narrow chord "Townend" type speed-ring cowling, to straighten air flows around the Lambert engine, but the "90 DeLuxe" around the Lambert engine, but the "90 De-Luxe" introduced a radical tunnel-type engine fairing that baffled the air-flow around the engine cylinders and boosted top speed by at least 15 m.p.h. Some cooling difficulties promoted changes in the cowling, so the model 90-A for 1935 introduced the normal N.A.C.A. type of engine cowling with streamlined humps over the engine cylinders, to hold down overall diameter and improve visibility over the front. First fitted with an "oleo" landing gear, as typical on the earlier "Monosport," by 1934 the gear was changed to incorporate two streamlined vees with central bracing of streamlined steel wire; wheel pants were part of the gear's configuration, the shock-cord snubbing mechanism was housed in the fuselage belly and this type of "gear" remained more or less standard for the balance of the series. The fabric-covered tail-group was built up of welded 1025 steel tubing, varying occasionally in shape and area; the horizontal stabilizer was adjustable in flight. Later versions had bungee-cord trim adjustment that changed the back and forth pressures on the "stick" for nose-high or nose-low conditions. Deep seat cushions were removable to accommodate seat-pack parachutes; weights of parachutes (20 lbs. each) were deducted from the baggage allowance. For earliest models, a Hartzell wooden propeller and 6.50 x 10 semi-airwheels were standard equipment; a metal prop, speed-ring engine cowling, engine starter and wheel pants were optional. On later

models, engine cowling (tunnel-type or N.A.C.A. type), metal propeller and wheel brakes were standard equipment; engine starter was optional. On the very latest models, a tail wheel, navigation lights, first-aid kit and fire extinguisher were also standard equipment. The next "Monocoupe" development after the "Ninety" of 1930 was the "Model 110" as described in the chapter for ATC #327 in this volume.

Listed below is a partial listing of Monocoupe 90 entries as gleaned from registration records:

NC-157K;	Model 90	(# 500)	Lambert 90.
NC-139K;	″	(# 501)	″
NC-166K;	″	(# 502)	″
NC-169K;	″	(# 503)	″
NC-170K;	″	(# 504)	″
NC-502W;	″	(# 534)	″
NC-504W;	″	(# 535)	″
NC-514W;	″	(# 536)	″
NC-505W;	″	(# 537)	″
NC-506W;	″	(# 538)	″
NC-507W;	″	(# 539)	″
NC-509W;	″	(# 542)	″
NC-510W;	″	(# 543)	″
NC-511W;	″	(# 544)	″
NC-512W;	″	(# 545)	″
NC-513W;	″	(# 546)	″
NC-515W;	″	(# 548)	″
NC-516W;	″	(# 549)	″
NC-522W;	″	(# 550)	″
NC-525W;	″	(# 551)	″
NC-526W;	″	(# 553)	″
NC-540W;	″	(# 554)	″
NC-535W;	″	(# 555)	″

Registration number for serial #503 unverified; serial #504 was first production airplane; NC-171K was ser. #505 and numbers ran consecutively to NC-199K which was ser. #533; for added interest we mention that ser. #618 was a model 90 mfgd. 3-1932 & ser. #A-765 was model 90-A mfgd. 8-25-37; this ATC also for 90-AF and 90-AL.

A.T.C. #307
(4-2-30)
FORD "TRI-MOTOR", 9-AT-A

Fig. 33. Ford "Tri-Motor" model 9-A shown here during test as X-7585; upon approval became NC-423H.

A good many people remember the colorful Ford "Tri-Motor," of course, and many thought they knew its lineage well, but most did not realize that the three-engined "Ford" was built in at least 7 basic models plus a good number of modifications. Apparently to save development time and to further confuse the issue for belabored historians, without intent, Ford developed the practice of modifying certain existing models into a range of different models that sometimes varied only in powerplants, or in wing configuration, or even a combination of both. Thus, the Ford model 9-A (also 9-AT-A) was actually converted into its new form from a model 4-AT-B (ser. # 4-AT-39) by replacing the 3 Wright J5 engines with 3 brand new Pratt & Whitney "Wasp Junior" engines of 300 h.p. each. The new "Junior" engine was a scaled-down version of the popular big "Wasp" and Ford was among the very first to use it. In fact, engine with serial #1 was installed in the nose of the 9-A, serial #3 was in the left wing nacelle and serial #4 was in the right wing nacelle. It makes one wonder who got away with #2! For development tests the prototype 9-AT-A bore its registration as NX-7585, but soon after approval tests were completed, the registration was changed to NC-423H and the serial no. was now changed to #9-AT-1. In June of 1930 there was an announcement that Pennsylvania Air Lines had purchased a brand new Ford 9-A that was to be used in scheduled service on their Pittsburgh to Cleveland run; this could have been the prototype ship that Ford was grooming but no further record substantiates this. No doubt due to the unhealthy atmosphere created by the business depression there was no demand for the 9-A, and no more examples were built beyond the prototype airplane. It is likely that the Ford Motor Co. kept "Nine-A" on for demonstration and other various chores around the place for several years, until 1934, when the ship seems to have been exported for duties in Honduras.

The Ford "Tri-Motor" model 9-A (also as 9-AT-A) was a three-engined high winged monoplane of all-metal construction and was basically similar to the earlier 4-AT-B (refer to chapter for ATC #87 of U. S. CIVIL AIRCRAFT, Vol. 1) except for the installation of different engines and a change in the wing nacelles which were dropped some 7.5 inches. Seating was arranged for two pilots and 11 passengers in comparable surroundings as featured in the various other Ford models. With a performance more or less comparable to the model 4-AT-E (refer to chapter for ATC #132 of U. S. CIVIL AIRCRAFT, Vol. 2), the 9-A was somehow capable of slightly more speed, but was 196 lbs. heavier in its empty state so this was detrimental to its payload capacity. Baggage space was 30 cu. ft. and all passenger seating was easily removable to provide some 400 cu. ft. of cargo area. In normal times and under better conditions, it is a good bet that the model 9-A would have been a popular model in the Ford "Tri-Motor" line-up, but as it was, its future fairly died in the bud and Henry Ford could have saved some money by letting this development die on the drawing board. The type certificate number for the Ford model 9-A was issued 4-2-30 and apparently only one example of this model was built by the Ford Motor Co. in its Aircraft Div. at Ford Airport in Dearborn, Mich.

Listed below are specifications and performance data for the Ford "Tri-Motor" model 9-A (9-AT-A) as powered with 3 "Wasp Jr." engines of 300 h.p. each; length overall 49'10"; height overall 12'8"; wing span 74'0"; wing chord at root 156"; wing chord at tip 92"; total wing area 785 sq. ft.; airfoil Goettingen 386; wt. empty 6863 lbs.; useful load 3267 lbs.; payload with 235 gal. fuel was 1325 lbs.; gross wt. 10,130 lbs.; max. speed 135; cruising speed 115; landing speed 58; climb 920 ft. first min. at sea level; climb in 10 min. was 7200 ft.; ceiling 14,000 ft.; gas cap. 235 gal.; oil cap. 24 gal.; cruising range at 45 gal. per hour was 570 miles; price at the factory field was $40,000.

The construction details and general arrangement for the Ford model 9-A were typical to that of the model 4-AT-E, as described in the chapter for ATC #132 of U. S. CIVIL AIRCRAFT, Vol. 2. The cabin interior measured 57" wide x 16'2" long x 67" high for a volume capacity of over 400 cu. ft.; pilots cockpit dimensions were 47" x 44" and the entrance to this station was through the main cabin only. The treadle type landing gear had a wheel tread of 201 in. using "Aerol" (air-oil) shock absorbing struts; wheels were 36x8 and brakes were standard equipment. All fuel was carried in the wing with the capacity divided among three tanks. Cabin dome lights, cabin ventilation and cabin heaters were fitted for passenger comfort; windows were of shatterproof glass throughout length of the cabin. Metal propellers, inertia-type engine starters and navigation lights were standard equipment.

Listed below is the only known example of the Ford model 9-A (9-AT-A) as gleaned from registration records:
NC-423H; Model 9-A (#9-AT-1) 3 Wasp Jr.

A fairly complete resume of the Ford "Tri-Motor" breed naturally starts with the model 3-AT, which was basically a single-engined 2-AT that was modified extensively to mount 3 Wright J4 engines. Impractical, unsatisfactory, and soon destroyed in a hangar fire, the 3-AT was quickly followed by the improved 4-AT series which bore more normal lines. The model 4-AT-A was the first model in this series and 14 examples were built under Group 2 approval 2-9 issued 9-25-28; the first 6 had 3 Wright J4 engines of 200 h.p. each, but the balance of these models had 3 Wright J5 engines of 220 h.p. each. Next, came the

improved model 4-AT-B with greater wing span, more allowable gross weight and a better performance with 3 Wright J5 engines. First approved on a Group 2 memo numbered 2-10, this was superseded by ATC #87 issued 11-19-28; the 4-AT-B was built in some 34 examples. The model 4-AT-C was a 4-AT-B (ser. #4-AT-47) that mounted a 450 h.p. "Wasp" engine in the nose and 2 Wright J5 engines under the wings. This model was modified on a Group 2 approval numbered 2-11, issued 10-27-28 and crashed to destruction early in its life.

The model 4-AT-D was a designation allotted to 3 ships and every one bore a difference. One of the models 4-AT-D (ser. #4-AT-24) was a model 4-AT-B with the larger 5-AT wing center-section and 3 Wright J4 or J5 engines. Another of the models 4-AT-D was a standard 4-AT-B (ser. #4-AT-37) with a 300 h.p. Wright J6 engine in the nose and Wright J5 engines in the wings. Still another of the models 4-AT-D was a 4-AT-B (ser. #4-AT-40) that was modified by installation of 3 Wright J6 engines of 300 h.p. each and might well be called the prototype for the popular model 4-AT-E. All 3 of the models 4-AT-D remained in the experimental category except for ser. #4-AT-24 which received Group 2 approval numbered 2-322 in 1-23-31 with 3 Wright J5 engines. The next model in line was the popular 4-AT-E that mounted 3 Wright J6 engines of 300 h.p. each; built in some 16 examples, it was approved under ATC #132 issued 3-30-29 with 4 slight revisions added to this certificate up to 9-23-31. Several of the 4-AT-E were built for the U. S. Army Air Corps as the C-3A. The model 4-AT-F in one example (ser. #4-AT-71) was allowed increased gross weight and incorporated several refinements, otherwise it was similar to the 4-AT-E. Approval for this model was on ATC #441 issued 8-10-31 and it was the last 4-AT model to be built. The model 4-AT-G, in one example only (ser. #4-AT-66), was a model 4-AT-E that had been modified by installation of 3 Packard Diesel engines of 225 h.p. each; this model was later redesignated the 11-AT but no more examples of this model were built.

With the models 4-AT-C and 4-AT-D showing some worthwhile increases in performance, Ford decided to go one better and developed the popular 5-AT series with 3 Pratt & Whitney "Wasp" engines of 420 h.p. each. The first in this series was the model 5-AT-A which was basically a 4-AT-B fuselage with an increase in wing span and the 3 "Wasp" engines; 3 examples of this model were on Group 2 approval numbered 2-32 issued 2-7-29. The model 5-AT-B was similar but carried 2 more passengers (15 place) and was allowed more gross weight. About 40 examples were built under ATC #156. The next model in this series, the 5-AT-C, had a longer fuselage to carry up to 17 and was also allowed increased gross weight. About 50 examples were built under

Fig. 34. Center engine of 9-A was first "Wasp Junior" built.

ATC #165. A seaplane version of this model, the 5-AT-CS, was built under ATC #296. The model 5-AT-D was perhaps the finest of the Ford "Tri-Motor" series and will be discussed in the chapter for ATC #409.

The model 6-AT-A was a model 5-AT-C that was modified by removing the 3 "Wasp" engines and replacing them with 3 Wright J6 engines of 300 h.p. each; an economical load-carrier, 3 examples were built under ATC #173 and a seaplane version of this model was on Group 2 approval numbered 2-80. The model 7-AT-A was a 6-AT-A that had the Wright J6 engine in the nose removed and replaced with a 420 h.p. "Wasp"; one example was built under ATC #246. The model 8-AT was an unusual version that was basically a 5-AT-C with all 3 engines removed and modified into a single engined air freighter; one example was built under Group 2 approval numbered 2-485 and both water-cooled and aircooled engines were tested in this configuration. The model 9-AT-A was discussed in this chapter (ATC #307); the model 10-AT was a four-engined design that was never built, the 11-AT has already been mentioned (4-AT-G) and the 12-AT was a 3-engined version of the 10-AT design that was not built either. This now leaves 3 of the Ford "Tri-Motor" yet to be discussed; the model 5-AT-D on ATC #409, the model 13-AT on ATC #431 and the model 4-AT-F on ATC #441. All will appear in U.S. CIVIL AIRCRAFT, Vol. 5.

A.T.C. #308
(4-2-30)
DETROIT-LOCKHEED "VEGA",
DL-1

Fig. 35. "Vega" DL-1 flown by Amelia Earhart to new speed records.

The Detroit-Lockheed model DL-1 was also a "Vega" type but it was an airplane of slightly different manner and breed. Though details were typical of the standard "Vega" model 5-B, the DL-1 was unusual in the fact that it was arranged with a metal monocoque fuselage newly developed at Detroit; the wing for this model was of the standard Lockheed all-wood cantilever construction. Unusual, too, was the fact that the metal fuselage was fabricated in the shops of Detroit Aircraft and then shipped to Lockheed in Calif., there to be mated with the wooden wing, the landing gear, the tail group; then other details of final assembly were performed. Of seven passenger capacity, similar in arrangement to the "Vega" 5-B (refer to ATC #227), the DL-1 was designed primarily as a small, speedy air-liner for lines offering fast and frequent schedules. With a promise of being not so fragile in service as the wooden-fuselaged "Vega", the metal-bodied DL-1 appealed to various air-lines because of its structure. The first 3 examples of the model DL-1, used first as demonstrators of Detroit Aircraft, were leased by the New York & Western Airlines for a run from New York City to Pittsburgh. Repossessed by DAC because of the lines early failure, the planes were leased and finally sold to Transcontinental & Western Air. Other examples of the DL-1 "Vega" went to Bowen Air Lines, Hanford Tri-State (becoming Mid-Continent Airlines), Wedell-Williams,

Varney Speed Lines & Varney Air Transport (becoming Continental Air Lines), Braniff Airways, Central Air Lines, and one went into Mexican service. Because of the failure of some lines to survive and because of stiff competition amongst carriers, most of the DL-1 were destined to shift from owner to owner. Some were owned privately and some became "company ships" for Standard Oil, General Tire & Rubber and others. Two of the basic DL-1, one more or less in standard configuration and one arranged as a "Speed-Vega," served in the Air Corps as the YIC-12 and the YIC-17. A few of the DL-1 had only short lives but several lasted 10 years or more.

Having the experience for metal airplane construction within its organization, the Detroit Aircraft Corp. designed an all metal cigar-shaped monocoque fuselage for the speedy "Vega" in 1929. Mated to a wooden wing shipped from Lockheed in Calif., the first model DL-1 was ready for tests early in 1930. Mostly as a promotional stunt, Amelia Earhart, best-known lady-pilot in the U.S.A., was engaged to fly the prototype (June-July 1930) in speed and load trials over a new course laid out near Grosse Isle, Mich. Though not quite as fast as the wooden "Lockheed," the metal-fuse-laged DL-1 performed well with varied loads on varied distances. The tousle-haired lady pilot set 3 new international records with one tour over the course at 181.18 m.p.h. Certifica-

Fig. 36. Army Y1C-17 was "Speed-Vega" version of DL-1; puny landing gear was worrisome.

tion for the Detroit-Lockheed was approved in April of 1930 and several more metal fuselages had been shipped to Lockheed Aircraft for mating with wooden wings. Stories heard since indicate that a new metal wing had also been built for use on the DL-1, but the extra weight of the new wing was prohibitive and money for further experimentation was not available either, so the project was dropped. Of the 10 metal "Vega" that were built, only 6 came under the provisions of ATC #308. One model DL-1A (also as DL-1 Special) was built for Lt. Comdr. Glen Kidston of London on Group 2 approval, numbered 2-316, and exported to England. A DL-1 was sold to the Air Corps and tested for cargo and personnel transport as the YIC-12, but Army pilots labeled it cranky and did not particularly like it. A model DL-1B, specially arranged as a "Speed-

Vega," topped 221 m.p.h. and was the fastest Army plane at that time. Designated the YIC-17, this version was flown by Capt. Ira Eaker on an attempted record transcontinental flight but cracked up in a forced landing before reaching its goal. The last "Metal-Vega" was built for Richard Von Hake on a Group 2 approval, numbered 2-448, in 1933 and sold to Morrell & Co., meat packers, for business transport and good-will promotion; this veteran "Vega," as some reports indicate, is probably still in storage somewhere and waits to be rebuilt for continued service.

The Detroit-Lockheed "Vega" model DL-1 was a high-winged cabin monoplane with seating for 7 and, except for its metal monocoque fuselage, was typical of the "Vega" 5-B or 5-C. Being somewhat heavier when empty (with allowance for more useful load and

Fig. 37. DL-1 used by General Tire Co. as "Miss Streamline."

operating with a greater gross load) caused a detriment to its all-round performance which was still a fair swap for the added reliability. Powered with the 9 cyl. P & W "Wasp" C1 engine of 450 h.p. the DL-1 was a bit slower than its "wooden sisters" and did not seem to fly as well in general, but the differences were easily overlooked in a period of extended service. Offered also with a special wire-braced "speed-gear" as the DL-1B, this version was a bit faster than the one with standard landing gear, but this spindly arrangement was not very hardy and was prone to occasional trouble. The prototype DL-1 was first introduced with slim high-pressure tires, a conical propeller spinner and no N.A.C.A. engine cowling; but most subsequent ships of this model had the N.A.C.A. engine fairing, low-pressure tires and did not use any propeller spinner. The prototype was soon modified in like manner. The type certificate number for the Detroit-Lockheed models DL-1 and DL-1B was issued 4-2-30 and some 6 examples of this model were manufactured by the Lockheed Aircraft Corp. at Burbank, Calif. The fuselages for this model were fabricated in the Detroit plant and shipped to Calif. for final assembly. As a division of the Detroit Aircraft Corp., Lockheed had Carl B. Squiers as general manager; chief engineer Gerry Vultee was succeeded by Richard A. Von Hake early in 1930; James Gerschler and Richard W. Palmer were project engineers. Veteran "Herb" Fahy was chief test pilot until April of 1930, succeeded by Marshall "Babe" Headle.

Listed below are specifications and performance data for the metal-fuselaged Detroit-Lockheed model DL-1 as powered with the 450 h.p. "Wasp" C1 engine; length overall 27'6"; height overall 8'6"; wing span 41'0"; wing chord at root 102"; wing chord at tip 63"; wing chord mean 81"; total wing area 279 sq. ft.; airfoil at root Clark Y-18; airfoil at tip Clark Y-9.5; wt. empty 2595 lbs.; useful load 1905 lb.; payload with 94 gal. fuel was 1075 lbs.; gross wt. 4500 lbs.; max. speed 178; cruising speed 153; landing speed 60; stall speed 60; climb 1275 ft. first min. at sea level; ceiling 19,750 ft.; gas cap. 94 gal.; oil cap. 12 gal.; cruising range at 22 gal. per hour was 600 miles; price at the factory was $21,985. raised to $22,000. in March of 1931. The following figures are for prototype before approval; wt. empty 2560; useful load 1707; payload with 94 gal. fuel 877; gross wt. 4270 lbs.; with these weight differences, performance did vary slightly from that listed above. With slight modification in wing frame structure, allowable gross weight was increased to 4750 lbs.; likewise, fuel capacity could be increased to 180 gal. and oil capacity to 15 gal.

The construction details and general arrangement for the model DL-1 were similar to that of the "Vega" 5-B as described in the chapter for ATC #227 of U. S. CIVIL AIRCRAFT, Vol. 3, except for the following: the fuselage framework was of metal monocoque construction; aluminum alloy annular rings, bulkheads and metal stringers formed the basic framework to which was riveted an "Al-

Fig. 38. DL-1 used by Braniff Airways; cities served shown on fin.

clad" metal skin. The cabin interior was arranged to seat 6 passengers, and window and door placement was typical; upholstery was padded pig-skin, and a 12 cu. ft. baggage compartment to the rear had allowance for 100 lbs. The all-wood wing framework was of the standard "Vega" cantilever construction with a fuel tank mounted in the wing flanking each side of the fuselage. An additional fuel tank of 50 gal. cap. could be mounted in the center portion of the wing forming the cabin roof or this area could be converted to an extra baggage compartment with allowance for up to 300 lbs. Total allowable fuel capacity was not to exceed 180 gals. The standard landing gear was of two long telescopic legs using air-oil shock absorbing struts; wheels were normally 9.50x12 semi-airwheels and Bendix brakes were standard equipment. Wheel streamlines were available for use over 32x6 wheels and high-pressure tires. The so-called "speed-gear" was of two wire-braced rigid legs using either 9.50x12 wheels without fairings or 32x6 wheels with tear-drop wheel pants. The tail-group was of the standard "Vega" all-wood construction as used on the models 5-B or 5-C. A tail wheel, N.A.C.A. engine cowling, electric engine starter, battery, cabin lights and navigation lights and metal ground adjustable propeller were standard equipment. Wheel streamlines, landing lights,

a chemical toilet, parachute flares, extra fuel tanks, oil cooling radiator and Goodyear "airwheels" were optional. The next and the last "Vega" development was the improved model 5-C as described in the chapter for ATC #384 in this volume.

Listed below are various Metal-Vega model DL-1, DL-1A and DL-1B entries as gleaned from registration records:

NC-497H;	Model	DL-1	(#135)	Wasp	450.
NC-483M;	”	DL-1B	(#136)	”	
NC-288W;	”	DL-1B	(#137)	”	
NC-8497;	”	DL-1B	(#154)	”	
G-ABGK;	”	DL-1A	(#155)	”	
NC-8495;	”	DL-1	(#156)	”	
NC-8496;	”	DL-1	(#157)	”	
AC.31-405;	”	DL-1	(#158)	”	
AC.31-408;	”	DL-1B	(#159)	”	
NC-12288;	”	DL-1B	(#161)	”	

Serial #135 later as DL-1B; serial #155 also as DL-1 Special on Group 2 approval 2-316, exported to England; serial #156 later as NC-239M; serial #157 later as XA-DAY in Mexico; serial #158 as YIC-12 for Army Air Corps; serial # 159 also as DL-1B Special, designated YIC-17 for Army Air Corps; serial #161 as DL-1B Special on Group 2 approval 2-448; approval for all models expired on 9-30-39.

Fig. 38A. Used by Braniff into 1937, ship was active at least 5 more years.

Fig. 39. Curtiss "Robin" model 4C-1A mounted improved "Challenger" engine of 185 h.p.

To prolong the popular "Robin" design just a little longer, by making it a more useful carrier and to meet competition in the cabin monoplane field (a field that was practically being over-run by the bargain-priced Stinson "Junior"), Curtiss-Robertson planned a few hurried changes here and there to remake the "Robin" into a competitive 4-seater. A previous token effort in this direction was a modified C-1 that was approved earlier on ATC #270 as the model 4-C, but this was more or less to study the actual changes that would be necessary to convert the 3-place "Robin" into a practical airplane for four. A more serious effort was then initiated later by Ralph Damon with project engineer E. M. Flesh and test-pilot H. Lloyd Child contributing greatly to the new design. It was planned in such a manner that would make possible the use of a big percentage of assemblies already on hand from the standard 3-place "Robin." Tested by H. Lloyd Child at St. Louis on March 3-4-15 of 1930 the 4-place "Robin" was pronounced quite satisfactory and quickly groomed for production and early delivery. The cabin interior was now 4 in. wider to give everyone aboard a little more room but the cabin entry doors were on the right side only which made it rather awkward for the pilot, that is, if he wanted to get out after the right front seat was occupied. The interior trim

was improved to appeal to those who would be looking for a family-type airplane, and the fuselage was now faired out much better to a soft and more rounded shape. Complimentary to the new shape, some examples had a new vee-type windshield and the latest series Curtiss "Challenger" engine of 185 h.p. was used to bloster up the performance. A stress analysis was made for the new 4C-1A by Alex Tsongas and several changes were made here and there to beef up the structure. In this manner then, the "Robin's" final shape was formed in the model 4C-1A and it soon after made its bow as the last of the familiar and very popular "Robin" monoplanes. Introduced first in 1928 the "Robin" spanned a production period of some two years and well over 700 were built in all; on the scene for a good many years, it is still well remembered by thousands.

As shown, the Curtiss-Robertson "Robin" model 4C-1A was a high-winged cabin monoplane with seating for four and was basically typical of previous "Robins", except for modifications to cabin arrangement which allowed extra seating. It had a cleaned up outward appearance for more buyer appeal. A quickly noticeable feature was the redesigned fin and rudder to provide more directional stability and better control at the lower speeds, especially in

Fig. 40. The bigger "Robin" was ideal for leisurely cross-country hops.

"spin recovery". Though not an aggravating short-coming, all "Robins" were more or less marginal in prompt spin recovery and the re-designed tail-group would have been welcome long before this. With its empty weight increased a good deal and its gross weight increased by some 250 lbs., the model 4C-1A was a "Robin" that waddled more than it hopped. This was acceptable in view of the fact that it was leveled specifically at the family man who flew for the fun of it, so the loss in snap and performance would not be so noticeable. Flying the 4C-1A presented no problems, and its behavior promoted what a good many called a nice ride. With a cruising range of nearly 500 miles one would guess that this bigger "Robin" would be ideal for cross-country flights that had no other purpose other than to be up there watching the scenery go slowly by. The type certificate number for the 4-place "Robin" model 4C-1A was issued 4-3-30 and some 11 or more examples of this model were manufactured by the Curtiss-Robertson Div. on Lambert Field in St. Louis, Mo. By Nov. of 1930, the Curtiss-Robertson Div. and the Travel Air Co. were merged to operate in St. Louis as the Curtiss-Wright Airplane Co. The rugged nature and the popularity of the various "Robins" naturally promoted longevity; it is no wonder that some 300 or more were still in active service 10 years later.

In reference to the "Robin" model 4-C, as described in the chapter for ATC #270 in U.S.

CIVIL AIRCRAFT, Vol. 3, it has since been learned that this first 4-seated version of the "Robin" was actually developed in Garden City under the direction of T. P. Wright and not at the Curtiss-Robertson plant as stated. H. Lloyd Child tested this first 4-place "Robin" model 4-C (X-8336) at Garden City, L.I., N.Y. on Sept. 4-5-6-9 of 1929. Using a standard "Robin" model C for the conversion, two seats were placed up front; due to the narrow fuselage at this point the added right hand seat was staggered back slightly to allow shoulder room for the extra occupant in front. Because of this arrangement it is assumed that controls were provided on the left hand side only. From this trial arrangement of seating four in a standard "Robin" (model 4-C) it was suggested by test-pilots H. Lloyd Child and Paul Boyd that the width of the fuselage in front be increased by at least 4 in. to allow normal side-by-side seating as in the later 4C-1A which is described and pictured in this chapter.

Listed below are specifications and performance data for the Curtiss "Robin" model 4C-1A as powered with the 6 cyl. Curtiss "Challenger" (R-600) engine of 185 h.p.; length overall 25'5"; height overall 8'0"; wing span 41'4"; wing chord 72"; total wing area 223 sq.ft.; airfoil Curtiss C-72; wt. empty 1811 lbs.; useful load 1039 lbs.; payload with 50 gal. fuel was 531 lbs.; gross wt. 2850 lbs.; max. speed 115; cruising speed (85%) 98; stalling speed 60; landing speed 55; climb 505 ft. first min. at sea level; climb in 10 min. was 4300 ft.; climb to 5000 ft. was 12 min.; climb to 10,000 ft. was 32 min. with climb rate still at 145 ft. per min.; service ceiling 11,200 ft.; gas cap. 50 gal.; oil cap. 5 gal.; cruising range at 10 gal. per hour was 430 miles; price at the factory was $7995. By this time in 1930, the price of the lovable "OX-5 Robin" had been lowered to $2495.

The fuselage framework was built up of welded chrome-moly steel tubing in a Warren truss form, faired to shape with rounded balsa-wood pads over the longerons, and wooden fairing strips down the sides; the completed framework was covered in fabric. Upholstery was of broad-

Fig. 41. Buxom 4C-1A had ample room to seat four.

cloth fabric, the windows were of shatter-proof glass and a small baggage compartment was behind the rear seat with allowance for 36 lbs. Front seats were of the bucket type and rear seat was of the bench type; two large entry doors were both on the right hand side. The windows slid open for ventilation and cabin heat was available. The semi-cantilever wing framework, in two halves, was built up of solid spruce spar beams and stamped-out "Alclad" metal wing ribs; the leading edges were covered with duralumin sheet and the completed framework was covered in fabric. Two fuel tanks were mounted one each in the root end of each wing half. The out-rigger landing gear of 96 in. tread was built into the wing bracing struts to form a rigid truss; shock absorbers were oleo-spring struts, wheels were 30x5 and Bendix brakes were standard equipment. All strut junctions and exposed fittings were carefully streamlined by formed metal covers and a low drag speed-ring engine cowling was available to shroud the "Challenger" engine. The fabric covered tail-group was built up of welded chrome-moly sheet and tubing into surfaces of larger area; the fin was ground adjustable and horizontal stabilizer was adjustable in flight. A Curtiss-Reed metal propeller, hand crank inertia-type engine starter, and a tail wheel, were also standard equip-

ment. Orange and cream was a popular color combination for the 4-place "Robin" but other colors were available too. The next Curtiss-Robertson (Curtiss-Wright) development was the twin-engined "Kingbird" described in the chapters for ATC #347 and #348 in this volume.

Listed below are Curtiss "Robin" model 4C-1A entries as gleaned from various records: NC-509N; Model 4C-1A (#702)

			— Challenger R-600
NC-510N;	"	(#704)	"
NC-512N;	"	(#706)	"
NC-514N;	"	(#708)	"
NC-516N;	"	(#710)	"
NC-563N;	"	(#712)	"
NC-564N;	"	(#714)	"
NC-565N;	"	(#716)	"
NC-566N;	"	(#718)	"
NC-567N;	"	(#720)	"
NC-606V;	"	(#730)	"

Serial #704 also listed as model 4C-2; serial #700-767-769 were model 4C-1 on Group 2 approval numbered 2-198 issued 4-3-30; these 3 were NC-508N, NC-625V, NC-554N. For a complete resume of all the "Robin" models refer to chapters for ATC #40-63-68-69-143-144-220-221-268-270-309 in U.S. CIVIL AIRCRAFT, Vols. 1-2-3-4.

A.T.C. #310
(4-5-30)
SPARTAN, MODEL C4-225

Fig. 42. Spartan C4-225 with Wright J6-7-225 engine was first of new monoplane series.

With the market for open cockpit biplanes now spread rather thinly amongst a host of manufacturers, Spartan Aircraft wisely decided to bolster their line-up for the new year with an attractive cabin monoplane. This was an airplane that was primarily developed as a practical family-type airplane and could double in duty as a comfortable and speedy air-taxi for the business man. Though of entirely different dimension and arrangement than the biplanes which were still coming off the line occasionally, the new cabin monoplane was definitely recognizable as a "Spartan" at very first sight. Faired lavishly to a plump, rounded figure and carefully streamlined at nearly every point — the "Spartan Monoplane" bore unmistakable signs of careful study and great care in its design, from inside out. Rex B. Beisel, chief of engineering at the Spartan factory had everyone enthused about this new airplane and plans were quickly laid out for a whole series of similar types to serve in a broader range of price and purpose. The first of these to roll off the line was the model C4-225, a handsome and well appointed vehicle seating four and powered with the 7 cyl. Wright J6 (R-760) engine of 225 h.p. Knowing full well they would have to compete with such already entrenched favorites as the Stinson, Monocoach, the Verville 104, and others, "Spartan" tried their level best to offer a craft that would be just about the best that good money could buy, paying particular attention to details and niceties which would appeal to the

hard to please. In the space of a year's time, the Spartan design staff stretched this new series to include 5 models with three different seating arrangements in the choice of three different powerplants. With measured caution, dictated by economic uncertainty in the country, only a few examples were built now and then; all of these were soon sold, but sales for a craft of this type were slow so the model C4-225 remained rather scarce in actual number.

The Spartan model C4-225 was a high-winged cabin monoplane with ample seating for four, and offered high quality accommodations arranged in good taste. Always catering more closely to owners who could well afford the best, Spartan spared very little expense in providing accessories and equipment which were usually considered deluxe. To further accommodate the clientele who would be buying this sort of ship, the model C4-225 was made easy to fly, pleasantly stable, and it did not require the maximum in piloting technique for satisfactory operation. This is not to imply that the C4-225 was a slow-minded dud, but only to point out that its nature was best suited to those who were going somewhere and would rather get there relaxed; or for those who were not going anywhere in particular but would rather be happily engaged in the appreciation of beautiful scenery, instead of constantly guiding an errant airplane. Powered with the Wright "Seven" (R-760) engine of 225 h.p., the C4-225 could be judged slightly under-powered according to the norm

Fig. 43. Spartan flying school used C4-225 for advanced flight training.

for a craft of this type, but this did not seem to hamper its ability to any great extent. Quite clean in the aerodynamic sense and of good proportion, the C4-225 was not left dangling behind by many, and whatever short-comings it did have in performance maximums were off-set by its hardy and pleasant nature. Very well equipped, the "Spartan" C4-225 owner was not plagued to buy the usual extras, and could oper-ate in the confidence that his ship was purposely stout for added safety; purposely large and pur-posely spacious for more useful operating utility. The type certificate number for the "Spartan" model C4-225 (as powered with the 225 h.p. Wright J6 engine) was issued 4-5-30 and later amended to include the improved Wright R-760 of 240 h.p. Some 5 examples of the model C4-225 were manufactured by the Spartan Aircraft Co. at Tulsa, Okla.

Listed below are specifications and perform-ance data for the "Spartan" model C4-225 as powered with the 225 h.p. Wright J6 engine; length overall 31'6" (32'3") length shown in brackets is for revised rudder shape; height overall 9'0"; wing span 50'0"; wing chord 80"; total wing area 299 sq.ft.; airfoil Clark Y (un-verified); wt. empty 2325 lbs.; useful load 1190 lbs.; payload with 85 gal. fuel was 470 lb.; pay-load with 60 gal. fuel was 620 lbs. which in-cluded 100 lbs. of baggage; gross wt. 3515 lbs.; max. speed 130; cruising speed 110; landing speed 50; climb 800 ft. first min. at sea level; ceiling 14,200 ft.; gas cap. max. 85 gal.; oil cap. 6 gal.; cruising range at 12.5 gal. per hour was 700 miles; price at the factory was $9750. Some of the later examples were powered with the improved Wright R-760 of 240 h.p. so perform-ance figures would be slightly better.

The fuselage framework was built up of welded chrome-moly and 1025 steel tubing, with steel gusset plates at all points of stress, extensively faired with wooden formers and fairing strips to a well-rounded shape, and fabric covered. The spacious cabin, 40" wide x 84" long x 51" high, seated four in deep cushion comfort and with ample leg room; the cabin was tastefully appointed in hardware and trim with provisions for heating and ventilation. A large door on either side with large convenient step provided unhampered entry, and a baggage compartment of 9.7 cu. ft. capacity with allow-ance for 100 lbs. was located behind the rear seat. The wing framework was built of solid spruce spars routed to an I-beam section with spruce and plywood truss-type wing ribs; the leading edges were covered with dural metal sheet and the completed framework was cov-ered in fabric. Two fuel tanks were mounted in the wing flanking each side of the fuselage. The landing gear of 120 in. tread was of the out-rigger type using oil-spring shock absorbing struts; wheels were 32x6 and Bendix brakes were standard equipment. Rather unusual were the landing gear "spats" that faired the U-type fork assembly to which the wheels were at-tached; the landing gear assembly was arranged into the wing bracing struts as a rigid truss. A swivel-type tail wheel was mounted on end of the fuselage with a cut-out in the rudder to pro-vide working clearance. The tail-group was a composite structure of wood and metal; the horizontal stabilizer was of wooden spars, dural metal ribs and formed dural sheet leading edge; the fin was of spruce spars and dural (duralum-in) ribs with elevators and rudder of dural tube spars and sheet dural metal ribs — all was cov-

Fig. 44. C4-225 was ideal family-type airplane for four.

ered in fabric. The horizontal stabilizer was adjustable in flight; both rudder and the elevators were aerodynamically balanced for light "stick" loads. A metal propeller, dual wheel-type controls, exceptionally complete instrument panel, wheel brakes, fire extinguisher, booster magneto, tail wheel, exhaust tail-pipe, Eclipse engine starter, and navigation lights were standard equipment. The next development in the "Spartan" cabin monoplane was the model C4-300

described in the chapter for ATC #383 in this volume.

Listed below are "Spartan" model C4-225 entries as gleaned from registration records:

NC-750N; Model C4-225 (#B-1) Wright R-760.
NC-751N;   "   (#B-2)   "
NC-752N;   "   (#B-3)   "
      ;   "   (#B-4)   "
NC-10483;   "   (#B-5)   "

Registration number for serial #B-4 unknown.

Fig. 45. Waco model RNF with 110 h.p. Warner engine; shown here over Troy, Ohio.

The new "Model F" was an interesting series of open cockpit biplanes that Waco Aircraft was indeed proud to announce. Though enjoying a tremendous success with various models built previously, "Waco" had learned by now that engine horsepower, though a sure bet to quicken the blood of any sportsman, was expensive to buy and rather expensive to maintain. To offer now an airplane of good performance with nominal horsepower that somewhat leaner pocketbooks could better afford, an intelligent redesign of old concepts was necessary. Such a design would have to provide a performance comparable to that to which operators were accustomed, a practical paying load, ample room and good comfort — all without resorting to a higher horsepower which would add useless weight, or an increase in overall bulk. In the new "Model F" this was all happily achieved almost to perfection. As introduced to the flying public in April of 1930, the new "Model F" appeared as a tidy combination that was fortunate to inherit every traditional feature of the dependable "Waco" design, plus the sensible selection of an efficient configuration that was

trimmed and condensed to contain nothing superfluous; nearly every pound of its make-up was a useful contribution, so available horsepower was not taxed to the point of over-work. Here now indeed, was an airplane that was the ideal for general-purpose service, as practiced by the average operator, and a craft that was entirely content to operate from anywhere. The short-field performance of the new "F" was a source of much amazement and demonstrations repeatedly proved that a good pilot could land it safely within the confines of a 100 ft. circle. Needless to say, news of the perky "Model F" traveled fast through flying circles and orders began pouring in at a delightful rate. As we recall, reception of the Waco F was so enthusiastic that scores of these craft were soon flying all over the country and some even abroad. In the parlance of later years, we might say that Waco Aircraft had surely hit the "jack-pot."

As shown here in varied views, the Waco model RNF was an open cockpit biplane, seating three, with lines typical and somewhat reminiscent of the past, but with the evidence that considerable study had gone into the basic con-

Fig. 46. Short-coupled RNF did better job with less airframe.

figuration and general aerodynamic arrangement. Almost to the point of being dainty, the new RNF was smaller and a good deal lighter, but left the impression that there was ample room for a practical load that could be handled easily with still a good margin of performance. Basically stable and very easy to fly, the RNF soon took on duties as a pilot-trainer in addition to other varied chores that would come up in general-purpose service. Stout of frame and very maneuverable, the landplane version was often used for teaching "acrobatics" in secondary phases of pilot-training. Powered with the 7 cyl. Warner "Scarab" engine of 110 h.p., all-round performance of the new Model F was surprisingly lively and good responsive control endeared it to many who flew their craft simply for sport. Just to prove what a good pilot could do with the RNF, Johnnie Livingston entered the "dead-stick" landing contests at the 1930

National Air Races held in Chicago. With several scores remarkably "close to the mark," Livingston came up with 2 first, 2 second, and 2 third place wins in the daily events. In the "balloon busting" contests he deftly wheeled the RNF to 3 first place wins with almost the grace of a ballet dancer. Also arranged to permit installation of several engines in the 100-125 h.p. range, the new Model F was soon seen in a variety of interesting versions. The type certificate number for the Waco model RNF, as powered with the 110 h.p. Warner engine, was issued 4-7-30, and estimates were reported that 150 or more examples of this model were manufactured by the Waco Aircraft Co. at Troy, Ohio through 1931. Inherent safety, popularity, and rugged character naturally promoted a long life for this model so it is not surprising that well over 100 were still in active service 10 years later, and it has become a very popular model for "restoration" even at this late date.

Listed below are specifications and performance data for the Waco model RNF as powered with the 110 h.p. Warner "Scarab" engine; length overall 20'8"; height overall 8'4"; wing span upper 29'6"; wing span lower 27'5"; wing chord both 57"; wing area upper 130.5 sq.ft.; wing area lower 111 sq.ft.; total wing area 241.5 sq.ft.; airfoil (N.A.C.A.) M-18; wt. empty landplane 1150 lbs.; useful load 747 lbs.; payload with 32 gal. fuel was 367 lbs.; gross wt. 1897 lbs.; max. speed 112; cruising speed 95; landing speed 35; climb 730 ft. first min. at sea level; ceiling 15,000 ft.; gas cap. 32; oil cap. 3 gal; cruising range at 7 gal per hour was 400 miles; price at the factory field was $4250. in 1930,

Fig. 47. Waco RNF on Edo floats.

with a hike to $4450. in April of 1931. The following figures are for the RNF as a seaplane on Edo model L twin-float gear; wt. empty 1421 lbs.; useful load 757 lbs.; payload with 32 gal. fuel was 377 lbs.; gross wt. 2178 lbs.; max speed 102; cruising speed 88; landing speed 45; climb 620 ft. first min. at sea level; ceiling 12,000 ft.; price with float-gear was approx. $5500. An amendment later allowed the installation of 125 h.p. Warner "Scarab" engine with allowable gross wts. boosted to 1938 lbs. for landplane and 2226 lbs. for seaplane; the increased power, only slightly cancelled out by the additional weight, provided small increases in general performance ability.

The fuselage framework was built up of welded chrome-moly and 1025 steel tubing into a rigid truss, deeply faired to shape with wooden formers and fairing strips, then fabric covered. Because of the close-coupled design, the passenger cockpit was placed well forward where the fuel tank would normally be and pilot's cockpit was immediately behind; both cockpits were roomy and well appointed. Early versions of the Model F had bulbous looking windshields of molded plastic, but because of some distortion, these were later changed to shields formed in the normal manner of sheet stock; a small baggage locker was in the dash-panel of the front cockpit with an allowance for 15 lbs. The wing framework was built up of solid spruce spars routed to an I-beam section, with spruce and mahogany plywood girder-type wing ribs; the leading edges were covered with duralumin sheet and the completed framework was covered in fabric. With no place in the fuselage for a tank, the fuel tank was mounted in the center-section panel of the upper wing; a direct-reading fuel gauge was in easy view. The fabric covered tail-group was built up of welded steel tubing; the elevators were provided with aerodynamic balance and the horizontal stabilizer was adjustable in flight. The robust landing gear of 76 in. tread was of the out-rigger type with oleo-spring shock absorbing struts; 6.50x10 low pressure semi-airwheels and wheel brakes were standard equipment. The seaplane version of the RNF as mounted on "Edo" floats was required to carry a 10 lb. anchor in the pilot's cockpit and an auxiliary fin was mounted on the under-side of the fuselage to provide added area. A slight modification was also required to give the seaplane ¾ in. more negative travel in adjustment of the horizontal stabilizer. A wooden Hartzell propeller, wiring for navigation lights and a swiveling tail wheel were standard equipment. Dual joystick controls, navigation lights, a Heywood air-operated engine starter, adjustable metal prop, skis, speed-ring engine cowling and wheel streamlines were optional. The RNF was available in color combinations of blue, green, or vermillion fuselage, with silver wings. The next

Fig. 48. Seen on pasture-airports everywhere, the RNF was popular for all-purpose use.

Fig. 49. RNF was forerunner in design to next decade of Waco biplanes.

Fig. 50. Various tests such as Waco 10 Special led to design of new F-series.

development in the "Waco F" series was the Kinner-powered model KNF as described in the chapter for ATC #313 in this volume.

Listed below is only a partial listing of Waco RNF entries as gleaned from registration records:

NC-677N; Model RNF (#3222) Warner 110.
NC-679N;   ”   (#3223)   ”
NC-685N;   ”   (#3237)   ”
NC-687N;   ”   (#3238)   ”
NC-695N;   ”   (#3240)   ”
NC-694N;   ”   (#3242)   ”
NC-692N;   ”   (#3244)   ”
NC-689N;   ”   (#3245)   ”
NC-834V;   ”   (#3249)   ”
NC-832V;   ”   (#3250)   ”
NC-835V;   ”   (#3251)   ”
NC-842V;   ”   (#3252)   ”
NC-836V;   ”   (#3253)   ”
NC-838V;   ”   (#3254)   ”
NC-839V;   ”   (#3255)   ”
NC-846V;   ”   (#3256)   ”
NC-849V;   ”   (#3259)   ”
NC-856V;   ”   (#3261)   ”
NC-850V;   ”   (#3262)   ”

NC-853V; Model RNF (#3263) Warner 110.
NC-848V;   ”   (#3264)   ”
NC-851V;   ”   (#3266)   ”
NC-855V;   ”   (#3267)   ”
NC-858V;   ”   (#3268)   ”
NC-854V;   ”   (#3269)   ”
NC-102Y;   ”   (#3270)   ”
NC-859V;   ”   (#3271)   ”
NC-101Y;   ”   (#3272)   ”
NC-800V;   ”   (#3273)   ”
NC-860V;   ”   (#3274)   ”
NC-857V;   ”   (#3275)   ”
NC-862V;   ”   (#3276)   ”
NC-107Y;   ”   (#3277)   ”
NC-863V;   ”   (#3278)   ”
NC-861V;   ”   (#3279)   ”
NC-866V;   ”   (#3280)   ”
NC-114Y;   ”   (#3281)   ”
NC-103Y;   ”   (#3282)   ”
NC-104Y;   ”   (#3284)   ”
NC-122Y;   ”   (#3288)   ”
   NS-31;   ”   (#3473)   ”

Serial #3223 also as model KNF; seaplane versions of RNF were serial #3318, #3335, and up; X-652N was prototype airplane for RNF series.

Fig. 51. Dressed up with cowling and pants, RNF was favorite of flying sportsmen.

A.T.C. #312
(4-8-30)
STINSON "JUNIOR", SM-8D

Fig. 52. Stinson SM-80 with 225 h.p. Packard "Diesel" engine.

The popular SM-8 series of the Stinson "Junior" was introduced a bit earlier with the 7 cyl. Wright J6 engine of 225 h.p. and the 9 cyl. Lycoming R-680 engine of 215 h.p., both of which stirred a considerable interest in the industry, and the value per dollar promoted a welcome surge of buying. With 3 models already available in the new "Junior" line, Stinson introduced the SM-8D version which was basically typical of other models in this series, but mounted the revolutionary Packard "Diesel" engine. This was the first certificated installation of the Packard DR-980 engine and both the "Junior" SM-8D and the "Packard" engine were formally introduced at the Detroit Air Show (April 5-13) for 1930. There was considerable interest in this new engine during show-time because of its novelty; many other aircraft manufacturers soon had standard models developed to use this new "diesel" powerplant. The Packard Motor Co. of Detroit had a prototype engine mounted for extensive flight testing in a Stinson "Detroiter" model SM-1DX since Sept. of 1928, so it was only logical that Stinson would also have the first approved installation to offer in its line-up for 1930. The "Junior" model SM-8D was a hit at the air show and provoked considerable discussion and interest because of the many practical features offered by the Packard diesel-type engine, but economic unrest in this nation caused most prospective buyers to take on a "wait and see" attitude. This was certainly not conducive to the craft's popularity and many prospective sales were thus side-tracked to "Junior" models mounting the more familiar gasoline engines; as a consequence, only a few examples of the SM-8D were built and it re-

mained a rare and relatively unknown version in the "Junior" line-up.

The Stinson "Junior" model SM-8D was a high-winged cabin monoplane with seating for 4, and was basically typical of either the SM-8A and SM-8B, except for its engine installation and the slight modifications necessary for this particular combination. The Packard "Diesel" engine (which was a joint effort by Capt. Lionel M. Woolson of Packard Motors and Prof. Dohner, formerly of Germany), was of an ideal power range for the "Junior" type airplane, and of a configuration suitable for a neat installation. Performance in this new engine-airplane combination was comparable to other "Junior models with bonus features added, such as better fuel economy, lower cost of fuel per gallon, far less fire hazard, and the elimination of intricate ignition systems for better reliability. On the other hand, the compression-ignition engine, because of its operating principle, developed harsh vibrations throughout the framework causing varying degrees of annoyance to apprehensive pilots and passengers. "Diesel" fuel was of a type similar to furnace oil and because of its peculiar properties was often tagged as "smelly & messy" so one gets the impression that the SM-8D was certainly not handed a rosy future to go on; the extras it did possess in economy and utility could hardly overcome the objections, some of which were real and some fancied. The type certificate number for the Stinson "Junior" model SM-8D was issued 4-8-30 and records indicate that probably no more than 2 examples of this model were manufactured by the Stinson Aircraft Corp. at Wayne, Mich.

Fig. 53. Efficient and cheap to operate, SM-8D "Junior" showed great promise.

Listed below are specifications and performance data for the Stinson "Junior" model SM-8D as powered with the 225 h.p. Packard "Diesel" engine; length overall 29'0"; height overall 8'9"; wing span 41'8"; wing chord 75"; total wing area 234 sq.ft.; airfoil Clark Y; wt. empty 2175 lbs.; useful load 1025 lbs.; payload with 50 gal. fuel was 505 lbs.; gross wt. 3200 lbs.; max. speed 128; cruising speed 108; landing speed 50; climb 800 ft. first min. at sea level; ceiling 14,000 ft.; fuel cap. 50 gal.; oil cap. 6 gal.; cruising range at 10 gal. per hour was 495 miles; price at the factory field was $8995.

The construction details and general arrangement of the model SM-8D were similar to that of the models SM-8A and SM-8B as described in the chapters for ATC #294 and #295 in U.S. CIVIL AIRCRAFT, Vol. 3. For interior soundproofing and insulation the "Junior" had its cabin walls lined with a 1 inch blanket of "Balsam Wool," a deadening material designed to screen out excessive noise and provide a barrier for extremes in temperature. Easy and un-

hampered entry was gained with a convenient step and a large door on each side, with comfortable seats upholstered in broadcloth or mohair fabrics; a baggage compartment of 7 cu.ft. capacity was located to the rear of the cabin area with allowance for 50 lbs. Access to baggage was from the outside. Dual wheel-type controls, wheel brakes, a ground adjustable metal propeller, tail wheel and electric engine starter were standard equipment. The next development in the Stinson "Junior" line was the model SM-7B as described in the chapter for ATC #329 in this volume.

Listed below are the only known Stinson SM-8D entries as gleaned from registration records:

NC-200W; Model SM-8D (#4036)
                    — Packard DR-980.
NC-227W; Model SM-8D (#4100)
                    — Packard DR-980.

Serial #4100 first listed as model SM-8A with Lycoming 215 engine, later converted to model SM-8D with Packard "Diesel" engine.

Fig. 54. Waco model KNF with Kinner K5 engine, shown on factory field during early tests.

Normal to practice for years, Waco Aircraft planned the new "Model F" series biplane with the specific intent of offering this airplane in several versions. In this case, the manufacture of model variants would present no problems because the identity of a particular version was wholly dominated by its engine installation. The basic airframe, from the firewall back, was the same for all types. Introduced as a companion model in this new series, the model KNF was powered with the 5 cyl. Kinner K5 engine of 100 h.p. and aimed specifically at those who would prefer a slightly cheaper airplane that could be operated at less cost. Though slight, the increase in economy added up to a sizeable amount during 100 hours of flying time, so most operators who conducted small, busy pilot-training schools would select the KNF for the extra profits. The slight reduction in power available, as compared to the previously described 110 h.p. RNF, actually caused very little difference in the performance of the model KNF and its general behavior was quite typical. It is believed that the first flyable prototype of the "Waco F" series was a Kinner K5 powered version but the

Warner powered version was almost a parallel development. In all respects the KNF did very well in sales but the Warner powered RNF was by far the most popular. The F-series of this particular period, which would include the RNF-KNF-INF-MNF, was enthusiastically received, outselling every open cockpit biplane in the country during the period of 1930-31.

The Waco model KNF was an open cockpit biplane seating 3, and its almost petite arrangement signified that careful study had been made to arrive at an airplane that would handle a good payload and still deliver a better than average performance on nominal h.p. Powered with the 5 cyl. Kinner K5 engine of 100 h.p. the KNF was certainly more than adequate for all the chores normally handled in general-purpose service, with yet enough spirit to appeal to the private-owner who flew just for the sport of it. Easy to fly, obedient and quite responsive, the KNF was used most by pilot-operators for training purposes and was very well liked by the student pilots. The extreme stagger between the upper and lower wings, in relations to the cockpits, offered the pilot excellent visibility and a good view of the ground on take-offs and landing; wheel brakes and a swiveling tail-wheel offered excellent maneuverability on the ground. The structure of the "Model F" was quite robust and could absorb an amazing amount of punishment whether in the air or on the ground; this, too, had great appeal to the average private owner or flying service operator. Later in the year Waco Aircraft introduced the model INF (which was also Kinner powered), but in a model that offered 25 more horsepower for those who wanted a little extra in performance. The type certificate number for the model KNF, as powered with the 100 h.p. Kinner K5 engine, was issued 4-12-30, and though no corrected tally is available it is believed that at least 20 examples of this model were manufactured by the Waco Aircraft Co. at Troy, Ohio.

Listed below are specifications and performance data for the "Waco" model KNF as powered with the 100 h.p. Kinner K5 engine; length overall 20'9"; height overall 8'4"; wing span upper 29'6"; wing span lower 27'5"; wing chord upper & lower 57"; wing area upper 130.5 sq. ft.; wing area lower 111 sq. ft.; total wing area 241.5 sq. ft.; airfoil (N.A.C.A.) M-18; wt. empty 1125 lbs.; useful load 747 lbs.; payload with 32 gal. fuel was 367 lbs.; gross wt. 1872 lbs.; max. speed 110; crusing speed 93; landing speed 35; climb 720 ft. first min. at sea level; ceiling

14,500 ft.; gas cap. 32 gal.; oil cap. 3 gal.; cruising range at 6.5 gal. per hour was 430 miles; price at the factory field was $4100. in 1930, with a hike to $4300. in April of 1931. An amendment later allowed increase in gross weight to 1938 lbs. which would alter listed performance figures to a slight degree.

The construction details and general arrangement of the "Waco" model KNF were typical of the RNF as described here previously in the chapter for ATC #311. The cockpits were deep and well protected with a large door on left side for easy entry to the front cockpit; entry to the rear cockpit was an easy step from the wing-walk. With the trailing edge of the upper wing projecting to about midway over the pilot's cockpit, two hand-holds were provided in the wing's trailing edge to offer the pilot something to hang on to for better exit or entry. A large cut-out in the trailing edge of the center-section panel offered excellent vision overhead. The Model F had ailerons on each wing panel that were connected together in pairs by a streamlined push-pull strut; the ailerons were hinged on a line that was canted to the line of flight for more responsive action. The outrigger landing gear of 76 in. tread was fitted with low pressure semi-airwheels of 6.50x10 dimension; wheel brakes and tail wheel were standard equipment. A baggage locker in dash-panel of front cockpit had allowance for 15 lbs. A Hartzell wooden

Fig. 55. Efficient design of KNF rendered excellent performance on nominal power.

Fig. 57. KNF was also favorite for all-purpose use; several served through next decade.

propeller, and wiring for navigation lights were also standard equipment. A metal propeller, Heywood air-operated engine starter and wheel streamlines were optional. The model KNF was offered with silver sings in combination with a blue, green, or vermillion fuselage. The next "Waco" development was the Packard diesel-powered model HSO as described in the chapter for ATC #333 in this volume.

Listed below are "Waco" model KNF entries as gleaned from registration records:

X-651N;	Model KNF	(#3203)	Kinner K5
X-653N;	"	(#3208)	"
NC-679N;	"	(#3223)	"
NC-681N;	"	(#3224)	"
NC-686N;	"	(#3239)	"
NC-690N;	"	(#3241)	"
NC-699N;	"	(#3243)	"
NC-698N;	"	(#3246)	"
NC-833V;	"	(#3248)	"
NC-852V;	"	(#3260)	"
NC-105Y;	"	(#3283)	"
NC-108Y;	"	(#3291)	"
NC-109Y;	"	(#3295)	"
NC-113Y;	"	(#3297)	"
NC-126Y;	"	(#3306)	"
NC-605Y;	"	(#3324)	"
NC-614Y;	"	(#3335)	"

Serial #3203 and #3208 were two of the original prototype airplanes; serial #3223 also as RNF in early test.

Fig. 56. Earlier tests such as Waco 10 with Kinner K5 engine led to F-series design.

REARWIN "KEN-ROYCE," 2000-CO

Fig. 58. Rearwin "Ken-Royce" model 2000-CO with 165 h.p. Continental A-70 engine.

Enthusiastically planned and honestly presented, the Rearwin "Ken-Royce" was a good example of the high performance biplane that harbored no super-duper gimmicks nor any claims to possessing anything revolutionary; it was a careful refinement of design and existing practice rather than a departure from the conventional type. As shown, the new model 2000-CO, as compared to the earlier model 2000-C (refer to ATC #232), was a companion model quite typical except for its powerplant installation and some minor modifications necessary to this new combination. Powered now with the 7 cyl. Continental A-70 engine of 165 h.p. the model 2000-CO was a somewhat lighter airplane that carried a bit more in payload with approximately the same performance. This particular engine-airplane combination was also of somewhat smoother nature because of the inherent smoothness of the new "Continental" engine; this tended to bring out some of the airplane's finer inherent attributes. Performance was certainly above the average for this amount of power and a good many pilots had commented that the "Ken-Royce", in either version, was one of the finest flying airplanes of this type. In short, its personality was very enjoyable and many expressed a genuine sorrow that this delightful machine was built in such small number. Because of the sagging market for this type of sport airplane not many examples of the 2000-

CO were built; the last example was most likely built sometime in 1932. Actually, these were the only biplanes ever built by Rearwin and though quite significant, they did not play a very large part in the overall company history. In 1931, Rae Rearwin, noted for his sound business policies, began viewing the apparent success of various flivver-plane manufacturers with pensive interest and decided then that, for this particular time at least, this was a very sensible approach to the vexing problem of marketing airplanes. In the pert Rearwin "Junior" monoplane (ATC #434), brought out later in 1931, the company launched an entirely new concept for their wares and formed a new beginning that led to the manufacture of some of the finest light airplanes in this country.

The Rearwin "Ken-Royce" as shown, was a 3 place open cockpit biplane of rather large proportion but of good aerodynamic arrangement; a craft that was especially planned for the busy flying-salesman or the active sportsman-pilot who would be making frequent short hops in and out of all sorts of landing places. Powered now with the 7 cyl. Continental A-70 engine of 165 h.p., the "Ken-Royce" 2000-CO offered utility and good performance somewhat beyond the average for this amount of power; consequently, it posed as a very good value for the money invested. Its lean and willowy frame, upheld by graceful outspreading wings, suggest-

Fig. 59. Excellent behavior and spirited nature of 2000-CO drew favorable comment from pilots.

ed faint feminine qualities but there was certainly no girlishness in the behavior of the "Ken-Royce"; flight characteristics were honestly pleasant, control was brisk and sharp and its rather effervescent nature gave enjoyment that must surely have been habit-forming. Miss Jean LaRene, famous aviatrix of various accomplishments and generally associated with "American Eagle" airplanes, owned and flew a "Ken-Royce" biplane for years and loved every inch of it. At the National Air Races for 1930 held in Chicago, C. B. Allen rounded the pylons to first place with a Ken-Royce 2000-CO in the sportsman-pilots event. This was a race (5 lap) for open ships of 650 cu. in. class and "Hoot" Gibson of cowboy-movies fame was his only threat in an Axelson-powered "Swallow" biplane. In another 650 cu. in. event, J. B. Story flew this same ship to third place, turning in lap times of better than 120 m.p.h. Flying the 2000-CO just a few weeks later in the National Air Tour for 1930, J. B. Story earned an 11th place spot with an average speed of over 119 m.p.h. for the route, goaded by some very determined competition. The type certificate number for the Rearwin "Ken-Royce" model 2000-CO, as powered with the 165 h.p. Continental A-70 engine, was issued 4-16-30 and some 2 or more examples of this model were manufactured by Rearwin Airplanes, Inc. on Fairfax Field in Kansas City, Kansas. Rae A. Rearwin was president; Fred Landgraf was chief engineer and genial George E. Halsey was chief test pilot who also engaged in sales promotion.

Listed below are specifications and performance data for the "Ken-Royce" model 2000-CO: length overall 25'0"; height overall 9'11"; wing span upper 35'0"; wing span lower 31'6"; wing chord upper 66"; wing chord lower 48"; wing area upper 186 sq. ft.; wing area lower 114 sq. ft.; total wing area 300 sq. t.; airfoil "Rhode-St. Genese"; wt. empty 1447 lbs.; useful load 912 lbs.; payload with 55 gal. fuel was 385 lbs.; gross wt. 2359 lbs.; max. speed 130; cruising speed 110; landing speed 45; climb 950 ft. first min. at sea level; climb to 10,000 ft. was 18 min.; ceiling 20,000 ft.; gas cap. 55 gal.; oil cap. 4 gal.; cruising range at 10 gal. per hour was 550 miles; price at the factory field was $6500.

The fuselage framework was built up of welded chrome-moly and low-carbon steel tubing, heavily faired with formers and fairing strips to a streamlined shape, then covered in fabric. The cockpits were deep and well protected, neatly upholstered in fine leather, and bucket-type seats were designed to fit a parachute pack. A small compartment was under the front seat for stowing baggage and this compartment could also be used for the installation of a battery; when a battery was installed, no baggage was allowed in this space. The fuel supply was carried in two tanks; a main tank of 35 gal. capacity was mounted high in the fuselage just head of the front cockpit and an auxiliary fuel tank of 20 ga. capacity was mounted in the center-section panel of the upper wing. The wing framework was built up of laminated spruce spar beams that were not routed out but left with

Fig. 60. Faired-out to reduce drag, 2000-CO shown was fast-stepper in pylon races.

solid section, with wing ribs of basswood webs and spruce cap-strips; the leading edges were covered with birch plywood sheet to preserve the airfoil form and the completed framework was covered in fabric. The split-axle landing gear employed "Rearwin" rubber-oil shock absorbing struts; wheels were 28x4 or 30x5 and Bendix brakes were standard equipment. The fabric covered tail-group was built up of welded steel tubing and steel channel section; the fin was ground adjustable and the horizontal stabilizer was adjustable in flight. The elevators were operated by a metal push-pull tube, the rudder was operated by braided steel cables and the large Friese-type ailerons, in the upper wing, were operated by a combination of torque-tubes,

bellcranks and flexible steel cable; dual joy-stick controls were optional. Navigation lights, wheel brakes, tail wheel and metal propeller were standard equipment; an inertia-type engine starter was optional. The next Rearwin development was the "Junior" flivver-type monoplane as described in the chapter for ATC #434.

Listed below are the only known "Ken-Royce" model 2000-CO entries as gleaned from registration records;
NC-400V; Model 2000-CO (#104) Cont. A-70.
NC-12579;   ”   (#   )   ”
Serial number for NC-12579 unknown but is likely to be #105.

A.T.C. #315
(3-16-30)
INLAND "SUPER SPORT", W-500

Fig. 61. Inland "Super Sport" model W-500 with 110 h.p. Warner "Scarab" engine.

The model W-500 "Super Sport" was easily the "star" of the Inland monoplane series. It was a capable and spirited airplane with enough power per pound to deliver a performance that commanded respectful notice at all the major functions where the airplane-people gathered. First brought to public attention at the National Air Races for 1929 held at Cleveland, O., the "Super Sport" (designated then as the model S-300-U) was flown by Wilfred G. Moore to 5th place in a 50 mile event averaging over 116 m.p.h. Flying then on a restricted license, the fleet "Inland" parasol monoplane showed itself well and proved the high caliber performance that was inherent in this new design. In a timed speed run over a 100 kilometer course the "Super Sport" averaged 127 m.p.h. and later set an altitude record for light airplanes by reaching 19,700 ft. Earning its government approval (ATC) by early 1930, the "Super Sport" continued to uphold its zest for mixing with the fast ones. During the National Air Races for 1930 held at Chicago, Ill., Wilfred Moore flew the W-500 to second place in the Brownsville, Tex. to Chicago Derby, just barely a minute behind hard-flying Johnnie Livingston in his speedy "Monocoupe". In a 25 mile event for women in open ships of 500 cu. in. class, Mae Haizlip, Vera Dawn Walker and Betty Lund each flew a "Super Sport" to finish 1-2-4 respectively. In a 25 mile event for men in open ships of 450

cu. in. class, these same 3 "Supers" were flown by Wilfred Moore, Wm. A. Ong and William "Bill" Green, to finish 1-4-5 respectively. Crowding into the open ship event of 650 cu. in. class, Wm. Ong and Wil Moore breezed in to finish 1-2. In the sportsman-pilots race (amateur category) for open ships of 450 cu. in. class, these same 3 "Super Sport" already well limbered up, were flown by Art Hardgrave (Inland's president), W. G. Houston and M. C. Meigs, to a rousing 1-2-3 finish respectively, capping off a very successful and profitable showing at America's premier air races. Nimble and fleet, the Inland "Super Sport" went on to establish itself as an ideal craft for the sportsman, whether seasoned or amateur, who enjoyed the lively performance available with such small effort on a nominal amount of power.

Pert and quite handsome, the Inland "Super Sport" model W-500 was a "parasol" type monoplane designed to give two people the practical companionship of side-by-side seating in the sporty atmosphere of an open cockpit. Of rather simple lay-out and quite rugged in body and character, the "Super" was well able to absorb the strain and hard knocks bound to appear in every-day amateur sport flying. Well planned, with good aerodynamic proportion, the flight characteristics of this little monoplane were delightfully nimble, positive and eager, but it did take just a little determination

Fig. 62. Wilfred Moore flew "Super Sport" to several records.

to become complete master of the ship. The original "Super Sport", as introduced in 1929, was at that time designated the model S-300-U; it operated in the restricted category, while engaged in promotion, until it was approved as the model W-500 on this certificate. The "Super Sport" was built in rather small number, probably because of the average person's inability to invest hard-earned money into a strictly "sport" airplane at this time, but the few that were built served usefully for many years. The type certificate number for the Inland "Super Sport" model W-500 was issued 3-16-30 (this may be an error in the records and it is more likely to be 4-16-30, according to certificates issued for ATC #314 and #316) and 6 or more examples of this model were manufactured by the Inland Aviation Co. on Fairfax Field in Kansas City, Kan. Arthur Hardgrave was the president; Milton C. Baumann was chief engineer with Allan Smith as his assistant; Wilfred G. Moore was general manager and sales manager; Bert Thomas conducted most of the flight testing, while Wm. A. "Bill" Ong, Wm. "Bill" Green and Charlie Dailey were sales representatives who did most of the promotion work in the field.

Listed below are specifications and performance data for the Inland "Super Sport" model W-500, as powered with the 7 cyl. Warner "Scarab" engine of 110 h.p.: length overall 19'4"; height overall 7'7"; wing span 30'0"; wing chord 60"; total wing area 144 sq. ft.; airfoil (N.A.C.A.) M-12; wt. empty 916 lbs.; useful load 574 lbs.; payload with 31 gal. fuel was 195 lbs.; gross wt. 1490 lbs.; max. speed 130; cruising speed 112; landing speed 45; stall speed 48; climb 1200 ft. first min. at sea level; climb in 10 min. was 9000 ft.; ceiling 19,650 ft.; gas cap. max. 31 gal.; gas cap. normal 24 gal.; oil cap. 3 gal.; cruising range at 7 gal. per hour was 470 miles; price at the factory field was $4985,

lowered to $4580 in April of 1931. All examples of this model were eligible for improved Warner engine of 125 h.p.; with this engine, certain performance figures would be proportionately increased.

The fuselage framework was built up of welded chrome-moly steel tubing in a rigid truss form, faired to shape with wooden formers and fairing strips, then fabric covered. There was a large cockpit door on the right side with a convenient step for exit or entry; the baggage compartment was a dural metal bin placed in the turtle-back section just behind the cockpit. Protected with a locking door panel this compartment had capacity for 5 cu. ft. and a maximum allowance for 25 lbs. The side-by-side seating was quite chummy but still comfortable for medium sized people; the interior was well upholstered and seat cushions were removable from the bucket seats to allow for parachute packs. A small sky-light of pyralin was provided over the cockpit from the rear wing spar back; dual joy-stick controls were optional. The semi-cantilever wing framework, in two halves, was built up of heavy-sectioned solid spruce spar beams that were taper-routed at various points for lightness; the truss-type wing ribs were built up of spruce diagonals reinforced with plywood gussets, the leading edges were covered in dural metal sheet and the completed framework was covered in fabric. The balanced ailerons were a welded steel tube framework covered with fabric. A 12 gal. gravity-feed fuel tank was mounted in the root end of each wing half and a 7 gal. auxiliary fuel tank was mounted forward in the fuselage. In later modifications to the engine mount, and landing gear chassis with improved brakes, the 7 gal. fuselage tank was removed to retain the approved gross weight; maximum fuel capacity was then 24 gal. Robust looking N-type struts braced the wing to the fuselage over a pyramidal cabane of

Fig. 63. "Super Sport" with speed-ring and wheel pants gained extra speed.

streamlined steel tubing, and the out-rigger landing gear of 87 in. tread was built into a rigid truss with the wing-brace struts; shock absorbers were oil-draulic struts, wheels were 26 x 4 and wheel brakes were standard equipment. Semi-airwheels of 6.50 x 10 dimension and Air Products wheel brakes were optional. The oleo tail-skid, mounted into lower end of the rudder, was steerable in a limited range to help in ground maneuvering. The fabric covered tail-group was built up of welded chrome-moly steel tubing; the fin was ground adjustable and the horizontal stabilizer was adjustable in flight. A metal propeller was standard equipment. Navigation lights, wheel pants and a speed-ring engine cowling were available as optional equipment. The next development in the Inland "Sport"

monoplane was the model S-300-E as described in the chapter for ATC #342 in this volume.

Listed below are Inland "Super Sport model W-500 entries as gleaned from registration records:

NC-8088:	Model W-500	(#W-505)	Warner 110
NC-252N;		(# W-512)	,,
NC-265N;	,,	(# W-514)	,,
NC-263N;	,,	(# W-)	,,
NC-448V;	,,	(# W-516)	,,
NC-262N;	,,	(# W-517)	,,
NC-258N;	,,	(# W-518)	,,

Serial #W-505 first as model S-300-U (#S-305); serial #W-512 first as model S-300 (#S-312); serial number for NC-263N unknown; serial #W-516 later modified to an "S-300 Special."

A.T.C. #316
(4-16-30)
VERVILLE "AIR COACH", 104-P

Fig. 64. Verville "Air Coach" model 104-P mounted revolutionary Packard "Diesel" engine.

The slender and trim looking Verville "Air Coach" was yet another airplane model offered with the newly introduced Packard "Diesel" air-cooled engine and was one of the more handsome in this combination. As evident upon closer study, it contained many original and worthwhile features in its general makeup to set it somewhat apart from others in its class. More safety, better economy, and more practical operation were the promises held forth in this new version of the "Air Coach" (104-P) but actually, these features fell somewhat short of promises and the average accustomed standards. Mind you, the Verville "Air Coach" 104-P was still capable to fly against the very best, but it was hampered by the many fancied and real problems in this version with the new "Diesel" engine. The basic "diesel" principle of compression ignition, however meritorious, created hammering and teeth-chattering jolts through the airframe. That was annoying to say the least and if reports from the field are true, almost everything shook loose in the airplane in just a short while. Added to this, was the rather unpleasant odor of the "diesel fuel" (somewhat like kerosene) that permeated the very core of the airplane, it seems, with a stench that lingered in spite of all efforts to remove it. All this generally caused repeated unkind comments and complaints. However objectionable, this interesting engine with its new principle for aircraft did have its merits and could have been developed to more complete satisfaction had there been more interest and more time. Iron-

ically enough, Capt. Lionel M. Woolson, who fathered the development of this revolutionary engine in this country, lost his life in a crash during a severe storm while flying in the Diesel-powered Verville 104-P; however, it was certainly neither the fault of the staunch airplane, nor the engine, which caused the fatal crash. Reports establish that it was the unusual severity of the storm. Enthusiasm for further development and refinement of the engine naturally waned after the death of Woolson, and it soon was dropped entirely. Recorded evidence points to only one example of the Verville 104-P but it also had been reported that one was sold in Italy for use by the air-attache in the embassy. The "Air Coach" model 104-P was part of the Verville line on display at the annual Detroit Air Show for 1930, held on April 5-13 in the new million dollar hangar on Detroit's "City Airport"; 81 colorful aircraft were on display and customer interest was rather high in spite of the sagging economic conditions.

The Verville "Air Coach" model 104-P as powered with the 9 cyl. Packard DR-980 "diesel" engine was a high-winged cabin monoplane, with seating for 4, in quarters of ample room and comfort. An ingenious design in the fuselage structure eliminated the need for all unsightly trusses in the cabin and window area, with interior coach-work that rivaled the spaciousness and beauty of the finest automobile. Designed carefully with good aerodynamic proportion, the "Coach" was light and easy on the controls and quite delightful in its graceful

Fig. 65. Graceful "Air Coach" shows careful attention to detail which resulted in top notch performance.

behavior aloft. With 225 h.p. on tap, performance was more than adequate for this type of airplane, with a good margin of reserve power. With the unusual economy actually possible in the Packard "Diesel" engine, operating expenses were greatly reduced per mile and the cruising range was extended by a bonus margin. It was the belief in some circles that had the "Packard" oil-burning engine been sufficiently developed to overcome most of its objectionable characteristics, it is likely that it would have found a high place for itself in general commercial aircraft operations. The type certificate number for the Verville model 104-P was issued 4-16-30 and no more than one or two examples of this model were manufactured by the Verville Aircraft Co. at Detroit, Mich.

Listed below are specifications and performance date for the Verville "Air Coach" model 104-P as powered with the 225 h.p. Packard DR-980 engine: length overall 28'8"; height overall 7'11"; wing span 44'0"; wing chord 81"; total wing area 270 sq. ft.; airfoil Clark Y; wt. empty 2300 lbs.; useful load 1100 lbs.; payload with 60 gal. fuel was 490 lbs.; gross wt. 3400 lbs.; max. speed 130; cruising speed 110; landing speed 50; climb 900 ft. first min. at sea level; climb in 10 min. was 7000 ft.; service ceiling 18,000 ft.; fuel cap. 60 gal.; oil cap. 7 gal.; cruising range at 10 gal. per hour was 640 miles; price upon introduction was $12,000 reduced to

$11,000 later in 1930.

The construction details and general arrangement of the Verville 104-P were typical to that of the model 104-C as described in the chapter for ATC #267 of U. S. CIVIL AIRCRAFT, Vol. 3. Because of the extra weight of the Packard "Diesel" installation, no baggage was allowed when carrying four people. Cabin area was 92 x 41 in. and all windows were of shatter-proof glass. Dual wheel-type controls were provided with right hand wheel quickly removable when carrying a passenger in right hand front seat; the pilot had access to an unusually complete instrument panel. Landing gear tread was 96 in. with 32x6 wheels; Bendix brakes were individually operated by stirrups on the rudder pedals. A full-caster tail wheel was provided for better ground maneuvering. Cabin dome and instrument panel lights, navigation lights, parachute flares and retractable landing lights were also provided for night flying. A metal propeller, electric inertia-type engine starter and battery were also standard equipment. The next Verville development was the "Sportsman" model AT sport biplane, as described in the chapter for ATC #323, in this volume.

Listed below is the only known example of the Verville "Air Coach" model 104-P as gleaned from registration records:
NC-70W; Model 104-P (# 6) Packard DR-980.

A.T.C. #317
(4-24-30)
DAVIS, D-1-66 (D-1-85)

Fig. 66. Davis model D-1-66 with 85 h.p. Le Blond engine; dainty beauty complimented by its playful nature.

The petite and comely Davis "parasol monoplane" was an enchanting, and strictly for fun airplane that enjoyed good solid popularity. Not specifically designed for pilot training nor any other form of commercial drudgery, it was kept neat and trim, quite girlish in its form and temperament, and very well suited for the private owner-flyer — one who enjoyed owning his own airplane and was willing to take time to pamper it a little. With a little special care came familiarity, better understanding and fuller appreciation of its friendly nature and willingness to respond. Enthusiasm and pride among owners always ran high for this airplane, in all of its versions, and its charm most likely laid in the happy balance of the responsive performance, durability of structure and its clean practical lines. Available up to now in the 60 h.p. model D-1 and the 100 h.p. model D-1-K, the new model D-1-66 (or D-1-85) was introduced as an in-between model that mounted the new 5 cyl. LeBlond "66" engine. The new "66" was actually the 85 h.p. model 5DF engine that was greatly improved in its make-up and performance, and had been groomed to replace the earlier engines. With the new 85 h.p. LeBlond 5DF, the model D-1-85 (D-1-66) was attuned more to the average sport flier's needs; it posed well to be the most popular model in the "Davis" line-up. Perhaps the most famous example of this version was the petted and pampered D-1-85 that was used for air-racing by Art Chester. Winning often and always placing well, Chester and the Davis D-1-85 came to be a combination hard to beat, whether racing around a course of pylons or vying for a win in "spot" landings. In the National Air Races for 1932, Art Chester flew the Davis D-1-85 to Division 1 first place in the Sohio Mystery Derby. Davis "mono" owners were a happy lot and never seemed to get disgruntled about anything, so it leaves one to wonder why a lot more examples of this little charmer were not built. In recent years several of the "Davis" monoplanes were rebuilt to fly again; they are now among the most sought-after airplanes in this particular class.

The Davis model D-1-85 (D-1-66) was an open cockpit "parasol monoplane" seating two in tandem, an arrangement that was perhaps just a bit cramped for room, but the being-hugged feeling seemed to help create an intimate understanding between man and airplane. Pure flying pleasure was built into this slender craft and its maneuverability or response was only limited by the pilot's mood or his ability. For the sport-flyer who flew alone a good deal of the time, a metal panel cover was available to close off the front cockpit and eliminate the disturbance in the air flow over this portion of the fuselage. For those in search of a little more speed, whatever the incentive, wheel fairings were available and also a Townend type "speed ring" engine cowling. Thus bedecked with a faired-over cockpit, wheel pants and engine fairing,

Fig. 67. NACA cowling blended into lines of D-1-85 for greater speed.

the D-1-85 was good for nearly 125 m.p.h. Only 15 lbs. heavier empty than the model D-1, with allowance for 31 lbs. more useful load, and a gross weight that amounted to only 46 lbs. more, the D-1-85 was a sprightly performer with a substantial increase in practical utility for the average owner. The type certificate number for the Davis model D-1-85 (D-1-66) was issued 4-24-30 and some 4 or more examples of this model were manufactured by the Davis Aircraft Corp. at Richmond, Ind. Walter C. Davis was president; H. C. Davis was V.P. and Geo. W. Davis was secretary, making it a cozy family operation. There is evidence that Walter Davis might have contemplated a production model with the 5 cyl. "Lambert 90" engine because the prototype for the model D-1-85 (ser. #301), only a month earlier, had been approved as the model D-1-L on a Group 2 approval numbered 2-195. However, this ship was converted to the model D-1-85.

Listed below are specifications and performance data for the Davis model D-1-85 (D-1-66) as powered with the 85 h.p. LeBlond 5DF engine; length overall 21'6"; height overall 7'3"; wing span 30'3"; max. wing chord 63", wing chord mean 56"; total wing area 145 sq. ft.; airfoil "Goettingen" 387 modified at max. thickness tápering to "Clark Y" at root and tip; wt. empty 854 lbs.; useful load 526 lbs.; payload with 20 gal. fuel was 217 lbs. which included 2 parachutes at 20 lbs. each or 47 lbs. baggage; gross wt. 1380 lbs.; max. speed 112; cruising speed 97; landing speed 40; climb 780 ft. first min. at sea level; ceiling 12,000 ft.; gas cap. 20 gal.; oil cap. 2.5 gal.; cruising range at 5 gal. per hour was 360 miles; price at the factory originally quoted at $3595, reduced to $3395 and lowered to $2795 in April of 1931.

The construction details and general arrangement of the Davis model D-1-85 were similar to that of the models D-1 and D-1-K, as described in the chapters for ATC #256 and #272 in U. S. CIVIL AIRCRAFT, Vol. 3. The windshields were larger, for better protection, with option

Fig. 68. Art Chester and his D-1-85 were well-known on the racing circuits.

Fig. 69. Arrangement of D-1-85 was favored for sport-flying.

to close off front pit with a metal panel. The baggage compartment was located in the turtle-back section behind the rear cockpit with allowance for 47 lbs. Fairing over landing gear vees and streamlined wheel pants were optional equipment; also eligible with 6.50 x 10 semi-airwheels with brakes. A Hartzell wooden propeller was standard equipment but a metal propeller was optional. A narrow-chord "speed ring" engine cowling was available in 1930-31 but later could be replaced by the deep-chord N.A.C.A. type engine fairing. A direct-reading fuel gauge, wiring for navigation lights, and a spring-leaf tail skid were standard equipment. Other Davis monoplane developments were built under various Group 2 approvals; this would include models such as the D-1-L on 2-195 and the D-1W with Warner 110 engine on 2-394.

Listed below are Davis model D-1-85 (D-1-66) entries as gleaned from registration records:

NC-855N;	Davis D-1-85	(# 128)	LeBlond 5DF.
NC-647N;	"	(# 301)	"
NC-150Y;	"	(# 302)	"
NC-13546;	"	(#)	"

Serial #128 modified to D-1-85 from earlier model D-1; serial #301 first as model D-1-L with Lambert 90 engine; serial number unknown for NC-13546; approval for model D-1-85 expired on 7-15-33.

Fig. 70. Alexander "Bullet" model C-7 with 165 h.p. Wright J6-5 engine; revolutionary design was ahead of its time.

To most everyone, the mention of Alexander "Eaglerock" brings quickly to mind a large and graceful biplane soaring slowly but serenely on lightly-loaded wings, wings that could carry it either high amongst the mountain peaks or let it down gently into some valley pasture floor. It is then odd to associate the name of "Eaglerock" with an airplane such as the speedy "Bullet", a craft that had the misfortune of being much too far ahead of its time. Though practical minded enough to build and sell an all-purpose biplane that was well attuned to the particular period, Alexander engineers (and one Al W. Mooney in particular), spent many hours looking into the future of aviation and envisioned exciting things to come; the visions slowly translated themselves into a certain pattern and from this the "Bullet" concept was finally born. Introduced proudly to the flying public at large, during the National Air Races of 1929, it was viewed on exhibit and also in action, creating an interest the likes of which is rarely seen; entered in contest the speedy "Bullet" soon proved that it was a harbinger of things to come. Flown by Errett Williams, the "Bullet" (as powered with the 165 h.p. Wright J6 engine), finished first in the Philadelphia to Cleveland Derby leading the straining field by nearly an hour, then finishing first in the 720 cu. in. race with an average of 134.6 m.p.h., finishing first in the speed event

for the Aviation Town & Country Club Trophy at 133.75 m.p.h., and then averaging 132.4 and 134.01 m.p.h. in two of the Australian Pursuit races, where faster ships were handicapped by starting off last. Edith Foltz, a comely aviatrix from the great northwest, flew a Kinner-powered "Bullet" (100 h.p.) to 7th place overall and to 2nd place in class during the Woman's Derby from Santa Monica, Calif. to Cleveland, Ohio. Later in Oct. of 1930, Mrs. Keith Miller a spunky aviatrix whose talent certainly belied her tiny size, had acquired the "Bullett" so ably flown by Errett Williams, and flew it cross-country from New York to Los Angeles in 25 hours, 44 mins.; 3 days later she flew back to New York in 21 hours, 47 mins. Ordinarily, performance such as repeatedly demonstrated by the fleet flying "Bullet", would bring a flood of inquiries and orders and an airplane's future would be well assured, but the future of the Alexander "Bullet" soon took on mysterious aspects that cannot be fully explained. Occasionally there would be talk of troubles with the retracting landing gear, a mechanism not fully developed as yet, then there would be talk of tricky behavior and of unpredictable nature; this was all in the form of hangar-talk so it necessarily had to be taken with a grain of salt. The "Bullet" has also been accused of dangerous spin characteristics and to some extent this was

Fig. 71. "Bullet" fuselage profile shows airfoiled design.

true at first, but continued research and testing finally eased this situation to the point where it was more or less acceptable in the face of knowledge that had been compiled about this phenomenon up to this time. Pondering the point for just a moment, it seems a pity that the reputation thus hung upon the "Bullet", caused so many to shy away from it. Its career, that had looked so promising and rosy at first, was left hanging for the "depression" to finish off.

As shown here in prototype, the original "Bullet" was designed by Al W. Mooney, a revolutionary design that promised to set an entirely new standard in utility and performance; like most all pioneering efforts it was beset with its share of problems. Exhibited for the first time at the Detroit Air Show in April of 1929, enthusiastic interest heartened its creators to continue its development in hopes of licking the problems which were mostly mechanical in nature. Two more examples of the "Bullet", (one powered with the 5 cyl. Kinner K5 engine of 100 h.p. and one powered with the 5 cyl. Wright J6 engine of 165 h.p.) already showed improvement in the landing gear mechanism and some slight changes in the aerodynamic arrangement; it was these two craft that were so ably flown by Edith Foltz and Errett Williams at the 1929 National Air Races. Errett Williams, who had flown an OX-5 powered combo-wing "Eaglerock" biplane for a long while, took to the new "Bullet" with enthusiasm, and the transition from the one craft to the other must have been exhilarating to say the least. Al Mooney, versatile creator of the "Bullet" concept, tried hard to reshape his brain-child in preparation for certification tests but he left the firm after Ludwig Muther (formerly with Junkers and Fokker in

Europe and Keystone here in the U.S.A.) was called in to help on the project. Herr Ludwig Muther, often called the co-designer of the "Bullet", was apparently also hard put to solve some of the inherent problems. Later he worked in close co-operation with Dr. Max M. Munk, noted aerodynamicist, to finally arrive at an acceptable conclusion. A Group 2 approval numbered 2-181 was issued for the "Bullet" on 2-18-30, a type certificate number (ATC #318) for the "Bullet" was issued on 5-6-30, and by that time Dr. Munk had already left Alexander's employ to open a consulting office in New York City. The Alexander "Bullet" had now finally earned its certification but it was indeed a shallow victory because the aircraft market was going into a terrific slump and all the heart-rending preparation had been almost in vain. Although the Alexander "Bullet" had not yet reached a stage of development that could be called a complete success, there is no doubt that it left an indelible mark in the science of aircraft design and surely influenced continued research by others in designs of similar configuration; the "ghost" of the "Bullet" certainly rides with many low-winged cabin airplanes of latter-day design. The "Bullet", as originally conceived by Al Mooney, was a form later projected into the "Mooney" model AX and A-1 of 1930, and to some extent into the Mooney-designed "Monoprep" and "Monosport" low-winged monoplanes as developed by the Lambert Aircraft Corp. (formerly Monocoupe Corp.), a design that years later became known as the Culver "Dart."

Developed late in 1928 the "Bullet" was exhibited at the Detroit Air Show for 1929 where it was continuously the center of attraction. As first proposed, the "Bullet" series were

Fig. 72. Prototype "Bullet" with 100 h.p. Kinner K5 engine carried four.

to be offered in several sizes and were to be designated in rifle "caliber" numbers such as the .22, the .32, the .45 and so on; the model .22 was to be a two-seated version in typical form with an engine of about 60 h.p. The model .32 was the 4-place model with 150-165 h.p. engines and the proposed model .45 was to be a deluxe four-place version mounting engines of 300 h.p.; had the model .45 actually been developed it stood a good chance of offering speeds in the neighborhood of 200 m.p.h.!! Due to various problems in development the "Bullet" series was finally narrowed down to the four-place version with engines in the 165 h.p. range, offering the Axelson B-150 and the

165 h.p. Comet 7-E as alternate powerplants to the 5 cyl. Wright J6 (R-540) engine. Discarding the rifle caliber designation, the "Bullet" became the models C-1, C-3, C-5 and C-7; the model C-5 was the type as used by Errett Williams in the 1929 races and the model C-7 was the final version as approved on ATC #318, the first low-winged cabin monoplane to earn the "ATC" approval.

The Alexander "Bullet" model C-7 was a low-winged cabin monoplane of a futuristic proportion and arrangement that had comfortable seating for 4 people and a dog, plus all their baggage. Designed primarily to combat parasitic resistance, the low-winged form was

Fig. 73. Retracting landing gear on "Bullet" C-5 added considerably to top speed.

Fig. 74. "Bullet" model C-7 was culmination of two years testing; its appearance on airports of today would not be out of place.

chosen to provide wells in which to house the retracting landing gear; the wheels folded neatly into the underside of the wing and this alone boosted top speeds by as much as 18 m.p.h. The fuselage was also shaped into an "air-foiled" contour for low drag and added lift, but apparently the aerodynamic geometry of early models produced ineffectiveness in the controls during some attitudes and was blamed for the so-called bad spin characteristics; this was finally corrected to some extent by a substantial lengthening of the fuselage arm and a slight redesign of the tail group. Retracting landing gears were certainly not a new development because "amphibian airplanes" used this type of gear for several years, but a retracting gear that would be completely housed in the wing to cancel out all of its drag was something still quite new and that is where some of the trouble lie. Most of Alexander's problems with this airplane were simply a matter of mechanics and the gear was finally developed into an acceptable form, but the "Bullet" was finally offered with a fixed cantilever landing gear too. The fixed "gear" was highly streamlined by large metal fairings to reduce drag as much as possible, and the retractable landing gear was offered as optional; the difference between the two installations being about 12 m.p.h. With a cabin area of generous proportions there was ample room for four big people in surroundings of rich appointments; the abundance of window area provided excellent visibility in just about any direction. Performance with the 165 h.p. engine was exceptional by any standards of comparison and its overall efficiency was not matched by too many airplanes for several years to come. The type certificate number for the Alexander "Bullet" model C-7, as powered with the 5 cyl. Wright J6 engine, was issued 5-6-30 and some reports vaguely state that as many as 20 examples of the "Bullet" had been manufactured

by the Alexander Aircraft Co. at Colorado Springs, Colo.; however, registration lists fail to verify this. Genial J. Don Alexander was the president; D. M. Alexander was V.P. and general manager; J. A. McInaney was sales manager; Ludwig Muther was chief engineer and Garland Powell Peed, Jr. was test pilot, with occasional assignment as project engineer. It was pilot Peed who initiated the "Eaglerock" biplane into the select circle of commercial airplanes that had performed the difficult and hazardous outside loop.

Listed below are specifications and performance data for the Alexander "Bullet" model C-7 as powered with the 165 h.p. Wright J6 (R-540) engine: length overall 26'10"; height overall 7'9"; wing span 36'0"; max. chord at root 84"; total wing area 208 sq. ft.; airfoil "Goettingen" 387; wt. empty 1708 lbs.; useful load 1082 lbs.; payload with 41 gal. fuel was 625 lbs.; gross wt. 2790 lbs.; max. speed 148; cruising speed 125; landing speed 48; climb 700 ft. first min. at sea level; ceiling 15,000 ft.; gas cap. 41 gal.; oil cap. 5 gal.; cruising range at 9 gal. per hour was 540 miles; price at the factory field was $7200 reduced to $6500 in Aug. of 1930.

For comparison, the following figures are for the earlier model C-5 with 165 h.p. engine; length overall 21'6"; height overall 8'3"; wing span 38'7"; max. chord at root 75"; total wing area 202 sq. ft.; wt. empty 1302 lbs.; useful load 1098 lbs.; payload with 40 gal. fuel was 653 lbs.; gross wt. 2400 lbs.; max. speed 150 plus; cruising speed 128; landing speed 45; climb 838 ft. first min. at sea level; ceiling 15,000 ft.; gas cap. 40-70 gal.; oil cap. 5-7 gal.; cruising range 550-900 miles; price at factory $8888.

The version powered with 100 h.p. Kinner K5 engine was similar to model C-5 above, except for the following figures; wt. empty 1228 lbs.; useful load 1065 lbs.; payload with 40 gal. fuel was 620 lbs.; gross wt. 2293 lbs.; max. speed 130;

Fig. 75. Kinner-powered "Bullet" flown by Edith Foltz in Women's Air Derby of 1929.

cruising speed 111; landing speed 42; climb 640 ft.; ceiling 11,000 ft.; gas cap. 40 gal.; oil cap. 5 gal.; cruising range 550 miles; price at factory $6666.

The fuselage framework was built up of welded chrome-moly steel tubing faired to shape with formers and fairing strips, then fabric covered. The cabin walls were lined with Fabrikoid insulating material and the interior dimensions provided ample room for all; the upholstery was of genuine leather with trim, hardware and appointments of matching taste. Dual joy-stick controls were provided, with adjustable front seats for proper leg-room. The window area was designed for excellent vision from the pilot's station; a sky-light in the forward cabin roof offered visibility overhead and also served as a quick-release emergency exit. A large cabin door offered easy entry into the 48 x 114 x 48 in. cabin, and baggage was stowed in a cabin compartment of 4 cu. ft. capacity. The cantilever wing framework was in 3 sections, built up of spruce and plywood box-type spar beams and spruce truss-type wing ribs; the leading edges were completely covered back to front spar with plywood sheet, and the completed framework was covered in fabric. The center portion of the wing, of constant chord and thickness, fastened directly into the underside of the fuselage and outer portions of the wing of curving chord and tapering thickness were fastened to the central wing stub. The wing was a full cantilever design and required no external

bracing. Fuel tanks were mounted in the center portion of the wing and landing gear wheels retracted into wells on the underside. The fixed landing gear, which was optional, had a tread of 78 in. with cantilever oleo-legs that were encased in streamlined metal fairings; wheels were 30 x 5 and Bendix brakes were standard equipment. The fabric covered tail-group was built up of welded chrome-moly steel tubing in cantilever form; to "trim" for all conditions, the horizontal stabilizer was adjustable in flight through a rather extreme range of travel. A metal propeller, navigation lights, and hand crank inertia-type engine starter were standard equipment. The next development by Alexander was the flivver-type "Flyabout" monoplane model D-1 as described in the chapter for ATC #439.

Listed below are the only known Alexander "Bullet" entries of all models as gleaned from registration records:

R-8228;	Bullet C-1	(# 2001)	Kinner K5-100.
R-8227;	" C-3	(# 2002)	J6-5-165.
NR-700H;	" C-5	(# 2004)	J6-5-165.
R-705H;	" C-5	(# 2005)	Kinner K5-100.
NC-309V;	" C-7	(# 2013)	J6-5-165.

Group 2 approval numbered 2-181 first issued 2-18-30, superseded 5-6-30 by ATC #318; model designations and serial numbers for R-8227 and R-8228 not thoroughly verified; no explanation for missing serial numbers up to #2013.

BELLANCA "SKYROCKET", CH-400

Fig. 76. Bellanca "Skyrocket" model CH-400; 420 h.p. "Wasp" engine unleashed exciting performance.

The efficient Bellanca "Pacemaker" mono-plane was certainly most appropriate in its standard form for the smaller air-lines, the fixed base operators, and most of the business houses; however, there was coming now a need for a basically similar craft with a somewhat higher performance, a performance that would not be held to sensible operating costs, nor anything else that would stifle its ability "to rare up and go". There was a limited buyer group, a small band who were craving a craft similar to the pop-ular "Pacemaker", but they wanted an airplane of much more sporting nature and were very willing to pay extra for what they had to have. It was for this group that Bellanca developed the powerful "Skyrocket" and in judging from the bouquets tossed about by enthusiastic owners, it must have been just what they wanted. The new "Skyrocket" (CH-400) was not necessarily a special design because the "Pacemaker" (CH-300) concept was entirely suitable to the need. They only added a little more muscle to the frame and more power to drive it, to come up with a combination that shared most of the at-tributes contained in the CH design, plus the bonus features that were possible only through the application of more horsepower. Of course, the addition of 120 extra horsepower did not show up greatly in more available speed, but take-off time was substantially shortened, climb-out was increased by an exhilarating amount,

and general behavior was sporting enough to warm the cockles of any good man's heart. George Haldeman, well-known promotional pilot for Bellanca, was so proud and so con-fident of the ships he was demonstrating on a flight that most often he would feel compelled to do a little something extra to prove their capa-bilities; usually he would "stall" a few times or let it "spin" a few turns, just to show effortless and prompt recovery. Often he added a few "loops", a few "barrel-rolls", or would even fly inverted for a time, to show the maneuverability and response that was built into the "Skyrocket" These were capabilities, perhaps never used, but it was quite comforting to know the ability was there if ever needed. Owners of the thun-dering "Skyrocket" were generally well-heeled businessmen and sportsman-pilots; that roster included the Bendix Research Div., Western Telephone of Nebraska, the Shell Petroleum Co., the May Co. of Los Angeles, A. O. Smith Corp., the National Battery Co., machinery companies, and such well known "sports" as Roger Wolfe Kahn, demure Ann Harding, and lovable Wal-lace Beery of the Hollywood films.

The Bellanca "Skyrocket" model CH-400 was a high-winged cabin monoplane of compact proportion with seating for 6 in ample comfort and tasteful surroundings. Ordinarily there would be very little to get excited about in a large cabin monoplane such as this, but the Bel-

Fig. 77. "Skyrocket" popular for sport; ship shown owned by Ann Harding of movies.

lanca "Skyrocket" was indeed an exception because in spite of its comfort and rather plush appointments, it had plenty of performance to please the owner who harbored sporty inclinations. More or less a custom built airplane, no reasonable expense was spared to make it the finest in its class, so it was liberally equipped with operating accessories and flying aids. The Colorado-Utah Airways had need for carriers of rugged nature and high performance because of the difficult terrain and high operating altitudes they were obliged to put up with; the highest point on the route was the 11,382 foot "Monarch Pass". It is interesting to note that they selected the Bellanca "Skyrocket" for the job. Powered with the 9 cyl. Pratt & Whitney "Wasp" C-1 engine of 420 h.p., or the SC-1 of 450 h.p., the "Skyrocket" was not particularly blessed with a whole lot of extra speed by comparison, but its shorter take-offs, steeper climb-out and plenty of pep and spice in its general behavior were the extras that made this ship such a favorite. With a pilot alone the "Rocket" could climb out at 1850 feet per min. and that's really getting up there in the blue. The CH-400 would step out at an honest 150 m.p.h. and this could be raised by some 5 m.p.h. if a "speed-ring" cowling was mounted over the engine; wheel pants accounted for at least another 8 m.p.h. increase so the 'Rocket, in its deluxe version, was stepping off 160 m.p.h. or more. Available also as a seaplane, on Edo K twin-float gear, the CH-400 "Skyrocket" was not greatly hampered by the bulky pontoons and performed very well in this combination also. Approval for the first "Skyrocket" was issued 4-30-30 on a Group 2 certificate numbered 2-205 and the type certificate number was

Fig. 78. "Skyrocket" Deluxe owned by Wallace Beery, movie-actor who was famous flying sportsman.

Fig. 79. "Skyrocket" on Edo floats was high-performing seaplane.

also issued 4-30-30 for serials #602 and up; 3 other CH-400 examples of special nature were also on Group 2 approvals and a few examples were purchased by the U.S. Marines air-arm as the XRE. A total of 32 examples of this model, in several variants, were manufactured by the Bellanca Aircraft Corp. at New Castle, Delaware.

Listed below are specifications and performance data for the Bellanca model CH-400 "Skyrocket" as powered with the 420 h.p. "Wasp" C-1 engine: length overall 27'10"; height overall 8'4"; wing span 46'4"; wing chord 79"; total wing area 273 sq. ft.; airfoil "Bellanca" (modified R.A.F.); wt. empty 2592 lbs.; useful load 2008 lbs.; payload with 120 gal. fuel was 1053 lbs.; gross wt. 4600 lbs.; max. speed 155; cruising speed 134; landing speed 55; climb 1250 ft. first min. at sea level; ceiling 20,000 ft.; gas cap. 120 gal.; oil cap. 8 gal.; cruising range at 22 gal. per hour was 670 miles; price at the factory field was $17,800 raised to $17,950 in April of 1931. The following figures are for the seaplane as fitted with Edo K twin-float gear; length overall 30'8"; wt. empty 3012 lbs.; useful load 1988 lbs.; payload with 120 gal. fuel was 1033 lbs.; gross wt. 5000 lbs.; max. speed 146; cruising speed 125; landing speed 60; climb 900 ft. first min. at sea level; ceiling 18,000 ft.; price at the factory was $21,000 raised to $21,150 in April of 1931. The CH-400, both landplane and seaplane, was also available with "Wasp" SC-1 of 450 h.p. so performance figures would be proportionately better.

The fuselage framework was built up of weld-

Fig. 80. "Skyrocket" in Navy service, labeled XRE-2.

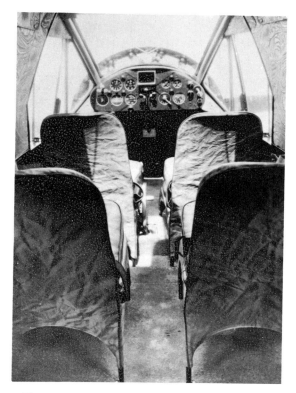

Fig. 81. Interior of Ann Harding's "Skyrocket" shows woman's touch.

ed chrome-moly steel tubing into an exceptionally rigid structure, faired to shape with spruce formers and fairing strips then fabric covered. Custom interiors were arranged on order and often included fancy curtains on the windows; windows slid open for ventilation and cabin heat was available. Entry to the cabin was gained through a large door and convenient step on either side; the 12 cu. ft. baggage compartment was to the rear with allowance of 208 lbs. for landplane and 188 lbs. for seaplane. The baggage was accessible from the inside or out. The wing framework in two halves, was built up of solid spruce spars that were routed to an I-beam section, with spruce truss type wing ribs; the completed framework was covered in fabric. The fuel tanks were mounted in the root end of each wing half, and the wings were braced to the fuselage by the distinctive "Bellanca" lift struts that contributed some 47 sq. ft. to the lifting area. The landing gear of 90 in. tread was fastened to a cantilever portion of the front wing strut using oleo-spring shock absorbing struts; wheels were 32x6 and Bendix brakes were standard

equipment. The fabric covered tail-group was a combination wood and steel tube structure; the horizontal stabilizer was adjustable in flight. A metal propeller, navigation lights, tail wheel, wheel brakes and inertia-type engine starter were standard equipment. The next "Skyrocket" development was the "Model D" as described in the chapter for ATC #480. The next development following the CH-400 was the model 300-W "Pacemaker" as for ATC #328 in this volume.

Listed below are various Bellanca "Skyrocket" model CH-400 entries as gleaned from registration records:

NC-179N; Model CH-400 (# 601) P & W Wasp.
NC-544V;   "   (# 602)   "
NC-545V;   "   (# 603)   "
NC-546V;   "   (# 604)   "
NC-547V;   "   (# 605)   "
NC-548V;   "   (# 606)   "
NC-10E;   "   (# 607)   "
NC-11E;   "   (# 608)   "
   ;   "   (# 609)   "
NC-10293;   "   (# 610)   "
NC-10294;   "   (# 611)   "
NC-751W;   "   (# 612)   "
NC-752W;   "   (# 613)   "
NC-753W;   "   (# 614)   "
NC-754W;   "   (# 615)   "
NC-778W;   "   (# 616)   "
NC-779W;   "   (# 617)   "
NC-780W;   "   (# 618)   "
NC-781W;   "   (# 619)   "
NC-10793;   "   (# 620)   "
NC-10794;   "   (# 621)   "
NC-10795;   "   (# 622)   "
NC-11661;   "   (# 625)   "
NC-12635;   "   (# 626)   "
NC-12636;   "   (# 627)   "
XB-XUY;   "   (# 631)   "

Serial #601 on Group 2 approval numbered 2-205; serial #603 unverified, may have later been CF-AOA in Canada; serial #605 on Group 2 approval numbered 2-226; serial #606 on Group 2 approval numbered 2-213; serial #607 on twin-float gear; registration number for serial #609 unknown; serial #621 later as NPC-24; registration number for serial #623-624 unknown; serial #625 unverified; registration for serial #628-629-630 unknown, may have been XRE with U.S. Navy or Marines; serial #631 (XB-XUY) operated in Mexico, was for sale in 1958 with 11,000 hours on airframe and still going strong; registration number for serial #632 unknown.

Fig. 82. "Fleetster" model 20 with 575 h.p. "Hornet" B engine, shown here as all-cargo version.

Already having some service experience with the smaller all-wood Lockheed "Air Express," and apparently quite intrigued with the capabilities of the "parasol wing" configuration, the N.Y.R.B.A. Line ordered several of the new "Fleetster" model 20 airplanes for service on their South American routes in Brazil and Argentina. By comparison, the Consolidated "Model 20" was a much bigger airplane than the "Air Express" type by Lockheed; it was longer, it was taller, the cigar-shaped fuselage was larger in diameter and it was heavier by nearly a ton. Actually easily done, the basic design of the "Fleetster" model 17 was used to make up the new "Model 20;" the internally braced cantilever wing was raised above the fuselage by about a foot, the pilot was moved from in front to an open cockpit further aft in the fuselage, and the area usually occupied by the pilot was then converted into a large cargo bin. The passenger compartment was still in the central portion of the body with plenty of room for 5, and all in all, this arrangement was a practical way to gain extra volume for heavier payloads without resorting to redesign of the fuselage structure. Capable of handling varying loads of cargo and passengers the "Fleetster" 20 was normally arranged in 3 different versions: one example — minus all seating — was fitted out as an all-cargo carrier; one example carried 3 passengers and over 800 lbs. of cargo; and one example (as eligible on this certificate) carried 5 passengers and 475 lbs. in cargo, a certain portion of which was allowed for passenger baggage. The "Fleetster" 20 was also convertible for operating on land or sea with

"float" fittings built into the fuselage structure; whether they were operated regularly on twin-float gear has not been verified. In 1931, the Pan American Airways System operated two of the "Fleetster" 20 they had bought previously from the defunct N.Y.R.B.A. Line, and later bought another from the factory. In more or less regular use up to some time in 1934, two of these high-winged "Fleetsters" were then reported scrapped and one was reported as sold to Aerovias. The fourth ship in this series was operating in Canada. Operating in neighboring countries, and being quite rare in number too, the early "Model 20" was not well known nor often seen in this country; however a later improved version called the model 20-A was used extensively here and became a familiar sight in various parts of the land.

As shown, the Consolidated "Fleetster" model 20 was more or less a "parasol" winged version of the flashing "Model 17," an arrangement that was somewhat better suited, with extra volume, for larger mixed payloads. In the version called "Model 20, Type 2," the "Fleetster" seated 5 passengers in the spacious cabin with provisions in a bin up forward for nearly 500 lbs. of cargo; rearrangement of the fuselage lay-out placed the pilot far aft in an open cockpit which was still preferred by a good many pilots over the closed-in cabin. As on the Model 17, the "Fleetster" 20 was equipped with a deep chord N.A.C.A. type low-drag engine cowl, and streamlined "wheel pants" were also available; thus fitted, the Model 20 was able to sustain a top speed of 175 m.p.h. with ease. Compared

Fig. 83. Cigar-shaped fuselage and cantilever wing of "Fleetster" kept drag at a minimum.

to the "cleaner" Model 17, there was some loss in maximum speed, due to the extra drag of the elevated wing and gaping open cockpit, but stability was slightly improved and the big "Parasol-20" was described as a pleasant airplane to fly. Powered with the big 9 cyl. "Hornet B" (R-1860) engine of 575 h.p., its all-round performance was rather impressive. Reserve power was available for those extra demands experienced over some of the difficult terrain in South America and in Canada. With the built-in capacity for extra paying loads, the "Fleetster" 20 was doing well in service but unsettled economic conditions and newer equipment finally forced it to the side-lines. The type certificate number for the "Fleetster" model 20, Type 2 was issued 5-7-30 and 2 examples were built in this version; 2 more of the Model 20 in a 1-4 place version called Type 1 were built under a Group 2 certificate numbered 2-231, manufactured by the Consolidated Aircraft Corp. at Buffalo, N.Y. Rueben Hollis Fleet was president and general manager; Lawrence D. Bell was V.P. & asst. gen. mgr.; I. M. Laddon, designer of the "Fleetster" series, was the chief engineer. Consolidated Aircraft had come a long way since its uncertain beginning in 1923 and now just 7 years later it was one of the largest aircraft manufacturing units in the U.S.A.

Listed below are specifications and performance data for the "Fleetster" model 20 as powered with the 575 h.p. Hornet B engine: length overall 31'9"; height overall 10'2"; wing span 45'0"; wing chord at root 106"; wing chord at tip 68"; total wing area 313.5 sq.ft.; airfoil "Goettingen" 398 modified; wt. empty 3439 lbs.; useful load 2461 lbs.; payload with 145 gal. fuel was 1325 lbs. (payload includes five passengers at 170 lbs. each and 475 lbs. cargo-baggage); gross wt. 5900 lbs.; max. speed (N.A.C.A. cowling & wheel pants) 175; cruising speed 148; landing speed 65; climb 1000 ft. first min. at sea level; climb to 10,000 ft. was 15 mins.; ceiling 17,500 ft.; gas cap. 145 gal.; oil cap. 12

gal.; cruising range at 32.5 gal. per hour was 625 miles; price at the factory was $30,413 fully equipped.

The fuselage framework was a semi-monocoque metal structure of circular section that was built up of duralumin annular rings and bulkheads, longitudinal stringers and stiffeners, to which was riveted a smooth aluminum alloy outer skin. The main cabin area for the seating of 5 was arranged with two full-width seats that could accommodate 3 across; the seats were in a "chummy" arrangement where the front seat faced backwards and the rear seat faced forward. Three passengers sat in front and two sat in back, all facing each other, very much like the arrangement in an old stage-coach. The pilot's open cockpit aft was deep and well protected with provisions for cockpit heat from the engine's exhaust system. A cargo compartment of 60 cu. ft. capacity was in front of the main cabin with allowance for up to 475 lbs. The main cabin section was 54" high x 62" wide x 80" long and all seating could be quickly removed for clear floor space to handle over a half ton of cargo. The cantilever wing framework, suspended above the fuselage by two heavy N-type struts, was built up of two heavy-sectioned box type spar beams and 2 auxiliary spar beams of spruce and plywood; spruce and plywood compression members and truss-type wing ribs completed the structure, which was then covered with plywood sheet. Two fuel tanks were mounted in the center portion of the wing and retractable landing lights were mounted out near the tips. The split-axle landing gear of 96 in. tread was of the long leg type using oleo-spring shock absorbing struts; wheels were 36x8 and Bendix brakes were standard equipment. The large fixed tail wheel was mounted deep in the fuselage and encased with a streamlined metal boot; large hand-rails on either side were provided to lift the tail around. The all-metal tail group was built up of dural box-type spars and channel former ribs covered with a

Fig. 84. Fleetster model 20 also used in Canadian service.

riveted aluminum alloy skin; the metal skin covering was stiffened with shallow corrugations at two-inch intervals. Of unsymmetrical shape, to offset engine torque, the metal fin was built integral to the fuselage and the horizontal stabilizer was adjustable in flight. A metal propeller, navigation lights, battery and generator, and electric inertia-type engine starter were standard equipment. Complete night-flying equipment, parachute flares, radio receiver and radio shielding, 35x15 low pressure wheels and tires, and wheel streamlines were available as optional equipment. The next Consolidated development was the rare Wright "Cyclone" powered "Fleetster" model 17-C as described in the chapter for ATC #369 in this volume.

Listed below are both versions of the "Fleetster" model 20 as gleaned from registration records:

NC-673M; Model 20, Type 1 (#1) Hornet B-575.
NC-674M;   ''   Type 1 (#2)   ''
CF-AIP;   Model 20, Type 2 (#3)   ''
NC-675M;   ''   Type 2 (#4)   ''

Serial #1-2 were built under Group 2 approval numbered 2-231; serial #1 later converted to Type 2; serial #3 operated in Canada; serial #4 later operated by Aerovias; all Model 20 also available with 575 h.p. Wright "Cyclone" engine but none recorded in this combination.

A.T.C. #321
(5-6-30)
STAR "CAVALIER", MODEL E

Fig. 85. Star "Cavalier" model E with 90 h.p. Lambert engine.

The Star "Cavalier" monoplane as introduced for 1930 was an example of some improvement over the earlier models. Building around a basic structure, already well proven in service, the new lines were fashioned into more rugged character with an eye especially for added utility and somewhat better performance. Now was the Model E, this new "Cavalier" was powered with the 5 cyl. Lambert R-266 engine of 90 h.p. and the healthy increase in horsepower was surely enough to alter its former personality. The new "Cavalier" was favored by private owners, and quite popular as a stand-by air taxi for various business people. Because of its better performance, it was especially popular in the oil-fields of the mid-west, where it could be seen operating out of fields that seemed nearly impossible to reach. Purposely void of extra fancy finery and other intricate trappings of small actual value, the "Cavalier E" was a practical craft that was able to perform all sorts of every-day duties, and as was once said about it, it didn't mind getting its face dirty; a willing and useful companion for people who had work to do.

Designed by Wm. D. "Billy" Parker and E. A. "Gus" Riggs the Star "Cavalier" had been in production since late in 1928; first introduced with the 55 h.p. Velie engine as the model B, about 15 examples were built under ATC # 138 (refer U. S. CIVIL AIRCRAFT, Vol. 2). About 6 months later a companion model was introduced as the model C; this version was powered with the 60 h.p. LeBlond 5D engine and about two of these were built under ATC # 255 (refer U. S. CIVIL AIRCRAFT, Vol. 3). One version, powered with the 4 cyl. Wright-Gipsy engine of 90 h. p., made its appearance at the National Air Races for 1929 held at Cleveland, O. This was an experimental craft and no production was planned for this particular type. Later in 1929 several manufacturers were showing an interest in the 5 cyl. Armstrong-Siddeley "Genet" (British) engine of 80 h.p.; a "Cavalier" version was developed, mounting this engine, as the Model D and 2 examples were built under Group 2 approval numbered 2-191. One example of the "Cavalier" was also powered with the 5 cyl. Warner "Scarab Jr." engine of 90 h. p. but this was also an experiment only, and no approval was planned for this version. The "Cavalier" model E was built in about 13 examples into 1931. The manufacture of this model must have been continued for several years afterwards because about 20 more examples were reported built by company sources, up to the time the approval (ATC # 321) had expired on 9-30-39.

The Star "Cavalier" model E was a small high winged cabin monoplane with side by side seat-

Fig. 86. Oil companies used "Cavalier" E for duty in the fields.

ing for two, similar to earlier models, but arranged in a new form of slightly more buxom proportions. Though basically typical of previous "Cavalier" versions, the model E was altered in fuselage form for better air-flow, the tail group was redesigned for better control (especially at the lower speeds), and a wide tread landing gear with "oleo" shock absorber legs was fitted to better cope with unimproved surfaces usually encountered on small pasture-airports. Powered with the 90 h. p. Lambert (R-266) engine, the "Model E" inherited plenty of eager get up and go that translated into good climb-out from short fields, and a fair cruising speed that allowed more ground coverage in a day's flying time. Most of the "Cavaliers" were based in portions of the mid-west, so many people did not have an opportunity to see one close. Despite its absence in other parts of the country, the "Cavalier" enjoyed a good reputation anywhere. Manufactured by a company that could hold itself together in spite of small sales volume, the "Cavaliers" were built unhurriedly, and ample time was alloted even to the smallest detail. The type certificate number for the "Cavalier" model E was issued 5-6-30 and some 13 examples of this model were built into 1931 by the Star Aircraft Co. at Bartlesville, Okla. John H. Kane was president; Wm. D. Parker was V. P., general manager and chief pilot; John Bain was secretary-treasurer, and Charles Peters had now replaced E. A. "Gus" Riggs as the chief engineer. The Star Aircraft Company's directors were also officers of the Phillips Petroleum Co., one of the largest independent oil companies in the U. S. A. Several "Cavalier" monoplanes were gainfully employed by "Phillips", and other oil companies in the area followed suit.

Listed below are specifications and performance data for the Star "Cavalier" Model E

as powered with the 90 h. p. Lambert engine; length overall 20'1"; height overall 6'8"; wing span 31'6"; wing chord 61"; total wing area 157 sq. ft.; airfoil Clark Y; wt. empty 919 lbs.; useful load 506 lbs.; payload with 22 gal. fuel was 185 lbs.; gross wt. 1425 lbs.; max. speed 115; crusing speed 98; landing speed 40; climb 850 ft. first min. at sea level; ceiling 16,000 ft.; gas cap. 22 gal.; oil cap. 2.5 gal.; crusing range at 5.5 gal. per hour was 355 miles; price at the factory was $3450. The following figures are for a later version with 142 sq. ft. of wing area, speed-ring engine cowling and wheel pants; max. speed 126; crusing speed 108; landing speed 45; all other figures remained more or less the same.

The construction details and general arrangement for the "Cavalier" model E were similar to the model B as described in the chapter for ATC #138 of U.S. CIVIL AIRCRAFT, Vol. 2, and the model C on ATC #255 of Vol. 3, except for the following changes; the welded chrome-moly steel fuselage was faired out to a deeper cross-section by wooden formers and fairing strips then fabric covered; window area was slightly increased for better vision and a larger door on the right side provided easy entry; a baggage shelf behind the seat had allowance for 15 lbs.; rudder pedals were dual and the single control-stick was provided with a cross-over extension for dual control. The wing framework was typical with an 11 gal. fuel tank mounted in each wing half, flanking the fuselage. Wing bracing struts were now of heavy-gauge steel tubes in a streamlined section instead of the round tubes encased in fairings. The long legged landing gear of 84 in. tread used oleo shock absorbing struts instead of spools of rubber bungee cord; wheels were 24x4 or 26x5 and Bendix brakes were standard equipment. A metal propeller, navigation lights, a speed-ring engine cowling, wheel pants and custom colors were offered as optional equipment.

changes: the welded chrome-moly steel fuselage was faired out to a deeper cross-section by wooden formers and fairing strips then fabric covered; window area was slightly increased for better vision and a larger door on the right side provided easy entry; a baggage shelf behind the seat had allowance for 15 lbs.; rudder pedals were dual and the single control-stick was provided with a cross-over extension for dual control. The wing framework was typical with an 11 gal. fuel tank mounted in each wing half, flanking the fuselage. Wing bracing struts were now of heavy-gauge steel tubes in a streamlined section instead of the round tubes encased in fairings. The long legged landing gear of 84 in. tread used oleo shock absorbing struts instead of spools of rubber bungee cord; wheels were 24x4 or 26x5 and Bendix brakes were standard equipment. A metal propeller, navigation lights, a speed-ring engine cowling, sheel pants and custom colors were offered as optional equipment.

Listed below are "Cavalier" model E entries as gleaned from registration records published through 1931:

NC-350V; Model E (# 121) Lambert 90.
NC-397V; Model E (# 122) "

Fig. 88. Buxom lines of "Cavalier" E harbored rugged frame for hard service.

NC-71W;	Model E (# 123)	"
NC-678W;	" (# 124)	"
NC-636W;	" (# 125)	"
NC-9E;	" (# 126)	"
NC-13E;	" (# 127)	"
NC-10359;	" (# 128)	"
NC-10535;	" (# 129)	"
NC-10583;	" (# 130)	"
NC-10585;	" (# 131)	"
NC-980N;	" (# 132)	"
NC-11007;	" (# 133)	"

Fig. 87. "Cavalier" E later used speed-ring cowling and wheel pants to improve performance.

A.T.C. #322
(5-9-30)
STEARMAN "SENIOR SPEEDMAIL",
4-EM

Fig. 89. Stearman "Sr. Speedmail" model 4-EM with 420 h.p. "Wasp" engine.

As pointed out in previous discussions on various models in the "Stearman 4" series, the "Junior Speedmail" was basically designed to serve as a high performance sport-plane, with inherent qualities easily adapted to various commercial uses. It is easy to see that the design was particularly slanted to possible use by air mail carriers and it was no great problem to convert the forward space, normally taken up by a passenger cockpit, into a hatch-covered cargo hold. A 600 lb. paying load was more than ample for the average mail-run, and the bonus performance that was available guaranteed slightly faster schedules with a utility and every-day reliability which assured profitable operations. Stemming directly from the sport model 4-E, as described in the chapter for ATC # 292 of U. S. CIVIL AIRCRAFT, Vol 3 the new model 4-EM (the letter M denoting coversion to mail-carrier) was basically typical in all respects, except for the fuselage modifications necessary to adapt it to the transport of mail; in this configuration the model 4-EM was labeled the "Senior Speedmail". According to scant records on this model that lead the historian to some reading between the lines, the 4-EM was built in only a few

examples; all of these were eventually operated in Canadian mail service. Powered with the 420-450 h.p. Pratt & Whitney "Wasp" engine, the "Stearman 4-EM" was capable of high performance from and over difficult terrain so its selection for operation in Canada, where the going was always rough, was quite understandable.

The Stearman model 4-EM "Senior Speedmail" was a fairly large open cockpit biplane, with flowing lines of beautiful proportions that were carefully engineered to take advantage of the "streamlining" properties in the large N. A. C. A. engine cowling. With seating for a pilot only, the space ahead of the cockpit was converted to two fire-resistant mail pits with locking metal hatch covers to protect the payload from the elements. With elimination of the drag-producing open cockpit as in the 3-place 4-E, the smooth forward fuselage of the 4-EM was often credited with a bit more speed. With 450 horses in the big "Wasp" engine up front, the 4-EM had plenty of power reserve to get off the ground quickly and a climb-out steep enough to warm the heart of any pilot who had to sweat out immovable obstacles directly in his course.

Fig. 90. Popular as high-performance sport plane, model 4-E was also popular for hauling mail.

Typical of earlier models, and especially those in the "Stearman 4" series, the 4-EM was blessed with amiable flight characteristics, and its general behavior was a boon to the pilot; the mail pilot usually had a difficult task before him and appreciated an airplane that would tend to help him in his chore. Of the 5 examples that are shown in records, 3 of the model 4-EM were converted from the 4-E sport and 2 examples were no doubt primarily built as the model 4-EM

Fig. 91. Rugged structure and high performance well-suited for hard service; 4-E series was popular in bush country.

for mail service in Canada. The type certificate number for the "Senior Speedmail" model 4-EM, as powered with the 450 h. p. "Wasp" engine, was issued 5-9-30 and it was manufactured by the Stearman Aircraft Co. at Wichita, Kansas.

Listed below are specifications and performance data for the Stearman model 4-EM as powered with the 450 h. p. "Wasp" engine: length overall 26'4"; height overall 10'2"; wing span upper 38'0"; wing span lower 28' 0"; wing chord upper 66"; wing chord lower 51"; wing area upper 204 sq. ft.; wing area lower 103 sq. ft.; total wing area 307 sq. ft.; airfoil Goettingen 436; wt. empty 2455 lbs.; useful load 1481 lbs.; payload with 106 gal. fuel was 600 lbs.; gross wt. 3936 lbs.; max. speed 160; cruising speed 134; landing speed 55; climb 1400 ft. first min. at sea level; ceiling 18,000 ft.; gas cap. 106 gal.; oil cap. 10 gal.; cruising range at 23 gal. per hour was 600 miles; price at the factory field was approx. $16,000. Models 4-EM with serial # 4010 and # 4028 were eligible with 3978 lbs. gross weight but this slight difference (42 lbs.) was hardly enough to alter the performance figures as listed above.

The construction details and general arrangement of the mail-carrying model 4-EM were typical to that of the sport model 4-E as described in the chapter for ATC # 292 of U. S. CIVIL AIRCRAFT, Vol. 3, except for the following notations. Being converted to the carrying of air-mail and cargo, the model 4-EM had two cargo holds forward of the pilot's cockpit that were metal-lined for fire resistance. Access to the cargo holds was by two locking type metal hatch covers; one hatch cover was immediately in front of the pilot's cockpit and the other hatch cover was forward of that in a position between the center-section struts. A 5 cu. ft. compartment in back of the pilot's cockpit was available for the pilot's personal effects. A Hamilton-Standard metal propeller, hand-crank inertia type engine starter, navigation lights, and wheel brakes were standard equipment. It is quite likely that an exhaust-type heater muff was also available to warm the pilot's cockpit during cold weather operations. A full complement of night-flying equipment was also available at option. The next development in the "Stearman 4" series was the model 4-CM as described in the chapter for ATC # 325 in this volume.

Listed below are Stearman model 4-EM entries as gleaned from various records:

NC-770H; Model 4-EM (# 4010) Wasp 420-450.
CF-AMB;   "   (# 4016)   "
CF-AMC;   "   (# 4017)   "
CF-ASE;   "   (# 4028)   "
CF-ASF;   "   (# 4029)   "

Serial # 4010-4016-4017 converted from model 4-E; Canadian registration number for serial # 4010 (Canadian Airways) unknown; this approval expired 9-30-39.

A.T.C. #323
(5-19-30)
VERVILLE "SPORTSMAN", AT

Fig. 92. Verville AT "Sportsman" with 165 h.p. Continental A-70 engine; beautiful lines typical of all Verville designs.

The trend for the development of the 2 place sport-trainer biplane, from a year or so earlier, continued steadily but was now taking a turn towards the more powerful craft, which was better capable of the performance that was singled out mainly for the sportsman. It needed to be better capable, too, of a maneuverability more attuned to requirements of flying schools that followed through with their pilot training into the more rigid and complex secondary phases. One of the first, and one of the finest examples to be designed in answer to this newer trend, was the Verville model AT, a craft meant specifically for the hard-flying sportsman and certainly ideal for the secondary phases of any flying school. Typical of "Verville" design, the "Sportsman" biplane was an airplane of beautiful proportion with well-rounded lines of feminine quality, a shapely grace that just barely disguised the vibrant muscle and robust nature underneath. While blessed with a classic beauty that would indeed kindle pride in the breast of any sportsman, the Verville AT was certainly no namby-pamby, and could deliver a rousing performance that was an equal to the very best. Equipped quite thoroughly, and appointed tastefully to suit the discriminating owner (one who usually bore the label of "sportsman-pilot"), the Verville "Sportsman" of necessity carried a rather high price tag for these lean times. It was only because of this that sales were fairly scant and the true potential of this versa-tile craft was never fully realized. Almost paralleling the development of the commercial "Sportsman" AT, was a similar tandem-cockpit biplane version that was specifically designed by Verville to government specifications as a primary trainer; the first of these became the YPT-10 and four were delivered to the Air Corps. The model YPT-10 was powered with the Wright "Five" (R-540) engine of 165 h.p.; one of these was converted to the YPT-10A powered with the new Continental A-70 (R-545) engine of 165 h.p. Another modification mounted the "geared" Wright 5 (GR-540) engine of 165 h.p. as the YPT-10B; still another modification mounted the 170 h.p. Kinner (R-720) engine as the YPT-10D. Three of the four original YPT-10 were later modified to mount the 180 h.p. Lycoming (R-680) engine and these became the YPT-10C. The trend of the higher powered two-place open cockpit biplane, designed specifically for the sportsman-pilot, continued in spasmodic success with the advent of fairly typical models from several different manufacturers; by 1934 this trend was all but ended by a new set of values and changing requirements set up by that relentless tide, the buying public.

Alfred Victor Verville who designed his first airplane in 1915, is perhaps best known and best remembered for his inventive genius that sponsored the design of the "Verville-Packard" and the "Verville-Sperry" racers — racing airplanes that won two of the fabulous "Pulitzer

Fig. 93. Spirited nature and high performance of Verville AT ideal for the sportsman.

Races" of the early "twenties". As chief of design for the Engineering Div. of the old Air Service, Verville was largely instrumental in the development of many aeronautical innovations that set the pattern for future designs to come. As a founder of the Buhl-Verville Aircraft Co. in 1925, Verville designed the "Airster" biplane that became the proud bearer of "Approved Type Certificate" #1. Blessed with a vision for the years to come, Verville left Buhl in 1927 to develop a line of low-winged "fighter ships" that featured retractable landing gears, a cantilever wing and other outstanding features definitely new to the "pursuit type" airplane for the armed services. But alas, all this was so far ahead of its time that it was received with jaundiced eye and skepticism so nothing ever came of it. Never one to rest on past laurels,

Verville re-entered the commercial aircraft business in 1928 to design and build the handsome "Air Coach" series monoplane, and now the "Sportsman" biplane and its related cousin the YPT-10. With aircraft sales steadily diminishing because of the economic distress in this country, Verville Aircraft was forced to close its doors by 1932 and Alfred Verville diverted his talents to designing house-trailers; it was of course a sad step, and a blow to the heart, but it was a way to keep the wolf from the door. In 1933 Verville joined the Bureau of Air Commerce (BAC) where he soon had charge of the section that "certificated" all commercial aircraft for civil manufacture. A capable engineer and always a visionary that could see beyond present needs, Verville served in various government agencies and was greatly in demand

Fig. 94. Beefy structure and rugged nature of AT ideal for advance flight training.

Fig. 95. Undressed AT reveals structure designed to withstand abnormal stresses.

up to the time of his retirement. For his significant and enduring contributions to the progress of aviation he was formally cited as an "Elder Statesman of Aviation".

The Verville "Sportsman" model AT was an open cockpit biplane of the sport-trainer type with seating for two, a craft that was specifically designed for the man who had a healthy zest for flying. Beautifully proportioned, with an exterior finish and detailed interior appointments that would please the most fastidious, the model AT was nevertheless a flying-machine first, and therein laid its greatest charm. Inherently stable and completely safe in any maneuver, the "Sportsman" possessed flight characteristics that were quite easy to master and a general behavior that was playful but friendly; maneuverability to any attitude was practically beyond any limit and its eager response was calculated to delight the man of action. Powered with the 7 cyl. Continental A-70 engine of 165 h.p. the Verville AT was not particularly blessed with speed, but its short field performance was very excellent and it operated happily from small grassy airports. Ground handling was easy and precise, visibility was excellent to a high degree and numerous other details of arrangement or equipment left very little to be desired. The type certificate number for the Verville "Sportsman" model AT, as powered with the 165 h.p. Continental A-70 engine, was issued 5-19-30. Some 10 examples of this model were manufactured by the Verville Aircraft Co. at Detroit, Mich. Alfred V. Verville was president and chief of engineering; R. S. Deering was V.P. and general manager; E. A. "Pete"

Goff was sales manager; and O. G. Blocher was chief engineer. Lee Gehlbach, crack-pilot who won the 1930 "Cirrus Derby" with the Command-Aire "Rocket", did a big portion of the test-flying and promotion on the "Sportsman"; several other "name pilots" also had a hand in the testing and development of the AT biplane.

Listed below are specifications and performance data for the Verville model AT "Sportsman;; as powered with the Continental A-70 engine: length overall 24'3"; height overall 8'9"; wing span upper 31'0"; wing span lower 31'0"; wing chord upper and lower 50"; wing area upper 128 sq. ft.; wing area lower 114 sq. ft.; total wing area 242 sq. ft.; airfoil was Clark Y-15; wt. empty 1562 lbs.; useful load 678 lbs.; payload with 35 gal. fuel was 226 lbs. (or crew wt. 390 lbs. and 31 lbs. baggage); gross wt. 2240 lbs.; max. speed 120; cruising speed 100; landing speed 50; climb 850 ft. first min. at sea level; climb in 10 mins. was 6500 ft.; ceiling 15,000 ft.; gas cap. 35 gal.; oil cap. 6 gal.; cruising range at 9 gal. per hour was 370 miles; price at the factory was $5250.

The fuselage framework was built up of welded chrome-moly steel tubing in a Pratt truss form, faired to shape with wooden formers and fairing strips, then fabric covered. The cockpits of 30x27 in. dimension were deep and well protected by 3-piece safety glass windshields and upholstered in auto-type leather; the deep-welled bucket seats were designed for a parachute pack and adjustable to height. A baggage compartment of 5 cu. ft. capacity with allowance for 31 lbs. was located in the turtle-back section

Fig. 96. Arrangement of AT offered easy access and good visibility.

behind the rear cockpit; a streamlined head-rest was provided for the rear cockpit. The wing framework was built up of heavy-sectioned solid spruce spar beams with spruce and plywood truss-type wing ribs; the leading edges were covered with dural metal sheet and the completed framework was covered in fabric. The upper wing was built in one continuous panel with the fuel tank mounted in the center portion; Friese-type ailerons were in the lower panels and built up of metal stampings fastened to a tubular metal spar. Interplane struts and interplane bracing was all of heavy gauge metal to withstand abnormal stresses that would be encountered in "acrobatic" flying. The excessive stagger of the lower wing, in relation to the upper wing, provided unhampered entry to the front cockpit and provided excellent visibility from either cockpit. A large cut-out with handgrips in the upper wing, plus wing-walks in the lower wing on either side, greatly enhanced entry, exit, or servicing. The engine compartment was vented by louvers of efficient and unusual design. The split-axle landing gear of 78 in. tread was of the rugged out-rigger type using APC oil-draulic shock absorbing struts; 7.50x10 balloon type tires with APC wheels and brakes were standard equipment. A full-caster tail wheel with a solid rubber tire and an oil-

draulic shock strut provided good control in ground maneuvering. The fabric covered tail-group was built up of welded chrome-moly steel tubing; the fin was ground adjustable and the horizontal stabilizer was adjustable in flight. A metal propeller, Heywood air-operated engine starter, navigation lights, a dry-cell battery, fire extinguisher, tool kit and first-aid kit were standard equipment. Also available as options were dual controls, duplicate instruments in the front cockpit, brake pedals in both cockpits, and custom colors.

Listed below are Verville "Sportsman" model AT entries as gleaned from registration records:

NC-450M; Verville AT (#1) Cont. A-70.
NC-451M;   ”   (#2)   ”
NC-452M;   ”   (#3)   ”
NC-453M;   ”   (#4)   ”
NC-454M;   ”   (#5)   ”
NC-455M;   ”   (#6)   ”
NC-456M;   ”   (#7)   ”
NC-457M;   ”   (#8)   ”
NC-458M;   ”   (#9)   ”
NC-459M;   ”   (#10)   ”

There were rumored reports of more examples built than are here listed but they do not show up in registration records. Approval for the AT expired 9-30-39 due to sale of design to White Aircraft Corp. of LeRoy, New York.

Fig. 97. New Standard model D-29-S was sport version of D-29-A trainer.

The model D-29-S pictured here has been thoroughly researched through all likely material and data sources, but aside from the information related here, it remains something of a mystery model in the development history of the various versions in the "New Standard" training plane series. From the clues garnered here and there, the picture of the D-29-S develops into that of a sport model version of the trainer; first from the "S" suffix letter in its model designation, and also because of the fact that 5 gals. more fuel was allowed for a greater cruising range. A weight comparison shows 30 lbs. added to the empty weight, 30 lbs. was added to the useful load, and this comes out with a 60 lb. increase to the gross weight. To complete the "sport model" picture of the D-29-S are the cockpit cowling panels which converted the "Sport" into a more conventional two-holer (2 separate cockpits) instead of the bath-tub type cockpit arrangement used on the D-29-A trainer. (See chapter for ATC #216 of U. S. CIVIL AIRCRAFT, Vol. 3). The 30 lbs. added to the empty weight was accumulated by such various items as the cockpit cowling panels, a fuel tank of greater capacity, low pressure semi-airwheels complete with brakes, a streamlined head-rest for the rear cockpit and metal fairing cuffs at the attachment points of the landing gear, the wing struts and the bracing

wires of the wing cellule. Of course the 30 lbs. added to the useful load was cancelled out by the extra 5 gals. of fuel allowed. This then added 60 lbs. to the gross weight and the D-29-S "Sport" must have felt the difference by falling slightly short in some of its performance ability, in comparison to that of the D-29-A trainer. No doubt the extra utility of a greater range, the added comfort of separate cockpits, and other fancy finery both inside and out were considered attractive enough to offset the slight performance loss. According to records compiled on the D-29 series, there is evidence of only one example in this particular model form. The only other entry of another D-29-S type was a D-29-A "Special" (later redesignated D-29-S) that was specifically arranged to the personal tastes and dictates of Charlie Day, chief engineer for New Standard Aircraft, who used it as his personal ship for business and occasional sport flying. This example of the D-29-S built for Day was built somewhat later in 1930 under a Group 2 approval numbered 2-272 and was quite typical of the model under discussion here. One must admit that to some extent the "Sport" model lost the look of "the ugly duckling" and was a fairly good looking airplane.

The New Standard D-29-S sport model was an open cockpit biplane with seating for two in tandem. It was arranged with appointments and accessories in keeping with crafts of this type. Extra protection was now provided by the deep individual cockpits and many other little changes were incorporated to set it apart from its workaday sister ship the D-29-A trainer. The streamlining of strut junctions and various other drag-producing fittings might have contributed a barely noticeable increase in speed, but most likely it was for a more finished-off effect than anything else. Extra fuel added miles to the range allowing longer hops on cross-country jaunts. Assuming that the D-29-S was more or less typical of the breed, it must have been pleasant and obedient, with ability attuned to the mood of the pilot. Even in this more attractive form, the D-29 "Sport" was not the type of airplane, in appearance or performance, that would foster excitement or quicken a man's pulse, as some of the other craft in this class would do; however the personality conceived into any airplane designed by Chas. H. Day was of such subtle flavor as to breed a companionship not very often equalled between man and machine. Purposely staunch and well behaved, the D-29 type was capable of most any task and prolonged intimate association with this friendly craft usually became a story of pure enjoyment. The type certificate number for the New Standard model D-29-S, as powered with the 5 cyl. Kinner K5 engine of 100 h.p., was issued 5-27-30 and it is likely that no more than one example of this model was manufactured by the New Standard Aircraft Corp. at Paterson, N. J. By mid-1930, business at New Standard was rather slow and there was some shuffling of company personnel; at this point, Chas. H. Day acted as president as well as the chief engineer and Geo. Daws was sales manager. Cloyd Clevenger, formerly test-pilot for Alexander and his "Eaglerock," was now distributor for "New Standard" aircraft in Mexico. With the initiation of 3 more designs (D-31; D-32; D-33) Charlie Day then resigned from the company later in 1931, to open an office in New York City as a consulting engineer. Shortly after the resignation of Day the company was liquidated; several concerns showed interest in purchasing assets of the company but no deals were formed at this time. The Barnard Aircraft Corp. of Syracuse, N. Y. was later associated with the D-29-A through D-33 series; there was a revival of the D-25 type by Ben Jones in 1935 and later by the White Aircraft Corp. of LeRoy, New York.

Listed below are specifications and performance data for the New Standard sport model D-29-S as powered with the 100 h.p. Kinner K5 engine: length overall 24'8"; height overall 8'9"; wing span upper 30'0"; wing span lower 30'0"; wing chord upper & lower 54"; wing area upper 126 sq. ft.; wing area lower 122 sq. ft.; total wing area 248 sq. ft.; airfoil "Clark Y"; wt. empty 1195 lbs.; useful load 655 lbs.; payload with 27 gal. fuel was 240 lbs.; crew wt.

allowed 190 lbs. for each occupant, 40 lbs. for 2 parachutes and 50 lbs. baggage; gross wt. 1850 lbs.; max. speed 102; cruising speed 85; landing speed 45; climb 730 ft. first min. at sea level; ceiling 10,800 ft.; gas cap. 27 gal.; oil cap. 3 gal.; cruising range at 6.5 gal. per hour was 340 miles; basic price at factory was $5000, plus the optional equipment.

The fuselage framework was built up of channel and open section dural metal members that were riveted and bolted together into a robust and simple structure, a structure that could be easily repaired with ordinary hand tools; the framework was now faired more liberally with wooden fairing strips then fabric covered. The individual cockpits of 35x23 in. dimension were deep and well protected by fairly large windshields; bucket type seats had deep wells for parachute packs. A streamlined head-rest was provided for rear cockpit, controls were dual of the joy-stick type, and a 2.5 cu. ft. locker provided allowance for 50 lbs. baggage, which included a 20 lb. tool-kit when carried. The robust wing framework was built up of laminated spruce spar beams with basswood and plywood built-up wing ribs; the leading edges were covered with plywood sheet and the completed framework was covered in fabric. Ailerons were on the lower wings only (the D-29-A trainer had 4 ailerons) and seemed to be of fairly large area. The gravity-feed fuel tank was mounted in the center-section panel of the upper wing and was actually shaped like a deep airfoil section. Interplane struts, interplane bracing and some fittings on the landing gear were faired by streamlined metal cuffs. The split-axle landing gear of 78 in. tread was of the out-rigger type using oil-draulic shock absorbing struts; 20x6 low pressure airwheels, complete with brakes, were standard equipment. The fabric covered tail-group was also built up of riveted and bolted channel and open section dural members, the vertical fin was ground adjustable and horizontal stabilizer was adjustable in flight. A metal propeller, navigation lights, and fire extinguisher bottle were standard equipment. An air-operated Heywood engine starter was optional. The next developments in the "New Standard" biplane were mostly for special purpose craft and prototypes of new models that were built under Group 2 approvals; the last airplane (Model D-33) was built late in 1931. Fifteen of the D-29 type were still operating actively in 1939.

Listed below is the only known example of the New Standard model D-29-S as gleaned from various records:
NC-171M; D-29-S (#1025) Kinner K5.
In the tabulation of D-29-A models for chapter on ATC #216, serial #1025 should not have been included; the model D-29-S (ser.#1025) was later modified into a model D-31 with 125 h.p. Kinner B5 engine on Group 2 approval numbered 2-276, issued 10-3-30.

A.T.C. #325
(5-27-30)
STEARMAN "SENIOR SPEEDMAIL," 4-CM (4-CM-1)

Fig. 98. Stearman "Sr. Speedmail" model 4-CM-1 with 300 h.p. Wright J6-9 engine.

No doubt the most successful version in the "Stearman 4" series was the model 4-CM and 4-CM-1, which was a mail-carrying biplane that was adapted to this job from the basic lay-out of the standard model 4-C. Though the model 4-C was primarily a 3 place sport-plane, its arrangement was definitely slanted to the adaption of the basic airframe to various commercial uses, so it was indeed no problem to install a mail-pit up forward instead of an open cockpit for passengers. With the ability to carry over 600 lbs. in payload at a performance rate slightly better than the average, a model 4-C was converted to the 4-CM configuration as a trial on Group 2 approval numbered 2-177. Coached by experience gained with previous mail-carriers, results in the 4-CM were very satisfactory and an interest displayed by air-line people prompted the production of several more examples which finally earned the approved type certificate. As a craft designed strictly to haul mail and cargo, the model 4-CM had an allowable payload of 770 lbs. and the attractive possibilities of this machine caused purchase of two for service on Canadian airways. In a progres-

sive redesign, with the incorporation of night-flying equipment and several other improvements, an amendment to the certificate allowed a boost in gross weight to take care of the additional equipment without suffering loss to the available payload. This improved version was the model 4-CM-1 and it posed as one of the finest mail-carriers in the country at that time. Having had previous experience with dependable "Stearman" equipment on portions of its far-flung system, American Airways ordered 12 of the model 4-CM-1 that were specially fitted to fly the mails at night. Company reports state they were used mostly on routes from St. Louis, Chicago, Evansville, and Atlanta with great success. The sale of 12 airplanes to American Airways may not sound like such a big deal, and it would not have been in more normal times, but the economic slump during 1930-31 caused a rapidly failing aircraft market and Stearman welcomed this quantity purchase with particular joy.

The "Stearman" model 4-CM (4-CM-1) "Senior Speedmail" was a fairly large open cockpit biplane with soft flowing lines of beau-

Fig. 99. 4-CM-1 used on American Airways routes in midwest.

tiful proportions that were carefully engineered around the big N.A.C.A. engine cowling. The bulky cross-section necessary to "fair out" the large diameter up front provided ample space to locate two good-sized cargo holds just forward of the pilot's cockpit. Well versed by now in the needs of the mail pilot, Stearman provided the model 4-CM with a well planned and comfortable cockpit that was equipped with two exhaust-manifold heaters for his cold-weather comfort. Well placed foot-steps and hand-holds also made the job of getting in and out a bit easier. Powered with the 9 cyl. Wright J6 (R-975) engine of 300 h.p., the 4-CM had a good range of performance with a reserve on tap for those unexpected situations when a maximum effort was required. Stemming from a sure-footed breed of airplane, the flight characteristics and general behavior of the 4-CM and 4-CM-1 were carefully planned to assist the pilot as much as possible. This points to the reason why pilots particularly enjoyed working with the 4-CM and 4-CM-1. The type of certificate number for the Stearman model 4-CM, as powered with the 300 h.p. Wright J6 engine, was issued 5-27-30 and later amended to allow changes as incorporated into the model 4-CM-1. A total of 15 examples in this version were manufactured by the Stearman Aircraft Co. at Wichita, Kansas.

Listed below are specifications and performance data for the "Sr. Speedmail" model 4-CM as powered with the 300 h.p. Wright R-975 engine: length overall 26'11"; height overall 10'2"; wing span upper 38'0"; wing span lower 28'0"; wing chord upper 66"; wing chord lower 51"; wing area upper 204 sq.ft.; wing area lower 103 sq.ft.; total wing area 307 sq.ft.; airfoil Goettingen 436; wt. empty 2285 lbs.; useful load 1651 lbs.; payload with 106 gal. fuel was 770 lbs.; gross wt. 3936 lbs.; max. speed 147; cruising speed 124; landing speed 55; climb

1000 ft. first min. at sea level; climb in 10 min. was 6900 ft.; ceiling 14,800 ft.; gas cap. 106 gal.; oil cap. 10 gal.; cruising range at 16 gal. per hour was 750 miles; price at factory field was approx. $12,500. The following figures are for the improved model 4-CM-1; wt. empty 2540 lbs.; useful load 1527 lbs.; crew wt. 190 lbs.; payload with 84 gal. fuel was 770 lbs.; gross wt. 4067 lbs.; max. speed 147; cruising speed 124; landing speed 57; climb 980 ft. first min. at sea level; climb in 10 min. was 6500 ft.; ceiling 14,500 ft.; gas cap. 84 gal.; oil cap. 8 gal.; cruising range at 16 gal. per hour was 600 miles; price at the factory field completely equipped was over $15,000.

The construction details and general arrangement of the model 4-CM were similar to that of the 4-C as described in the chapter for ATC #304 in this volume. Being expecially arranged for the transport of mail and package cargo, the model 4-CM had two cargo holds forward of the pilot's cockpit that were metal-lined for fire resistance. Access to the cargo space was by two locking type metal hatch covers; one hatch cover was immediately ahead of the cockpit and the other hatch cover was forward of that in a position between the center-section struts. Fuel tanks were mounted in the center-section panel of the upper wing. A 5 cu.ft. compartment in back of the pilot's cockpit was provided for a tool-kit and personal effects. A parachute of 20 lbs. was part of the 190 lb. crew weight allowance. The improved model 4-CM-1 was primarily arranged for operation at night, so it was equipped with a full complement of night-flying gear; this was to include navigation lights, cockpit lights, two retractable landing lights, parachute flares, Very pistol, a battery and radio receiver. Other equipment included two fire extinguishers, a hand crank inertia type engine starter, cockpit heaters and engine tool kit. Available also were 9.50x12 low pressure semi-airwheels, wheel streamlines (wheel pants), and a 5.00x4 tail wheel. As pictured, some of the later model 4-CM-1 were equipped with sliding side-plates and a modified head-rest to partially close in the pilot's cockpit for better cold weather protection. The next development in the "Stearman 4" series was the model 4-DM as described in the chapter for ATC #326 in this volume.

Listed below are Stearman model 4-CM and 4-CM-1 entries as gleaned from registration records:

NC-770H;	Model 4-CM	(#4010)	Wright R-975.
NC-772H;	"	(#4012)	"
CF-CCG;	"	(#4013)	"
CF-CCH;	"	(#4014)	"
NC-482W;	Model 4-CM-1	(#4030)	"
NC-483W;	"	(#4031)	"
NC-484W;	"	(#4032)	"
NC-485W;	"	(#4033)	"
NC-486W;	"	(#4034)	"
NC-487W;	"	(#4035)	"
NC-488W;	"	(#4036)	"
NC-489W;	"	(#4037)	"
NC-490W;	"	(#4038)	"
NC-11721;	"	(#4039)	"
NC-11722;	"	(#4040)	"
NC-11723;	"	(#4041)	"

Fig. 100. *Various views of 4-CM show classic lines of "Stearman" design.*

Serial #4010 converted to 4-CM from model 4-E and 4-EM; serial #4012 first as 4-C converted to 4-CM on Group 2 approval numbered 2-177; serial #4014 first as model 4-C; serial #4035 later converted to model 4-DM-1; serial #4032 later as "Special" with Ranger V-12 engine (X-484W).

Fig. 101. *4-CM-1 improvements included better pilot comfort and night-flying gear.*

Fig. 102. Stearman "Sr. Speedmail" model 4-DM mounted 300 h.p. "Wasp Junior" engine.

Another interesting version in the "Stearman 4" series was the mail-carrying model 4-DM, a cargo-hauling adaptation of the open cockpit model 4-D sport-plane that was powered with the 300 h.p. Pratt & Whitney "Wasp Junior" engine. As mentioned several times before, the 4-series were basically arranged for sport flying and business, but the design was definitely slanted to allow conversion that would fit various commercial uses. Whether in use for sport or business, or faithfully hauling the nation's mail, the "Stearman 4" in all its variants was one of the finest airplanes in this country and probably the most handsome airplane that Stearman ever built. The model 4-DM "Senior Speedmail" bears the added distinction of being the last production version in the "Stearman 4" series and also as the last mail-carrying airplane that Stearman Aircraft ever built. Well

noted for their earlier mail-planes which included such stars as the C2MB, the C3MB, the big M-2 and the LT-1, and now 3 distinct mail-plane versions in the 4-series, "Stearman" also produced popular models for business and sport. These could be easily identified as the classic C3B, the C3R, and 3 other versions in the distinctive "Stearman 4." This line-up of popular models had been the mainstay of Stearman production for 5 years but with the market for all craft of this type slowly dwindling away, the company was forced to look to new horizons for its future existence. It was logical that they should look to the armed services both here and abroad for a continuation of at least some business. For entry into this field "Stearman" leaned heavily on past experience, and developed a line of training airplanes which were to launch a program and new career that lasted for a span of nearly 15 years.

The Stearman "Senior Speedmail" model 4-DM was an open cockpit biplane bearing the flowing lines of beautiful proportions that were characteristic of every model in this particular series; patterned after the sport model 4-D, make-up of the 4-DM varied only in the substitution of a protected mail-pit in place of the passenger's open cockpit. The rounded and bulky cross-section necessary to fair out the large diameter of the engine fairing provided more than ample space to locate two fair-sized cargo holds just ahead of the pilot's cockpit. To complete the transformation, several other modifications were added to insure the pilot's comfort and to assure help in his rather difficult task; the mail-pilot's chore at this stage of the game was not an easy one and he was certainly grateful for a machine's willing cooperation. Powered with the 9 cyl. "Wasp Junior" (R-985) engine of 300 h.p., the model 4-DM had a good range of performance with a reserve on tap to bring into play when the chips were down; attuned to react to the pilot's every whim with a nature that was eager but still obedient, the "Stearman 4" was certainly a proud addition to any operators flight-line. The type certificate number for the "Senior Speedmail" model 4-DM was issued 5-27-30 and later amended to allow changes as incorporated into the model 4-DM-1. Only 1 example each of the 4-DM and 4-DM-1 series were manufactured by the Stearman Aircraft Co. at Wichita, Kan.; a division of the United Aircraft & Transport Corp. Walter Innes, Jr. was now the president; Mac Short was V.P. and chief engineer; and J. E. Schaefer was secretary and

Fig. 104. Revealing view of 4-DM shows front end details.

manager of sales. Lloyd C. Stearman, founder and president of the company since its inception back in 1926, had left with rather heavy heart to ride out the economic depression and try to establish himself in another line of business, for a time at least.

Listed below are specifications and performance data for the Stearman model 4-DM as powered with the 300 h.p. "Wasp Junior" en-

Fig. 103. 4-DM used also on airmail routes of American Airways.

gine: length overall 26'11"; height overall 10'2"; wing span upper 38'0"; wing span lower 28'0"; wing chord upper 66"; wing chord lower 51" wing area upper 204 sq. ft.; wing area lower 103 sq. ft.; total wing area 307 sq. ft.; airfoil Goettingen 436; wt. empty 2326 lbs.; useful load 1610 lbs.; payload with 106 gal. fuel was 729 lbs.; gross wt. 3936 lbs.; max. speed 147; cruising speed 124; landing speed 55; climb 1000 ft. first min. at sea level; climb in 10 min. was 6900 ft.; ceiling 14,800 ft.; gas cap. 106 gal.; oil cap. 10 gal.; cruising range at 16 gal. per hour was 750 miles; price at factory field was approx. $12,500. The following figures are for the improved model 4-DM-1; wt. empty 2581 lbs.; useful load 1486 lbs.; payload with 83 gal. fuel was 750 lb.; gross wt. 4067 lbs.; max. speed 147; cruising speed 124; landing speed 57; climb 980 ft. first min. at sea level; climb in 10 min. was 6500 ft.; ceiling 14,500 ft.; gas cap. 83 gal.; oil cap. 8 gal.; cruising range at 16 gal. per hour was 600 miles; price at the factory field completely equipped was near $15,000.

The construction details and general arrangement of the model 4-DM were much like those of the models 4-C and 4-D as described in the chapters for ATC #304 and #305 in this volume. Being especially arranged for the transport of mail and package cargo, the model 4-DM had two cargo holds forward of the pilot's cockpit that were lined with metal panels for fire resistance. Access to the cargo space was by two locking type metal hatch covers, adjacent to one another. To provide clear area in the fuselage, the fuel tanks were mounted in the center-section panel of the upper wing. A 5 cu. ft. compartment in back of the pilot's cockpit was provided for a tool kit and pilot's personal effects. The crew weight was 175 lbs. and a parachute at 20 lbs. was calculated as part of the payload. The improved model 4-DM-1 was primarily arranged for operation at night so it was equipped with a full complement of night-flying gear. This was to include navigation lights, cockpit lights, two retractable landing lights, parachute flares, a battery and radio receiver. Other equipment included two fire extinguishers, a hand crank inertia-type engine starter, cockpit heaters and engine tool kit. Available also were 9.50x12 low pressure semi-airwheels, wheel streamlines, and 5.00x4 tail wheel. Some of the later model 4-CM-1 were equipped with sliding side-plates and a modified head-rest to partially close in the pilot's cockpit for better weather protection; this may have been available for the 4-DM-1 also. The next development in the "Stearman" biplane was the model 6-A trainer as described in the chapter for ATC #364 in this volume.

Listed below are Stearman model 4-DM and 4-DM-1 entries as gleaned from registration records:

NC-774H; Model 4-DM (#4011) Wasp Jr.
NC-487W; Model 4-DM-1 (#4035 Wasp Jr.
Serial #4011 converted from a sport model 4-D; serial #4035 converted from a model 4-CM-1.

A.T.C. #327
(6-16-30)
"MONOCOUPE 110"

Fig. 105. The "Monocoupe 110" with 110 h.p. Warner "Scarab" engine.

As a basic development from the earlier "Monosport" series, the "Monocoupe 110" was designed especially as a sporting type airplane for the sportsman-pilot, one who would best enjoy a small cabin monoplane with some of the extras in speed and performance. The standard 90-series was actually the basis for this new development which was typical, except for the installation of the 7 cyl. Warner "Scarab" engine of 110-125 h.p. By the time almost 40 examples of the "Model 90" had been built, 2 prototypes of the "Model 110" were rolled off the line for flight test. Dainty and light, with not a useless pound in her frame, the "One-Ten" proved to be a fast-stepper with a fairly frisky nature. Company pilots were of course jubilant about its speed and sharp handling, so its creators smiled with satisfaction, — together they planned for things ahead. Pent-up enthusiasm could be tethered no longer so the "Monocoupe 110," in several examples, was carefully groomed to a shiny perfection and flown off to the arenas of up-coming air races. The 1930 National Air Races held at Chicago proved to be a sweeping victory for "Monocoupe" in general and the "Model 110" had somewhat of a field day; it was the winner in at least 4 events and placed high in many other events. Class for class, the "One-Ten" was practically unbeatable and even nosed out ships with more than twice the horsepower. With the

sweet taste of victory in its innards, the "Model 110" was hard to keep from the races but not all of them spent their time happily rounding the pylons; several served as fast air taxis in various fields of business. Tom Colby for one, genial promoter of "Berryloid" airplane finishes, managed a busy schedule over most of the nation with the help of his gayly painted "One-Ten." It was fairly evident that "Monocoupe" had a star on hand that was a real trouper, one quite able to handle the lead no matter what the part.

Facing baptism in the air-races of 1930, and coming through on top with flying colors, the Model 110 was soon the basis for much experiment in developments that were pointed at more and more speed. Variously shaped low-drag engine cowlings boosted speeds by many miles per hour, spindly low-drag landing gears were next in order and then came the mania for chopping off wing area and tail area, to further increase the speeds. The standard "Monocoupe 110" as such, almost petered out by 1933 but it did become the basis for the "110 Special" which was a "hairy" clipped wing version with 145 h.p., a craft that could flick off 185 m.p.h. with no apparent strain. By 1934 the 145 h.p. model "D-145" was introduced to production as a version combining many of the features that had been heretofore reserved for only the racing models. Because of its na-

Fig. 107. "One-Ten" was darling of the sportsmen but popular for business use also.

ture, the D-145 was certainly no airplane for putt-putting around the airport on a busy weekend! As to the Model 110 it had been filling the need for sporty transportation in a vehicle of nominal seating, but was not selling in any large number because of the many other makes of airplanes available at bargain prices during those depressed times. Had the times been more normal, it is certainly likely that many more examples would have been built and sold. Stout of frame and quite reliable, the "One-Ten" wore well as the years went by and at least 28 were still in active service by 1939.

The "Monocoupe" model 110 was a small high-winged cabin monoplane with side by side seating for two, in chummy but comfortable surroundings. Typical of the basic series introduced early in 1930 the "One-Ten" bore soft, rounded lines that reflected a girlish beauty and a step ahead in aerodynamic efficiency. Powered with the 7 cyl. Warner "Scarab" engine of 110-125 h.p. and with weight kept down to a sporting trim, the performance of this model was far beyond what one would expect for a ship of such nominal power. Inheriting all basic features of the "Monocoupe" line, the

Fig. 106. Early versions of 110 had oleo landing gear; speed-ring cowl was added to boost speed.

Fig. 108. 1932 model 110 shown at factory in Robertson, Mo.

Model 110 was of fairly good nature but expected the pilot to stay in firm command at all times. As short-field performance and climb-out from take off were very good, this little craft was an ideal companion with which to make calls to out-of-way places; a craft heartily welcome at any field, the "One-Ten" was always the source of great interest. The type certificate number for the "Monocoupe 110," as powered with the 110 h.p. Warner engine, was issued 6-16-30 and at least 20 or more examples of this model were manufactured by the end of 1931, according to the available records. The first few examples of the "One-Ten" were built by the Mono Aircraft Corp. at Moline, Ill. but the bulk of these models were later built by the Mono-

coupe Corp., which was a reorganization of the former company. Under the new banner, the company had quarters on Lambert Field in Robertson, Mo., Don A. Luscombe, president of the firm, flew a "One-Ten" on his personal travels but it was Johnnie Livingston and Vern Roberts who brought it to public notice and ever-lasting fame.

Listed below are specifications and performance data for the "Monocoupe 110" as powered with the 110 h.p. Warner "Scarab" engine: length overall 20'4"; height overall 6'11"; wing span 32'0"; wing chord 60"; total wing area 132 sq. ft.; airfoil Clark Y; wt. empty 991 lbs.; useful load 620 lbs.; payload with 30 gal. fuel was 247 lbs.; gross wt. 1611 lbs.; max.

Fig. 109. Modified often, this "One-Ten" was well-known on racing circuits.

speed 133 [142]; cruising speed 112 [120]; landing speed 45; climb 1050 ft. first min. at sea level; ceiling 16,000 ft.; gas cap. 30 gal.; oil cap. 3.5 gal.; cruising range at 7 gal. per hour was 450 [480] miles; figures in brackets are for "One-Ten" with speed-ring cowling and wheel pants; price at the factory field was $4500., raised to $4750 in 1931. The following figures are for the later models with noted changes; wt. empty 1020 lbs.; useful load 600 lbs.; payload with 28 gal. fuel was 239 lbs.; gross wt. 1620 lbs.; all other figures remained more or less the same, except when fitted with 125 h.p. Warner engine, in this case, a general increase in performance would be noticeable.

The following record might well be noted as the evolution of the "110 Special" or the famous "Clipped-Wing Monocoupe" as it was sometimes called. Johnnie Livingston, top-notch pilot in any type of competition, took delivery of his standard "Monocoupe 110" in the summer of 1930; it was strictly a stock machine, but it was powered by the improved "Scarab" series engine of 125 h.p. and topped off at about 135 m.p.h. Test pilot Vern Roberts had a similar ship, both craft were fitted with narrow-chord Townend type speed-ring engine cowlings for the 1930 National Air Races held at Chicago. Vern Roberts won many of the events and John Livingston usually hung on for a close second, except in the Class B "Pacific Derby" from Brownsville, Tex. to Chicago, which he won handily. Determined to win more than just occasionally, Livingston eyed his "Coupe" critically and noticed places for many improvements to increase his top speed. The following

months were spent in careful grooming; a new faired landing gear was installed, streamlined "cuffs" were fitted at all strut junctions, wheel pants now shrouded the wheels, the carburetor air-scoop was redesigned and all protruding nuts, bolts and fittings were covered over to cause the least disturbance to the air flow. A special engine fairing was also tested and installed and by the time of the 1931 National Air Races, Livingston's "Coupe" was doing 186 m.p.h. and winning. Happy with the results obtained, Livingston had further plans for 1932. Calculating for bare minimum wing area needed, a 23 foot wing (as compared to the standard 32 ft. wing) was ordered from the "Mono" factory along with a new tail-group of less area and span. A 145 h.p. Warner engine was also installed. He used smaller wheels, made a major change in the windshield shape, and continued experiments on engine cowling. Careful attention was paid to streamlining at all points and every surface was highly polished. Employing an extra boost by using a jet-type engine exhaust system, the "Coupe" was now doing 220 m.p.h. The evolution for speed had finally reached its peak at this point, and other racing craft were becoming faster than the Livingston "Coupe," but the lessons learned in its development were successfully used later on the famed "110 Special."

The construction details and general arrangement for the "Model 110" are typical to that of the "Model 90" as described in the chapter for ATC #306 in this volume. The wing bracing struts were vees of streamlined steel tubing, faired at the fuselage and wing

Fig. 110. Monocoupe 110 Special was design fostered by previous modifications to gain speed; 185 m.p.h. was not unusual.

junctions. The landing gear on earlier versions was typical of that used on the "90," but later versions of the "One-Ten" had faired vees on each side with streamlined steel tie-rods as center bracing in conjunction with a shock absorbing mechanism that was housed in the fuselage; this type of gear was usually fitted with wheel pants. Earlier versions used "Monoil" shock absorbing legs and later versions used a spool of rubber shock cord. The robust wing, though installed as a one-piece section, was actually built up in halves that were spliced together in the center; ailerons were of the plain unbalanced type. The baggage compartment behind the seat had ample room with an allowance for up to 74 lbs. A metal propeller and 6.50x10 semi-airwheels were standard equipment. Optional equipment included speed-ring engine cowling, wheel pants, Heywood engine starter, a battery, navigation lights and tail wheel; with these accessories added, the payload was reduced by the amount of weight actually added. All examples of the standard "One-Ten" were also eligible as the "110 Special" when modified to specifications of Group 2 approval numbered 2-452. The next "Monocoupe" development was the rare "Model 90-J" as described in the chapter for ATC #355 in this volume.

Listed below is a partial tally of "Monocoupe 110" entries as gleaned from registration records:

NC-508W; Model 110 (#5W41) Warner 110-125.

NC-501W;	Model 110	(#5W47)	Warner 110-125.
NC-517W;	"	(#5W57)	"
NC-518W;	"	(#5W58)	"
NC-521W;	"	(#5W67)	"
NC-524W;	"	(#5W68)	"
NC-527W;	"	(#5W69)	"
NC-533W;	"	(#5W70)	"
NC-542W;	"	(#5W90)	"
NC-544W;	"	(#5W91)	"
NC-10730;	"	(#5W92)	"
NC-546W;	"	(#5W95)	"
NC-10746;	"	(#5W98)	"
NC-570W;	"	(#5W99)	"
NS-29;	"	(#6W00)	"
NS-30;	"	(#6W01)	"
NC-573W;	"	(#6W05)	"
NC-589W;	"	(#6W08)	"
NC-12027;	"	(#6W12)	"
NC-12312;	"	(#6W13)	"
NC-112V;	"	(#6W22)	"
NC-114V;	"	(#6W23)	"
NC-115V;	"	(#6W24)	"
NC-477W;	"	(#6W33)	"
NC-478W;	"	(#6W34)	"
NC-12344;	"	(#6W38)	"
NC-12349;	"	(#6W43)	"
NC-12345;	"	(#6W47)	"
NC-12385:	"	(#6W56)	"
NC-12359;	"	(#6W58)	"

This approval for serial #5W40 and up; registration number for serial #5W40 unknown; a few of the standard "Model 110" were built in 1937-39.

A.T.C. #328
(6-19-30)
BELLANCA "PACEMAKER," 300-W

Fig. 111. Bellanca "Pacemaker" model 300-W mounted 300 h.p. "Wasp Junior" engine.

One of the lesser known but equally capable models in the famous "Bellance Monoplanes" line-up was the model 300-W, a "Pacemaker" that was typical in most all respects except that it was powered with the new Pratt & Whitney "Wasp Jr.", a 9 cyl. air-cooled radial engine of 300 h.p. P & W engines amassed a world-wide reputation for delivering dependable power over a long period of time and a number of buyers would prefer this engine, so Bellanca Aircraft developed the 300-W "Pacemaker" to fit this particular segment of the aircraft market. With only those modifications necessary to the assembly of this combination there was no tangible improvement to speak of; this is not meant to slight this new version because the "Pacemaker" was many times judged the best of its type and the 300-W was certainly as good as any of the best of them. Coming at a rather bad time because of the upset in the nation's economy, it did not sell in any large number, but had the money been easier to come by it would have sold quite well, that is sure. The popular CH-300 (refer ATC #129) had been selling well for about a year but it too, was falling off in sales. Bellanca was looking to newer models to help keep that mythical wolf from the factory door. The new 420 h.p. "Skyrocket" had been well accepted and was doing well, and the new "Airbus" which showed drawing-board promise of being the most efficient airplane Bellanca

had built, was nearing completion by April of 1930. An occasional "special" airplane was, of course, ordered for some new record-breaking flight. In reality, Bellanca Aircraft was getting by fairly well in these lean times and a lot better than most.

The Bellanca "Pacemaker" model 300-W was a high-winged cabin monoplane with ample space for six in a roomy interior that offered comfort, the latest in styles and a generous payload. Efficient performance of the basic "Pacemaker" design had been proven in the field of contest a good number of times. It speaks well for this new version to know that it shared all these inherent qualities, plus many little improvements that were continually added as airplanes were coming off the assembly line. Guiseppe Bellanca spent most of his time at the factory, hovering nearby like a happy mother-hen, and it was not unusual for him to suggest slight changes and slight improvements every day, probably making each "Bellanca" that came off the line a little bit better than the one before it. Powered with the "Wasp Junior" engine of 300 h.p., the model 300-W had a range of performance that allowed it to operate from most any of the smaller fields and its general behavior was above the average for a ship of this type. "Bellanca's" flight characteristics and general nature had a tendency to adjust to the driving mood and its pilots were

Fig. 112. 300-W warming up for test at Bellanca plant, shown here in prototype.

generally known as a happy lot, happy in the knowledge that this craft, like its many sister-ships, was always ready to meet the worst and come out on top of it all. The type certificate number for the model 300-W, both as land-plane and seaplane, was issued 6-19-30 and probably some 7 or more examples of this model were manufactured by the Bellanca Aircraft Corp. at New Castle, Delaware. Guiseppe M. Bellanca was president and chief engineer; R. B. C. Noorduyn, formerly as "Tony" Fokker's right-hand man, was now the vice president and engineering assistant.

Listed below are specifications and performance data for the Bellanca model 300-W "Pacemaker" as powered with the 300 h.p. "Wasp Jr." (R-985) engine; length overall 27' 10"; height overall 8'4"; wing span 46'4"; wing chord 79"; total wing area 273 sq. ft.; airfoil "Bellanca" (modified R.A.F. 15% thickness); wt. empty 2465 lbs.; useful load 1835 lbs.; payload with 112 gal. fuel was 930 lbs.; gross wt. 4300 lbs.; max. speed 145 (with speed-ring cowling); cruising speed 122; landing speed 50; climb 1000 ft. first min. at sea level; ceiling 18,000 ft.; gas cap. 112 gal.; oil cap. 8 gal.; cruising range at 16 gal. per hour was 730 miles; price at the factory field was $15,350 raised to $15,550 in April of 1931. The following figures are for seaplane on Edo K twin-float gear; wt. empty 2922 lbs.; useful load 1835 lbs.; payload with 112 gal. fuel was 930 lbs.; gross wt. 4757 lbs.; max. speed 136; cruising speed 114; landing speed 55; climb 700 ft. first min. at sea level; ceiling 16,000 ft.; price at factory was $18,400 raised to $18,600

in April of 1931; all other specifications and data typical to the listing above.

The construction details and general arrangement for the "Pacemaker" model 300-W was typical of that of the model CH-300 as described in the chapter for ATC #129 of U. S. CIVIL AIRCRAFT, Vol. 2. The fuselage framework was faired to shape with spruce fairing strips and fabric covered; a baggage compartment of 12 cu. ft. capacity was to the rear of the cabin with allowance for 83 lbs. and was easily accessible from the inside or out. Passenger seats were welded steel tube frames upholstered in fine fabrics and quickly removable to provide floor area for hauling cargo; the cabin interior had a minimum of 57 inches headroom. The robust wing framework was of solid spruce spar beams that were routed out for lightness, with spruce and basswood truss-type wing ribs; the leading edges were covered with ply-wood sheet clear around to the front spar and the completed framework was covered in fabric. The landing gear of 90 in. tread was fitted with "Alemite" grease fittings at all points of wear and the oleo-spring shock absorbing struts had a 6 in. travel to absorb even the most severe loads; wheels were 32x6 and Bendix brakes were standard equipment. A 4x5 tail wheel was now fitted for better ground handling. A metal propeller, Eclipse hand crank inertia-type engine starter, a wet-cell battery and navigation lights were standard equipment. Seaplane equipment also included water-proof engine cover, tool kit, mooring ropes, anchor and a paddle. The next Bellanca development was the big "Airbus" as described

Fig. 113. 300-W was rare version in "Pacemaker" series.

in the chapter for ATC #360 in this volume.

Listed below are "Pacemaker" model 300-W entries as gleaned from registration records:

NC-76W;	Model 300-W	(#301)	Wasp Jr.
NC-353W;	"	(#302)	"
NC-354W;	"	(#303)	"
NC-355W;	"	(#304)	"
NC-11694;	"	(#305)	"
NC-12617;	"	(#306)	"
;	"	(#307)	"
NC-33M;	"	(#147)	"

Serial #303 later with Noel Wien in Alaska; serial #305 and #307 later as model CH-400-W on Group 2 approval numbered 2-392; registration number for serial #307 unknown; serial #147 modified to 300-W from an earlier model CH-300.

Fig. 114. Stinson "Junior" model SM-7B mounted 300 h.p. "Wasp Junior" engine; fancy trimmed ship shown was owned by A. Felix Dupont, Jr.

In the new and varied line-up for 1930, Stinson Aircraft planned to concentrate more heavily on the attractive "Junior" cabin monoplane with several offerings in two different series; the SM-7 series, available in two versions, were of higher horsepower and carried bigger price tags. Knowing well from past experience that certain customers would be interested in an airplane of the 4-place cabin type but with a promise of more verve and better performance, Stinson singled out the model SM-7 "Junior" as the airplane best suited to that type of clientele. The model SM-7 was the "Junior" version designed specifically for the sportsman-pilot, the well-to-do family man or the business executive, who certainly wouldn't hesitate to pay a little more for a much better airplane. True, an airplane of this type would not bring on a rush of orders because the price tag (quite high for these times) would turn many away, but aircraft manufacturers sadly watched their market diminishing as the depression gained momentum; however, if only 5 examples or so were sold of a particular model, it helped to keep the builders in business. One of the oddities in the airplane business was the fact

that perhaps in the whole country there would be no more than 10 potential customers for a certain type of airplane, yet several manufacturers would feel the urge to vie for this limited business. The model SM-7 "Junior" found its small band of customers quite readily and was soon seen in widely scattered parts of the country. Walter Varney, aviation pioneer who had a zealous penchant for sponsoring air-lines, used the high-performance SM-7B on a "Varney Air Service" run from Oakland to Marysville in California. With small loads carried on a frequent schedule, this type of service was well suited to the abilities of this 300 h.p. "Junior." General Electric, the Auburn Automobile Co., and A. Felix DuPont, Jr. among others, found the spirited SM-7B also to their liking, a craft with just that extra bonus in performance that made the difference. Years later, at least one of the SM-7B was reported in service (as L-12-A) with the U.S.A.A.F., as a light utility transport and trainer.

The Stinson "Junior" model SM-7B was a high winged cabin monoplane that seated four nicely in a style and comfort that was quite plush and fairly equal to the best anyone had to

Fig. 115. High performance utility of SM-7B popular for business flying.

offer. Of a proportion and arrangement similar to the cheaper SM-8 series, the SM-7B was offered as a craft with more allowable gross weight. This added allowable weight was to permit more fuel load for a greater cruising range and yet have useful load enough to allow the addition of any number of desirable aids and accessories — such as speed-ring engine fairing, wheel streamlines, and perhaps even a radio, without cutting into the available payload too much. Powered with the 9 cyl. Pratt & Whitney "Wasp Jr." of 300 h.p. (as a companion model to the SM-7A with 300 h.p. Wright J6 engine) there was plenty of good performance available, with flight characteristics and general behavior of a nature that had made the Stinson "Junior" a country-wide favorite for many years. It has quite often been said that there were airplanes that were certainly more handsome, airplanes with more speed and airplanes that did excel in one way or another, but there were very few airplanes of this time that could even match the Stinson "Junior" as an all-round good and honest airplane. Versatile as to chore, with a nature well adaptable to temperament, the "Junior" fitted itself to the job quickly and fairly revelled in the master's touch. The type certificate number for the Stinson "Junior" model SM-7B as powered with the 300 h.p. "Wasp Jr." (R-985) engine, was issued 6-20-30 and 8 or more examples of this model were built by the Stinson Aircraft Corp. at Wayne, Mich.; a division of the Cord Corp. Business at Stinson was holding its own, production lines were humming and by the year's end· of 1930, several departments were scheduled to work 7 days a week in order to keep abreast of sales. Nearly 13% of all cabin-type airplanes in this country were built by

Stinson, certainly a record of outstanding achievement during those troubled times.

Listed below are specifications and performance data for the Stinson "Junior" model SM-7B, as powered with the 300 h.p. "Wasp Junior" engine: length overall 29'11"; height overall 8'9"; wing span 41'8"; wing chord 75"; total wing area 234 sq. ft.; airfoil "Clark Y"; wt. empty 2312 lbs.; useful load 1188 lbs.; payload with 90 gal. fuel was 428 lbs.; gross wt. 3500 lbs.; max. speed 142; cruising speed 120; landing speed 60; all speeds listed are with speed-ring engine cowling; climb 1000 ft. first min. at sea level; ceiling 18,000 ft.; gas.cap. 90 gal.; oil cap. 7 gal.; cruising range at 18 gal. per hour was 550 miles; price originally quoted at $11,195 , lowered to $10,695 and then reduced to $8995 in 1931.

The construction details and general arrangement of the model SM-7B were similar to the models SM-7A, the SM-8B and SM-8A as described in the chapters for ATC #298-294-295 in U. S. CIVIL AIRCRAFT, Vol. 3. Because of its higher price the SM-7B was upholstered in somewhat richer fabrics and the interior was bedecked with accessories which were available only as extra equipment in the cheaper SM-8 series. For that "extra" in speed and performance the SM-7B was available with a low drag speed-ring (Townend type) engine fairing and wheel streamlines (wheel pants); together they boosted the top speed to nearly 150 m.p.h. All cabin windows were of shatterproof glass and the cabin was fitted with heat and ventilation. The 40x54x48 inch cabin interior could be quickly cleared of 3 seats when there was cargo to be hauled. A baggage compartment behind the rear seat was of 7 cu. ft. capacity and had allowance for 50 lbs.; this compartment when fitted with radio gear cancelled out the baggage allowance. The out-rigger landing gear with a tread of 115 inches, used Stinson-made oleo shock absorbing struts; wheels were 30x5 with Kelsey-Hayes brakes. A ground-adjustable metal propeller, Eclipse electric engine starter, a battery, engine exhaust collector ring, wheel brakes, emergency parking brake, tail wheel, dual wheel-type controls, and navigation lights were part of the standard equipment. A speed-ring engine cowling, wheel streamlines, landing lights, parachute flares, generator, radio, skis, and 8.50x10 semi-airwheels were optional equipment. The next Stinson development was the tri-motored model SM-6000 as described in the chapter for ATC #335 in this volume.

Listed below are "Junior" model SM-7B entries as gleaned from registration records:

NC-217W;	Model SM-7B	(#3003)	Wasp Jr.
NC-933W;	"	(#3006)	"
NC-944W;	"	(#3007)	"
NC-936W;	"	(#3009)	"
NC-947W;	"	(#3010)	"
NC-11116:	"	(#3012)	"
NC-10812;	"	(#3013)	"
NC-11151;	"	(#3014)	"

Serial #3000 - 3001 - 3002 - 3004 - 3005 - 3008 - 3011 were "Junior" model SM-7A on ATC #298.

Fig. 116. Boeing "Monomail" model 200; its majestic flight was a prophecy of things to come.

Almost as a prophet, the Boeing "Monomail" emerged in mid-1930 as a sign in the skies of things to come — like the harbinger of an era that was to foster a new concept in transport aviation and be directly responsible for drastic changes in military aviation too. As an entirely new form in large transport aircraft, the new "Monomail" model 200 harbored such innovations as an all metal semi-monocoque fuselage with one of the earliest applications of "stressed skin" metal coverings, a full cantilever all-metal wing mounted in the low position, and a retractable landing gear to eliminate the drag imposed by this bulky assembly. Practically the only feature still present to mark it as a craft of this early decade was the pilot's open cockpit, a hold-over which seemed a habit hard to shake by old-line pilots. No doubt about it, this was a revolutionary design with a daring for new principles hardly akin to what one would expect from the "Boeing" shops; the "Monomail" was tangible proof that engineers were thinking and taking a good look into the near-by future. Introduced first as an experiment in all-cargo transport (Model 200), the "Monomail" was also arranged in a version fitted out as a passenger transport.(Model 221) with ample seating for 8. All preliminary testing was conducted by Boeing's test-pilot, Ed T. Allen. Erik Nelson, sales manager for Boeing and a top-notch pilot, was anxious to test the promise of this new design so he flew the prototype on a shake-down trip from Seattle to New York in Nov. of 1930. Soon after being tested (in actual service on Boeing's own air-line system), the "Monomail" provided much interesting data and an incentive of ex-

Fig. 117. View shows relative size of Model 200; engine was 575 h.p. "Hornet" B.

citing drive to study quickly all the possibilities that this basic design now uncovered. As a direct result from "Monomail" studies, Boeing very soon was able to announce the twin-engined YB-9 bomber, a revolutionary craft that was the fastest military bomber in the world at that time. Only a few years later, Boeing introduced the "Model 247" twin-engined passenger transport which undeniably influenced a whole new era in commercial aviation. Getting back to the "Monomail" of 1930-31 we are concerned here with the "Model 200," which was the first of the type, an all-cargo version carrying up to 2300 lbs. of miscellaneous air-freight at speeds of up to 160 m.p.h. Dispensing with experiments and tests by July of 1931, the "Monomail" took its place in active service on the San Francisco to Chicago leg of Boeing's system, which soon was to become the United Air Lines.

The Boeing "Monomail" model 200 as shown here in varied views, was a large low winged single-place monoplane of all-metal construction with an enclosed cargo hold in the forward portion for air-mail and package air-freight. Due to the preference voiced by seasoned flyers, the pilot still sat far aft in an open cockpit. Bristling with innovations both aerodynamic and structural, a most interesting feature on the "200" was, of course, the landing gear which retracted into wing wells to eliminate most of its parasitic resistance. Not fully retracted, the slightly protruding wheels still allowed a safe landing of sorts with "wheels up" in case of mal-

function to the extending mechanism, or in case the "pilot just forgot." Sensing that the "Monomail" was just the beginning of much greater possibilities for air-carriers along this line of concept, the Boeing system did not stock up on this model, but used only a single example of the cargo version as a test-bed for serious planning into the future of air transport. Powered with the 9 cyl. Pratt & Whitney "Hornet B" engine of 575 h.p., the available performance was rather high for a ship of this type and proved conclusively that increases in speed and utility were quite possible without resorting to brute horsepower. Subsequent testing in transcontinental service brought on various revisions to the "Monomail." Testing also proved that engine and propeller designs at this date were incapable of permitting the "200" to deliver its fullest potential. By the time the "Monomail" was fitted with a controllable pitch propeller to more fully utilize its inherent performance, it was slated to be replaced by newer equipment that it had inspired. The type certificate number for the Boeing "Monomail" model 200 all-cargo version was issued 6-24-30; only one example was manufactured by the Boeing Airplane Co. at Seattle, Wash.

Listed below are specifications and performance data for the Boeing "Monomail" model 200 all-cargo version, as powered with the 575 h.p. Hornet B engine: length overall 41'2"; height overall 12'6"; wing span 59'2"; wing chord at root 12'4"; wing chord at tip 7'6";

Fig. 118. Retractable landing gear and low-mounted cantilever wing of Model 200 was new concept in large transports.

total wing area 535 sq.ft.; airfoil "Boeing" 106; wt. empty 4626 lbs.; useful load 3374 lbs.; payload with 135 gal. fuel was 2280 lbs.; gross wt. 8000 lbs.; max. speed 158; cruising speed 137; landing speed 57; climb 720 ft. first min. at sea level; climb to 10,000 ft. was 21.5 min.; average takeoff run 800 ft.; ceiling 14,700 ft.; gas cap. 135 gal.; oil cap. max. 12 gal.; cruising range at 32 gal. per hour was 540 miles; no price was announced.

The semi-monocoque fuselage was built up of square duralumin tubing for the longerons, the longitudinal stiffeners and the oval bulkheads, bolted together in a framework that was then covered with smooth dural metal skin. The mail-cargo area up forward, with 220 cu.ft. capacity, was divided into 3 pits with access to each by a metal hatch cover on top. Behind the cargo compartment was the pilot's cockpit which was deep, well protected, and slightly elevated for better vision. The pilot's seat also had 7 inches of up and down adjustment for better visibility over the nose on take-offs and landing. Convenient steps built into the fuselage offered access to the open cockpit which was heated, and offered all the latest in instruments; the wheel brakes were operated from the rudder

pedals and a hand brake was provided for parking. Other appointments in the cockpit were a lighted dash panel, wheel-position indicator, fire extinguisher bottle, first-aid kit and the control for the parachute flares. Radio sending and receiving equipment was stored in a compartment aft of the pilot's cockpit. The cantilever wing framework was of two outer panels that were bolted to the center-section panel, an assembly built integral to the fuselage. The outer wing panels were built up of truss-type spar beams of bolted dural tubing, with truss-type wing ribs also of dural tubing; the completed framework was covered with smooth dural metal skin. The integral center-section panel of the wing housed two fuel tanks and also the retracting landing gear, which folded back into wheel wells; the wide tread gear used "Boeing" oleo type shock absorbers, and wheel brakes were standard equipment. A full-caster tail wheel with oleo shock absorber aided in ground maneuvering. The "Hornet" engine was mounted on a welded steel tube frame and was shrouded with a low-drag "speed ring" cowling; a collector ring with long tail pipe supplied heat to the carburetor and the pilot's cockpit. The large tail group was a dural tube structure cov-

Fig. 119. Single place "200" was all-cargo version.

ered with dural metal sheet; the fin was ground adjustable and the horizontal stabilizer was adjustable in flight. All movable control surfaces were of the over-hanging hinge type, for aerodynamic balance, to lighten the "stick loads." When pressed into regular service on Boeing's line, a complete set of night-flying equipment was installed on the Model 200. An adjustable metal propeller, and a combination hand-electric inertia type engine starter were also standard equipment. The next development in the "Monomail" was the Model 221, a passenger carrying version as described in the chapter for ATC #366 in this volume.

Listed below is the only example built of the "Monomail" model 200, as verified by company records:

X-725W; Model 200 (#1153) Hornet B-575.

Serial #1153 later converted into passenger carrying model 221-A.

Fig. 120. "Kari-Keen" model 90 with 90 h.p. Lambert engine.

All of the approved offerings in the small cabin airplane, for the past two seasons, had more than one thing in common, that is, they were mostly strut braced high-winged monoplanes of more or less comparable aerodynamic geometry and of a configuration that varied only to a small degree. The sweep of a line, the curve of a wing tip or the angle of a strut, was perhaps the only distinguishing mark between them all. But the pert "Kari-Keen," also a small cabin airplane for the seating of two, matched its purpose in somewhat typical manner but with the introduction of lines and features that set it well apart from all the rest. Of course, the prime distinguishing feature first seen in the Kari-Keen monoplane was its internally braced cantilever wing, a stout wooden framework of thick and robust proportion, the likes of which had never been used before on a high-wing monoplane so small. With the wing fastened directly to the top-side of the fuselage frame and no bracing struts hindering the view, one had the slight feeling of being suspended from some sort of sky-hook, and not the feeling of being encased in a cage-like framework as in most other small aircraft. Blessed with good performance, even in its earlier form, the new "Kari-Keen" as improved for 1930-31, incorporated many small modifications and the engine horsepower was upped a good bit to keep pace with the trend for higher performance. Standing to be judged on its own merits,

the "Kari-Keen" 90 had much to offer the private owner-pilot, both in practical utility and in performance, but enthusiastic acceptance of a new airplane in a field already over-crowded and with a market that was slowly dwindling to almost nothing, was just too much to expect.

The Kari-Keen cabin monoplane, first introduced just two years earlier by an auto-accessory firm of the same name, was at that time powered with a 5 cyl. Velie M-5 engine and known in this version as the "Coupe 60." Fairly well known in the middle-west and quite popular in that section, the "Kari-Keen 60" was built in only 24 examples, and the whole operation finally succumbed to the early days of the business depression in late 1929. In the hands of a receiver for some time, refinancing revived the company again and a concerted effort was made to promote the "60 Coupe" by way of heavy advertising and a display at the Detroit Air Show of early 1930; plans had also been laid for an improved model. The new model, the Kari-Keen 90, briefly enjoyed a spirited sales campaign but the small operating budget apparently did not permit this enthusiastic ballyhoo for very long. The "Kari-Keen 90," as redesigned by Stanley LaSha and Richard Gazely (engineering consultants of Washington, D.C.) had been more or less in development since late 1929 and was formally introduced in June of 1930. It is commendable that they stuck to the basic design, as laid out by

Fig. 121. Uncluttered lines of "Kari-Keen" 90 offered high performance on nominal power.

Swen Swanson, which was excellent in itself and only introduced such features as would be normal to the progressive improvement of a good design. Now powered with the 5 cyl. Lambert R-266 engine of 90 h.p., with moment arms slightly lengthened to improve the ride, the "Ninety" was shaplier and cleaner looking, with a performance that was more closely attuned to the needs of the average pilot operating from any of the small grassy airports. Flight characteristics were very pleasant, with repeated remarks that it flew exceptionally well and handled easy; with a design load factor of 6.5 positive and almost that in negative, its rugged nature tended to offer peace of mind during rough handling. With the cantilever wing suspended high off the ground, several desirable features automatically came into play. Walking around the airplane was easily done without jockeying around and ducking under to miss a wing strut; likewise, entry or exit was quick and easy through a large door without the practiced gymnastics usually required just to find the step. The pendulum stability inherent in the high wing design was perhaps slightly less in this cantilever configuration, but hardly enough to be noticeable. The visibility was, of course, excellent and the clean uncluttered design naturally paid off in some performance increases.

The "Kari-Keen 90," as already described in part, was a high-winged cabin monoplane with side-by-side seating for two in ample room and comfort, a chummy type of airplane that was quite often described as rather easy to like. The powerplant for the new "Model 90" was the Lambert engine of 90 h.p., an engine that was developed from the classic "Velie M-5"; the added horsepower (30) in this new Kari-Keen uncovered many of its better features. Knowing the airplane through slight acquaint-

ance and impressed by it to the point of admiration, it would be a distinct pleasure to be able to say that the "Ninety" was built in several hundred examples, to be found in all corners of the country, but actually this was not the case. The model 90 was built in only a very few examples and all too soon the company floundered again for a while until it was forced to go under for the second time. Revived again briefly in late 1931 as the "Sioux Coupe" by the Sioux Aircraft Corp., the "Kari-Keen" design made one more valiant try for some fame and a small profit, but this brief venture was to be its last. The type certificate number for the "Kari-Keen 90" as powered with the 90 h.p. Lambert R-266 engine was issued on 6-25-30 and 6 examples of this model were manufactured by the Kari-Keen Aircraft, Inc. at Sioux City, Iowa. Ryal Miller was president and general manager; E. Alden was V.P.; Arney Stensrud was sales manager with a possibility that Merle F. Cornell replaced him at a later date when Stensrud was promoted to factory superintendent in charge of production; engineering was by Stanley LaSha and Richard Gazely; Ole Fahlin was manager of the propeller division. Kari-Keen Aircraft, Inc. in its first reorganization, also absorbed the Dakota Propeller Co. with Ole Fahlin in charge of this division. Most people will remember Ole Fahlin as "the little ol' prop maker" but his diversified talents saw him doing far more than just that; when he wasn't busy carving propellers, he flew on test-flights, demonstration flights and assisted in sales promotion. Dr. B. H. Sprague, the former president, had left for California to set up a west coast sales branch.

Being a quiet, peaceful and extremely modest man, not much has ever been revealed about Swen S. Swanson so we feel it is quite appropriate to dwell here in some length on some

Fig. 122. "Kari-Keen" 90 was slight improvement over earlier Swanson design.

of his many accomplishments. Avidly interested in aeronautics, Swanson designed his first airplane in 1915 while still in his teens; the Swanson "Sport" with a 2 cyl. Lawrance engine was designed and built in 1922 while he was still in college. Swanson came to Lincoln-Standard Airplane Co. (formerly Nebrasks Aircraft Co.) in 1923, shortly after graduating from his aeronautical engineering training in Vermillion, So. Dakota, and became their chief engineer. At Lincoln-Standard in 1924, he teamed up with Harold K. Phillips, superintendent of maintenance at the L-S shop. They designed the single-place Lincoln "Sport" (largely based on designs for the earlier Swanson "Sport") — a diminutive biplane of 18 foot wing span with a 3 cyl. Anzani (French) engine of some 35 h.p. Outstanding performance was soon established, the word got around, and inquiries came from far and near. Construction kits were assembled with parts and materials for the home-builder and the frisky little "Sport" was soon to be seen flying in all parts of the country. Working happily in the repair shop most of the time, Swanson helped to recover the fabric on a Fokker D-7 "pursuit ship" of W.W.I vintage and was so intrigued with the internally braced cantilever wings used on the "Fokker" that they gave him inspiration for several new designs. He began toying with the idea of designing a small sport biplane with the cantilever wings. The well known "Arrow Sport" of 1926-27 was Swanson's first design to utilize the cantilever wing structure and after that time every one of his designs was patterned around the internally braced wing. Joining Kari-Keen Aircraft early in 1928, Swanson designed the Kari-Keen "Coupe" which was a high winged cabin monoplane that employed the cantilever wing, a craft which reflected a design style partially borrowed from the "Arrow Sport." Organized in June of 1928, Kari-Keen built 24 of these little craft through 1929, then was forced to

close its doors at the onset of the depression. The plant was still closed in the spring of 1930 but Swanson had already laid plans for an improved "Coupe." Fretting over the inactivity, Swanson left Kari-Keen and began developing the beautiful Swanson "Coupe" model W-15, a craft that was later built in Virginia. Meanwhile, the Kari-Keen plant was refinanced and reorganized later in 1930 and engineers LaSha and Gazely designed the "Kari-Keen 90" which was a modification of the earlier Swanson (Kari-Keen 60) design. Failing to achieve any success with his new "Coupe" model W-15, Swanson returned to the mid-west to design two more outstanding airplanes (the Swanson-Fahlin "Coupe" and the "Plymacoupe") before he succumbed to a very bad case of pneumonia. Swen S. Swanson passed away as a young man, but he did earn and leave behind the reputation among his fellow-men of being a quiet, modest man, with a lot of foresight, and he certainly possessed that rare genius of being an outstanding aircraft designer and engineer.

Listed below are specifications and performance data for the "Kari-Keen 90", as powered with the 90 h.p. Lambert R-266 engine: length overall 23'2"; height overall 7'0"; wing span 30'0"; wing chord at root 90"; wing chord at tip 54"; total wing area 150 sq. ft.; airfoil "Eiffel" 385; wt. empty 1014 lbs.; useful load 534 lbs.; payload with 25 gal. fuel 199 lbs.; gross wt. 1548 lbs.; max. speed 115; cruising speed 100; landing speed 45; climb 850 ft. first min. at sea level; ceiling 14,000 ft.; gas cap. normal 25 gal.; gas cap. max. 30 gal.; oil cap. 2.5 & 3.5 gal.; cruising range at 5.5 gal. per hour was 420 miles; price at the factory with Goodyear airwheels and brakes was $3365.

The robust fuselage was of specification 1025 steel tubing welded into an exceptionally rigid and sturdy framework, faired to shape with wooden fairing strips and fabric covered; aluminum panels covered the front section to the forward door hinge. Seating on a bench-type seat was side-by-side for two and a small baggage shelf with allowance for 25 lbs. was behind the seat; dual controls were of the joy-stick type. A large rectangular door and a convenient step on the right side afforded easy entry to the cabin which had ample width for plenty of shoulder room; a sky-light in the cabin roof offered vision overhead. Ample window area on all sides offered good visibility and the vee-type windshield in front was built up of flat sections to avoid distortion. The cantilever wing, fabricated into one continuous panel, tapered both in planform and section, was built up with box-type spar beams of heavy spruce flanges and 1/8 in. birch plywood sides; the truss-type wing ribs were of spruce, cap-strips and diagonals that were gusseted with mahogany plywood. Ailerons were of similar wooden construction and the completed wing framework was covered in fabric; the wing attached directly to the top longerons of the fuselage. A gravity-feed fuel tank of up to 30 gal. capacity was built into the leading edge of the wing within the cabin area; the fuel level gauge was visible in the cockpit. The split-axle landing gear of fairly wide tread and built up of

welded chrome-moly steel tubing, was of the cross-axle layout using oleo and spring shock absorbing struts; low pressure airwheels with brakes were standard equipment. The fabric covered tail-group was built up of welded 1025 steel tubing and the horizontal stabilizer was adjustable in flight; all models were also eligible with revised tail surfaces of more area. A wooden propeller and wiring for navigation lights were also standard equipment. The defunct Kari-Keen Aircraft, Inc. was reorganized again later in 1930 by the Sioux Aircraft Corp. offering the models 90-A and 90-B on ATC #413 and #414.

Listed below are examples of the Kari-Keen "Model 90" as gleaned from registration records:
NC-10190; Kari-Keen 90 (#L-301) Lambert 90.
NC-10380;   "   (#L-302)   "
NC-10447;   "   (#L-303)   "
NC-10448;   "   (#L-304)   "
NC-10543;   "   (#L-305)   "
NC-10544;   "   (#L-306)   "
Serial #L-306 later modified into model 90-A with 90 h.p. Brownback "Tiger" engine; approval for Kari-Keen 90 expired on 7-1-33.

A.T.C. #332
(6-25-30)
OGDEN "OSPREY," MODEL PC

Fig. 123. Ogden "Osprey" was baby tri-motor with 3 American "Cirrus" engines.

Another interesting entrant in the "baby tri-motor" field was the Ogden "Osprey," a medium-sized craft of quite plain and proper arrangement that promised only to offer safer air transportation with the maximum in reliability. A magic-sounding formula for safety in air travel, at least during these particular times, was the tri-engined airplane which promised the ability to continue flying even though there was failure in one of the engines. The "Osprey" even promised marginal flight capabilities on one engine only, a feat that would enable the craft to at least stagger sluggishly to some nearby haven of refuge for engine repairs. This promise of relative safety, especially in a craft suitable for business people or as a small air-liner, was an attraction that stimulated a speculative interest and would have resulted in a healthy rate of buying; however the introduction of the small tri-motor unfortunately came at a time when the economy of this nation was literally falling apart at the seams. Potential customers for this type of airplane generally found it nearly impossible or inadvisable to invest in the advantages of owning such an airplane. In spite of the unfavorable atmosphere for sales of aircraft in volume, several examples of the "Osprey" were built and nearly all were flown in a useful service of one kind or another. One "Osprey" was being readied for the All-American "Cirrus Derby" of 1930 but there is no verified record of its participation. Another "Osprey" was later bedecked with paraphenalia by a concern that called it a "sky-sign" and a "broadcaster," an arrangement that was used in the development (signs and loud-speakers) of air-borne advertising. In Jan. of 1933, after

weathering about 3 years of the so-called "Depression," four of the "Osprey" tri-motors were offered for sale at clearance prices by Henry Ogden. It is not known how he fell heir to the quartet of airplanes but it has been suggested that this was his share of the business. One of the ships had been used for a total of about 900 hours, two of them had about 200 hours of usage and one was still brand new. The new ship was offered for $5,000 or all four could have been bought for $10,000, with a large inventory of spare parts thrown in to boot. This offer poses immediately as a terrific buy but it might just as well have been a million dollars because airplane people just didn't have that kind of money in 1933.

Henry H. Ogden, noted Army flyer of 1924 round-the-world fame, left Kreutzer Aircraft about May of 1929 where he had been test-pilot during the development of the tri-motored "Air Coach" series. Either disgruntled or perhaps just looking for greener pastures of opportunity, Ogden was said to have instigated the development of the "Osprey" monoplane which was designed and engineered by Frederick G. Thearle, an engineer previously associated with the veteran Otto Timm. Henry Ogden test-hopped the prototype example later in 1929 and a second ship was quickly readied for show display at the St. Louis Air Exposition; this ship was tested in March of 1930 for approval on a Group 2 certificate. Some reports describe the very first experimental version of the "Osprey" as having three 5 cyl. Kinner K5 engines of 90-100 h.p. each, and performance data was even released for this example, but the engines must have been replaced by 3 "Cirrus" engines ra-

Fig. 124. Neatly cowled in-line engines of "Osprey" were innovation for multi-motored craft.

ther early in the development and the "Osprey" became the first tri-motor with small aircooled in-line engines. According to the comparison of data for these two different prototype installations there were definite advantages to be gained by use of the in-line engines; smaller frontal area and easier streamlining accounted for a generous boost in speed, and even an improvement in the craft's general behavior. Organized as the Ogden Aeronautical Corp. in May of 1929 at Inglewood, Calif., the company secured a modest plant with 5000 sq. ft. of floor space and tooled-up in mid-1930 to manufacture about one airplane per month. The "Osprey" was eventually built in three versions that differed mainly in the powerplants that were installed; the later versions coming off, used in-

verted in-line "Menasco" engines of 95 h.p. or 125 h.p. each.

The standard Ogden "Osprey" model PC (also listed as Model C) was a 6 place tri-motored cabin monoplane of the high winged type, a configuration that was quite conventional in most respects except that it was powered with three 4 cyl. "American Cirrus" engines of 90 h.p. each. Built to carry a substantial payload with good performance, the "Osprey" was primarily designed for the business-executive or the small feeder air-lines. Flight tests were conducted by Henry Ogden for over three months and performance was proved to be excellent in all respects. The big "Osprey" flew quite safely on any two of its three engines and was even reported to hold its own nicely in a lightly loaded condi-

Fig. 125. "Osprey" shown was fitted with loud-speakers for aerial broadcasting.

tion, with only the center engine operating. Top speeds of 130 m.p.h. were consistently recorded and economical cruising speeds were attained at 100 m.p.h. The "Osprey," though seemingly quite plain at first glance, had several unique features in its design and construction. This was the first "tri-motor" to use small in-line aircooled engines, and head resistance was thereby kept to a minimum because of the excellent cowling of the engines and streamlining of the engine nacelles. A then novel method of engine cooling was developed whereby the air entered the front aperture in controlled quantity and was baffled and deflected around each cylinder; the hot air was sucked out at the rear of the engine. This baffling and shrouding system insured even cooling around the cylinders and ideal operating temperatures at all times. Normally flown with two pilots at the helm, the control column had a swing-over wheel quickly accessible to either pilot; the individual wheel brakes were operated by a hand-brake lever located between the two front seats. This lever applied the desired braking tension and the rudder pedals transmitted the braking power to either one or both wheels. The large and roomy cabin area was 13 feet long and all seating could be easily removed for hauling cargo. Other cabin arrangements included deep plush seats and a small lavatory to the rear; even fancy curtains were available. The type certificate number for the Ogden "Osprey" model PC, as powered with 3 "Cirrus" engines, was issued 6-25-30 and 5 or more examples of this model were manufactured by the Ogden Aeronautical Corp. at Inglewood, Calif. L. Lindsay was president; Henry H. Ogden was V.P., general manager and chief pilot; Frederick G. Thearle was chief of design & engineering; Edward Donahue was factory superintendent.

Listed below are specifications and performance data for the Ogden "Osprey" model PC as powered with 3 A.C.E. "Cirrus" engines of 90 h.p. each: length overall 34'6"; height overall 9'3"; wing span 50'0"; wing chord 84"; total wing area 312 sq. ft.; airfoil Goettingen 398; wt. empty 2898 lbs.; useful load 1650 lbs.; payload with 90 gal. fuel and 1 pilot was 900 lbs., 735 lbs. with 2 pilots; gross wt. 4548 lbs.; max. speed 128; cruising speed 102; stalling speed 65; landing speed 60; climb 650 ft. first min. at sea level; climb in 10 min. was 5000 ft.; ceiling 15,000 ft.; gas cap. 90 gal.; oil cap. 6 gal.; cruising range at 18 gal. per hour was 450 miles; price at the factory was $18,900, lowered to $16,000, in May of 1931.

The fuselage framework was built up of welded chrome-moly steel tubing in a Warren truss form, faired to shape with wooden formers and fairing strips, then fabric covered. The cabin walls were insulated from the extremes in wea-

ther by "Insulite" and sound-proofed with blankets of "Dry-Zero," permitting conversation in near-normal tones; interior cabin lighting was fitted, there were adjustable louvers for ventilation and cabin heating was also available. Passenger seats were welded steel tube frames upholstered with wicker and leather trim and the cabin interior was upholstered in fine fabrics; to the rear of the cabin was a small lavatory, a 12 cu. ft. baggage compartment with allowance for 50 lbs. and an emergency exit door. Entry to the 154x45x55 inch cabin area was through a large door on left side rear with a large step for easy access; all windows were of shatter-proof glass and those at the pilots station could be rolled down. The husky wing framework, consisting of a center-section and two outer panels, were built up of spruce and plywood box-type spar beams with spruce truss-type wing ribs reinforced with mahogany plywood gussets; the leading edges were covered with spruce plywood sheet to the front spar and the completed framework was covered in fabric. Two 48 gal. fuel tanks of terne plate and dural metal sheet were mounted in the center-section panel, one each side of the fuselage. The wing was braced to the fuselage by two parallel struts with the engine mountings and landing gear fastened to the wing brace struts to form a strong and rigid truss. The outrigger landing gear of 168 in. tread used oil-draulic shock absorbing struts; wheels were 32x6 and Bendix brakes were standard equipment. A swiveling steerable tail wheel, mounted on the extreme end of the fuselage, was provided to improve the ground maneuvering. The fabric covered tail-group was built up of welded chrome-moly steel tubing; the rudder was notched to clear the tail wheel and the horizontal stabilizer was adjustable in flight. Metal propellers, 3 hand-crank Eclipse engine starters, tail wheel, Bendix brakes, and navigation lights were standard equipment. The next "Osprey" development was the Menasco-powered model PB as approved on a Group 2 approval numbered 2-295.

Listed below are Ogden "Osprey" model PC entries as gleaned from registration records:
X-187N; Osprey PC (#101) 3 Amer. Cirrus.
NC-398V;   "   (#102)   "
NC-121W;   "   (#103)   "
NC-122W;   "   (#104)   "
NC-149W;   "   (#106)   "
Serial #101 was experimental prototype and not on this approval; serial #102-103 first on Group 2 approval numbered 2-197 (issued 3-26-30) which was superseded by ATC #332; serial #105 (X-150W) was powered with Menasco B-4 engines on Group 2 approval numbered 2-295, issued 10-30-30; all examples also eligible with 3 A.C.E. "Cirrus" engines of 100 h.p. each.

Fig. 126. Model HSO was popular "Waco" straight-wing with a 225 h.p. Packard "Diesel" engine; shown here during National Air Tour.

By this particular time the quite numerous and very popular Waco "straight-wing" had almost run its course of normal development, and shiny new models looming on the horizon, relegated the SO-series to finish off the balance of its colorful existence by acting as test-beds for newly introduced powerplants. As an example in point, the "Waco" model HSO was specifically arranged to mount the new Packard "Diesel" engine for continued development in rigorous flight-test and to further exploit the popularity that accompanied this engine's introduction to the aircraft market. The model HSO was the third certificated installation of this engine, and soon several more manufacturers were to follow suit — in an effort to be ready for the flood of buying that was really expected to materialize as a result of promises held forth by this revolutionary powerplant. A Packard "Diesel" powered "Waco" biplane (HSO) was selected as the "official ship" of the 1930 National Air Tour to ferry the tour's manager around the course, enabling him to keep a watchful eye on the daily proceedings. Piloted by the venerable Walter E. Lees, who as a service engineer already had considerable experience in the development of this engine, and carrying genial Ray Collins the manager of the tour, this craft with its oil-burning powerplant created

enormous interest during the tour's many stops. Carrying these two men with all their baggage and maintaining a 100 m.p.h. cruising average, the "Waco" burned only 437 gals. of cheap furnace-oil on the 4935 mile course and averaged less than one cent per mile for fuel. Besides the economy of operation demonstrated time and again, the furnace-oil burning "Diesel" was fire-safe under most any conditions, which in itself was a great point of advantage. Though a somewhat controversial subject throughout later years, it might well be said that in the adaption of the "diesel" principle to a radial type air-cooled aircraft engine, the Packard Motor Car Co. had contributed at least a new and revolutionary chapter to aeronautical progress.

In the diesel-type engine, the real point of difference from the conventional system was the type of fuel ignition that was used. In the gasoline burning engine an electric spark is used to ignite the combustible mixture for a power stroke, while the "diesel" engine drew fresh air into the cylinders, and this air alone was compressed in the combustion chambers under very high pressure; the rather high compression pressure naturally squeezed and heated the air to a high degree. When atomized fuel, such as furnace-oil, was injected into the combustion chambers, the fuel was ignited as soon

Fig. 127. Airframe of HSO also tested in taper-wing configuration; shown here on Packard testing grounds.

as it came in contact with the super-heated air to cause a combustion or power stroke. This system known as "compression ignition" did not need all the devices used on gasoline-burning engines and therefore laid claim to simplicity and reliability; absence of an electrical "ignition system" also avoided causes of static and interference in aircraft radios. Numerous aircraft manufacturers introduced models powered by the Packard "Diesel" engine and several more were in various stages of planning and construction; it certainly posed a bright future for the "diesel" airplane engine. After the untimely death of Capt. Lionel Woolson (one of the designers greatly responsible for the development of this engine), and due to the unfortunate sag in aircraft buying, the Packard Motor Co. lost enthusiasm in its further development and promotion, choosing to shelve the project for the more favorable conditions of another time. Unfortunately, the project was never revived.

The "Waco" model HSO was an open cockpit biplane with seating for 3 and was more or less typical of previous models, such as the BSO and COS in most all respects, except for its power-plant installation and other slight modifications necessary for this particular combination. Power in this case was the new 9 cyl. Packard "Diesel" (DR-980) engine of 225 h.p. rated at 1950 r.p.m. Slightly heavier than the model CSO when empty, because of the heavier "Diesel" engine, the model HSO was nevertheless comparable in its all-round performance and general characteristics of behavior, suffering

only in some loss of useful load. For performance tests conducted on the sprawling Packard proving grounds, the motor company selected the tried and true "Waco" straight-wing for numerous trials on this new engine; because of the easy conversion from one configuration to the other, the HSO airframe was also tested in the "Taper-Wing" version known generally as the HTO. Either version performed exceptionally well and forecast the promise of a good future for the economical "Diesel" engine in craft of the general-purpose type. As far as can be remembered or determined from available records, this was the only example of the "Waco" biplane powered with the Packard "Diesel" engine. The type certificate number for approval of the standard model HSO "straight-wing" was issued 6-27-30 and the tapered wing version of this same airplane remained in the experimental or restricted category.

Listed below are specifications and performance data for the "Waco" model HSO, as powered with the 225 h.p. Packard DR-980 engine: length overall 22'4"; height overall 9'2"; wing span upper 30'7"; wing span lower 29'5"; wing chord both 62.5"; wing area upper 155 sq.ft.; wing area lower 133 sq.ft.; total wing area 288 sq.ft.; airfoil "Aeromarine" 2A modified; wt. empty 1814 lbs.; useful load 786 lbs.; payload with 50 gal. fuel was 251 lbs.; gross wt. 2600 lbs.; max. speed 128; cruising speed 108; landing speed 45; climb 1100 ft. first min. at sea level; service ceiling 20,000 ft.; fuel cap. 50-60 gal.; oil cap. 8 gal.; cruising range at 10 gal. per hour was 490-585 miles; price at the factory was $7500.

Fig. 128. Waco "Taper-Wing" and Packard Diesel were good combination but not slated for production.

Construction details and general arrangement for the model HSO were more or less typical of the model CSO as described in the chapter for ATC #240 in U.S. CIVIL AIRCRAFT, Vol. 3. Because of the increased empty weight and resulting loss in payload, the model HSO should be normally considered as a two-place airplane. Parachute packs at 20 lbs. each were considered as part of the payload so with pilot and one passenger at 165 lbs. each, wearing parachutes, baggage allowance was held to 50 lbs. A special metal propeller, an Eclipse engine starter, Bendix wheels and brakes, and dual joy-stick controls, were standard equipment. Navigation lights, Goodrich semi-airwheels, and windshields built up of shatter-proof glass panels were added equipment. The next "Waco" development was the model QSO as described in the chapter for ATC #337 in this volume.

Listed below is the only known example of the "Waco" model HSO as gleaned from various records:

NC-4N; Model HSO (#X-3101) Packard Diesel.

Registration number X-4N and NR-4N was the same airplane fitted with "tapered" wing panels.

A.T.C. #334
(6-27-30)
AEROMARINE-KLEMM,
AKL-26-B

Fig. 129. Aeromarine-Klemm model AKL-26-B mounted 85 h.p. Le Blond 5DF engine.

As discussed previously in several other chapters of U.S. CIVIL AIRCRAFT, the broad-winged Aeromarine-Klemm monoplane was an unusual light airplane — an airplane that was actually born of a motored glider in Germany some years back, and started out here in America as a 2-seated craft with an engine of 40 h.p. As originally conceived, it was a sport-craft offering flying in its strictest economy with a minimum of repairs and maintenance; a delightful combination full of surprises that delivered an amazing performance. Bear in mind that it was designed mainly for those who enjoyed flying for the sheer fun of it, within those limits imposed by a very inexpensive operating arrangement. The flight characteristics and general behavior of this 40 h.p. "flivver plane" were certainly pleasant and enjoyable, but if you were contemplating a long cross-country hop hoping to maintain high average speeds, you'd have done well to forget it. Most likely this is one of the reasons that "Aeromarine" finally gave in to the snow-balling trend of simply adding more power to get an overall increase in performance; thus, the next version of the AKL series sported a bigger engine of some 60 h.p. This added power was bound to bring out a little more verve and increased performance in general but not to a proportionate degree, and it was a clearly forseen fact that another power increase would soon be in order. We might agree that this was the right way and the best way to go because the average owner-pilot fairly worshipped the thrill of higher performance, but in the process of attaining that,

the simple uncomplicated reason for the little AKL-40 in the first place was now ignored and lost sight of. So now the latest offering by Aeromarine-Klemm was the model AKL-26-B, a typical version that held the line on damaging weight increases to get the most out of its 85 h.p. engine. Again there were some slight increases in all-round performance, the "85" fairing better in this respect than previous models, but of course operating expenses went up too; we strongly believe that this to some extent, actually turned away the potential customers it was supposed to attract. In spite of the fact that this discussion might sound a bit detrimental to the new AKL-26-B (also AKL-85) it must be conceded that by average standards used as a "yard-stick" for comparing airplanes in this country, the new model was a good airplane, and perhaps the finest ship that Aeromarine-Klemm ever built; it's a pity it was to be the last.

The Aeromarine-Klemm model AKL-85 (AKL-26B) was an open cockpit low-winged monoplane with seating for 2 in tandem; although still a typical light-plane in basic weights it was of surprisingly large dimension and was hardly a typical light airplane in size. The AKL inherited much of its rugged character and good solid feel from a robust vibration-absorbing and shock-deadening all wood construction. Of very good aerodynamic proportion in a simple and uncluttered design, the new AKL was very stable, practically flew by itself, and was often laughingly accused of causing student-pilots to day-dream while lazily viewing the pano-

Fig. 130. Increased performance of AKL-85 ideal for sport or pilot training.

rama of beauty below. Powered with the new 5 cyl. LeBlond 5DF engine of 85 h.p. the model AKL-85 did have much more verve in its character aloft than the previous lower-powered models of the AKL series, but it was quite easily managed and still one of the best training airplanes in this country. Fortified now with extra horsepower; the AKL-85 showed eager response and was well able to perform all the advanced maneuvers required in secondary pilot-training courses. More expensive now and slightly more costly to operate, this model was not the bargain that earlier models had been, but it had much in its favor and still managed to pay off well in profit and other dividends. The type certificate number for the model AKL-26-B (AKL-85) was issued 6-27-30 and 12 or more examples of this model were manufactured by the Aeromarine-Klemm Corp. at Keyport, N.J. with executive offices in New York City. Inglis M. Uppercu was president and C. L. Zakhartchenko was the chief engineer. By late 1931 business at A-K had fallen off drastically and preparations were

being made for an early plant shut-down; by mid-1932 approval for manufacture had expired on all models of the AKL. There was a half-hearted attempt in 1937 to revive the AKL design by Horace Keane, pioneer airplane builder, with the power of a Ford V-8 auto engine — but this offering met with no success in spite of its merits.

Listed below are specifications and performance data for the Aeromarine-Klemm model AKL-26-B, as powered with the 85 h.p. LeBlond 5DF engine: length overall 23'6"; height overall 7'0"; wing span 40'2"; wing chord at root 79"; total wing area 194 sq.ft.; airfoil Goettingen 387 modified; wt. empty 1016 lbs.; useful load 565 lbs.; payload with 29 gal. fuel was 206 lbs., allowing 170 lbs. for passenger and 36 lbs. for baggage; gross wt. 1581 lbs.; max. speed 97; cruising speed 85; stall speed 42; landing speed 40; climb 750 ft. first min. at sea level; ceiling 12,000 ft.; gas cap. 29 gal.; oil cap. 2 gal.; cruising range at 5.5 gal. per hour was 385 miles; price at the factory was $3700.

Fig. 131. Rugged nature of AKL popular with private-owner flyers.

Fig. 132. Aeromarine-Klemm testing unusual amphibious gear.

The boxy fuselage framework was built up of spruce longeron members with plywood bulkheads and formers; the completed framework was covered with plywood sheet to form a stiff semi-monocoque shell. The cockpits were now somewhat larger and deeper to offer better protection and easier entry or exit while wearing bulky clothing. The engine mount was a welded steel tube structure easily detachable from the framework; a baggage compartment of 6 cu. ft. capacity with allowance for 36 lbs. was in the turtle-back section of the fuselage behind the rear cockpit. The cantilever wing framework, in 3 sections, was built up of spruce and plywood box-type spar beams with plywood wing ribs that were lightened with suitable cut-outs; the completed framework was covered with laminated plywood sheet. The outer wing panels were of tapering chord and thickness. A gravity-feed fuel tank of 15 gal. capacity was mounted high in the fuselage just ahead of the front cockpit; an extra 14 gal. fuel tank was mounted in the right hand side of the wing center section. The split-axle landing gear of 85 inch tread was an inverted tripod structure on each side using oil-draulic shock absorbing struts; wheels were 20x9 inch "airwheels" and brakes were optional. The fixed surfaces of the tail-group were built up of spruce spars and plywood ribs, covered with thin plywood sheet; the movable surfaces were also built up of spruce spar beams and plywood rib-formers but covered with fabric. The

vertical stabilizer and the horizontal stabilizer were adjustable on the ground only and this applied to all of the AKL models; dual controls of the joy-stick type were provided. A tail-bracing strut of streamlined steel tubing was later added to dampen vibrations in this structure. An "Aeromarine" wooden propeller was standard equipment with wheel brakes, a speed-ring engine cowling, navigation lights and engine starter were listed as optional.

Listed below are Aeromarine-Klemm AKL-26-B (AKL-85) entries as gleaned from registration records:

	AKL-26B		LeBlond 5DF-85
-92N;	"	(#)	"
NR-387N;	"	(#2-62)	"
NC-393N;	"	(#3-68)	"
NC-394N;	"	(#3-69)	"
-863W;	"	(#3-74)	"
NC-864W;	"	(#3-75)	"
NC-753N;	"	(#3-86)	"
NC-866W;	"	(#4-79)	"
NC-869W;	"	(#4-80)	"
NC-870W;	"	(#4-81)	"
NC-868W;	"	(# -82)	"
NC-780Y;	"	(# -83)	"

Serial number for -92N unknown; serial # 2-62 listed also as L-27; serial # 3-74 listed also as L-27; serial #3-75 listed also as # 75; serial # -82 and -83 may have been # 4-82 and 4-83; serial # 3-73 and # 3-82 also eligible with heavier model L-27 center-section panel; approval expired 7-1-32.

STINSON "AIR LINER", SM-6000

Fig. 133. Stinson "Tri-Motor" model SM-6000 was powered by three 215 h.p. Lycoming engines.

Fortified with liberal financing that came about through a merger with the huge Cord Corp., "Stinson" launched an ambitious program of aircraft manufacture for 1930-31 and utterly amazed those in the industry with its many bargain offerings. Of course Stinson's prices were geared to volume selling and it was not long before manufactured units reached a total large enough to justify the nearly ridiculous prices. With a handful of different "Junior" models now available and older "Detroiter" models still on the available list, Stinson looked eagerly to the possibility of developing a relatively low cost tri-motored airplane for use by the nation's air-lines. Careful studies dictated a certain course to take in the design of a cheaper tri-motor and the SM-6000 "Air Liner", as shown and described here, was the end result. Introduced early in 1930 as a 10 place craft with 3 "Lycoming" engines for $23,900, the new "Air Liner" immediately classed itself as a bargain buy, with interested and unbelieving inquiries coming from all sides. After receiving certification later in 1930, Stinson announced a price reduction on the new SM-6000 to the ridiculously low figure of $18,000, mostly to entice the air-lines into buying, since several of them were known to be about ready for some new or larger equipment anyhow. This move had no immediate effect on any of the established carriers but it did prompt the formation of some new lines that could now offer a specialized service. One of the first of these was the New York-Philadelphia-Washington Airways which was popularly known as the "Ludington Line,"

a million-dollar company headed by C. T. Ludington. The line, buying the first batch to come off the production line, operated 7 of the SM-6000 liners between New York and Washington, D. C., giving one hour service on a schedule of 11 round trips daily. "Every hour on the hour" became their operating slogan. Business soon flourished to the rate of 150 passengers daily, and in the space of a few months, thousands of round trips had been recorded. One would certainly think that a tri-motored airplane in this price range would necessarily be a stripped-down vehicle devoid of normal comforts and conveniences but surprisingly this was not the case with the Stinson "Air Liner"; it was well arranged and well appointed in coach-style comfort for the shorter trips. Because of the nature of this frequent service to New York and Washington, the crew was limited to a pilot only and he was required to do everything. He would take tickets at the plane's door, see that passengers were seated properly and comfortably, fly the ship to its destination, supervise the unloading, and repeat this routine for the return trip. "Ludington Line" pilots were surely kept busy but none seemed to mind being glorified aerial bus-drivers. With increasing business the line added more equipment and the operating fleet soon comprised 10 of the Stinson (SM-6000) "Air Liner".

Though shrouded in some mystery because of the span of time involved, it has been generally conceded that the Stinson "Tri-Motor" was a redesigned development of the "Corman" Tri-Motor. The Corman company was a trade-name

Fig. 134. Corman "Tri-Motor" was basis for Stinson SM-6000 design.

arrived at by mating the names of E. L. Cord and L. B. Manning who were president and V. P. of the Cord Corp. As the Corman Aircraft Corp. of Dayton, O., the company built one or two examples of the "Corman 3000" and "Corman 6000", then turned over the task of redesigning this craft for quantity production to the engineers at Stinson Aircraft. The basic "Corman" as a 7 place craft was powered with three Wright "Whirlwind" J5 engines of 220 h.p. each, but Stinson had better access to Lycoming engines now so these were used in the SM-6000 series. Regardless of the fact that the SM-6000 was supposed to be a development of the earlier "Corman" design, it shows clearly that many great changes took place in the redesigning, and the new "tri-motor" configuration came out looking just as "Stinson as Stinson could be".

The Stinson "Air Liner" model SM-6000 was a large high-winged cabin monoplane of the tri-motored type with seating for eleven in its spacious interior. As developed in the first batch of this series, the "Air Liner" was arranged for coach-type service on the shorter routes; later versions (SM-6000-A and SM-6000-B) in this series were available in several arrangements, including a "Custom Club" special. Early examples of the SM-6000 had bare engines but low drag "Townend-ring" fairings were soon available, as were wheel streamlines over the large wheels to boost cruising speeds a little higher. Powered with 3 nine cyl. Lycoming R-680 engines of 215 h.p. each, the big SM-6000 delivered a performance that belied its well apparent bulk; a fully loaded take off required only some 700 feet and the landing roll hardly ever exceeded 400 feet. It is this short-field performance plus the ability to carry a sizeable payload that prolonged its useful life, so we find it not surprising that at least 25 of the high-wing "Stinson Tri-Motors" were still flying in active service up to 10 years later. The type certificate number for the 11 place model SM-6000 was issued 7-10-30 and at least 10 examples of this particular version were manufactured by the Stinson Aircraft Corp. at Wayne, Mich. "Eddie" Stinson, master pilot, was the

president of the company but he was rather hard to catch at his desk because he spent most of his time in the shop or flying around the countryside promoting "Stinson" airplanes.

Listed below are specifications and performance data for the Stinson "Air Liner" model SM-6000, as powered with 3 Lycoming R-680 engines of 215 h.p. each: length overall 42'10"; height overall 12'0"; wing span 60'0"; wing chord 105"; total wing area 490 sq. ft.; airfoil Goettingen 398; wt. empty 5575 [5625] lbs.; useful load 2825 [2875] lbs.; payload with 120 gal. fuel was 1815 [1865] lbs.; baggage 165 [215]lbs.; gross wt. 8400 [8500] lbs.; figures in brackets as eligible with radio installation; max. speed 138; crusing speed 115; landing speed 60; climb 1000 ft. first min. at sea level; climb to 10,000 ft. was 25 min.; ceiling 15,000 & 14,500 ft.; gas cap. 120 gal.; oil cap. 15 gal.; cruising range at 36 gal. per hour was 345 miles; normal price at the factory was $23,900 with special price announced at $18,000 for short time only.

The fuselage framework was built up of welded chrome-moly steel tubing, faired to shape with formers and fairing strips then fabric covered; the whole forward section to a point just behind the pilot's station was covered in dural metal sheet panels. The pilot's compartment had seating for two side by side and the main cabin was arranged for the seating of ten; a large entry door to the main cabin was on the right side to the rear. The wing framework, in two halves, was built up of chrome-moly steel tube spar beams that were welded into Warren truss girders with wing ribs riveted together of square duralumin tubing; the leading edges were covered with dural metal sheet and the completed framework was covered in fabric. Gravity-feed fuel tanks were mounted in the root ends of each wing half. The wing was braced by two parallel struts on each side and the engine nacelles were mounted to a truss connected to the wing braces; from this extended the landing gear which was of unusually wide tread. The husky landing gear used "Aerol" (air-oil) shock absorbing struts, wheels were 36x8 and Bendix brakes were standard equipment; a swivel-type tail wheel was mounted far aft for better ground maneuvering. The fabric covered tail-group was built up of welded chrome-moly and 1025 steel tubing; the rudder was acrodynamically balanced and the horizontal stabilizer was adjustable in flight. Metal propellers, exhaust collector-rings, electric inertia-type engine starters, wheel brakes and a tail wheel were standard equipment. The next development in the Stinson "Tri-Motor" series was the improved model SM-6000-A as described in the chapter for ATC # 367 in this volume.

Listed below are Stinson model SM-6000 entries as gleaned from registration records:

NC-974W; SM-6000 (# 5005) 3 Lyc. R-680
NC-975W; " (# 5006) "
NC-976W; " (# 5007) "
NC-977W; " (# 5008) "
NC-978W; " (# 5009) "
NC-979W; " (# 5010) "
NC-429Y; " (# 5011) "
NC-429Y; " (# 5012) "
NC-497Y; " (# 5013) "
NC-498Y; " (# 5014) "

This certificate for serial #5005 through #5014; approval expired 7-1-32.

A.T.C. #336
(7-11-30)
AMERICAN-MARCHETTI, S-56-B

Fig. 135. Savoia-Marchetti amphibian with 125 h.p. Kinner B5 engine; S-56-31 shown was one of last to be built.

In nearly every city and hamlet having the advantage of nearby water, there were fairly wealthy owners of motor-cars and speed-boats who were often chafing at the bit, so to speak, complaining because of the week-end road congestion and bother of surface travel to popular resorts and playgrounds. These were the people singled out by "American-Marchetti as likely prospects for their "baby amphibian". These were the people who would be ready and willing to take to the air in order to travel swiftly and unhampered in a small, easy-to-handle, amphibious aircraft expecially designed for private flying. The Savoia-Marchetti series S-56, being a small craft of simple and rugged construction was just about the ideal, and well suited for this type of all-purpose flying; it actually would cost little more to operate and maintain than the average speed-boat. First introduced here in late 1929 (see chapter for ATC # 287 of U. S. CIVIL AIRCRAFT, Vol. 3), the little S-56 was greeted with interest and good acceptance. By Feb. of 1930 at least 15 were being put into service and orders were logged for a good many more. With 100 h.p. tucked into its upper wing, the model S-56 had an average performance to be expected for a craft of this type but, as is always the case, there were those who clamored for more performance and were ready to pay the price. By mid-year 1930 the new S-56-B was introduced with an extra 25 h.p. to placate the

"sports" who were doing the most clamoring. The conversion was really quite simply done by just replacing the 100 h.p. Kinner K5 engine with the new Kinner B5 engine of 125 h.p. As a consequence, the model S-56-B had much more muscle in its frame, that translated into quicker take offs, a better climb-out and a slight increase in all-round speeds. Several of the earlier S-56 type were converted into the more powerful S-56-B version, and a newer improved version was offered as the model S-56-31. Although fairly expensive by comparison with similar craft, the S-56 series sold surprisingly well and quite a few were still in active service 10 years later.

The wooden-hulled model S-56-B was a small biplane of the classic flying boat type with the addition of a simple hand-operated retracting landing gear. The seating for three in two open cockpits might have lacked the plush appointments of formal comfort but it surely did make up for it in flying fun and practical utility. Equally at home on land or water, the perky S-56-B maneuvered well into docking or parking; take offs were clean with about an average run, and landings were never too much of a problem. Powered now with the 5 cyl. Kinner B5 (R-440) engine of 125 h.p., which was mounted under the upper wing in a "tractor" fashion, the model S-56-B had good performance for a versatile craft of this type with maintenance and

Fig. 136. Peter Talbot flying S-56-B stayed aloft for 22 hours; note large gas tank behind pilot.

operating costs still held to a sensible minimum. The newer model S-56-31, although typical, incorporated many slight changes as to structure and detail arrangement but none were of any great significance; all changes were for either convenience or extra strength. As a consequence, the S-56-31 topped off with a gross weight that was 18 lbs. ((2218 lbs. gross) heavier than the S-56-B, but this had little or no effect on overall performance. The only flight of any national interest made by the S-56-B type was a non-refueled endurance record set by Peter Talbot in Aug. of 1930 for light amphibian aircraft in this class; nursing his fuel as best he could, Talbot stayed aloft for 22 hours. Zachary Reynolds (notedtobacco-fortune heir and avid sportsman-pilot) owned and personally flew a special one-place version of the S-56 (S-56-C) that had been modified especially for his particular purpose, but later this craft was converted to a standard model S-56-31. The type certificate number for the model S-56-B was issued 7-11-30 and this was later amended to include the model S-56-31; some 10 or more examples were manufactured by the American Aeronautical Corp. who had a modest factory and a flight-ramp on 12 acres of shore front property in Port Washington, Long Island, N. Y. Ugo V. D'Annunzio was president; P. G. Zimmermann was V. P.; Albert Kapteyn was secretary-treasurer and Peter Talbot was company test pilot. Some records show that in 1931 the Dayton Airplane & Engine Co. of Dayton, Ohio acquired a controlling interest in the

American-Marchetti company but this has not been verified. Production of all models was more or less halted by late 1931 but all activity had not yet ceased and the company continued with a token force several years afterwards.

Listed below are specifications and performance data for the Savoia-Marchetti model S-56-B, as powered with the 125 h.p. Kinner B5 engine: length overall 25'0"; height on wheels 10'1"; wing span upper 34'1"; wing span lower 30'1"; wing chord upper 67"; wing chord lower 52"; wing area upper 179 sq. ft.; wing area lower 107 sq. ft.; total wing area 286 sq. ft.; airfoil "Marchetti"; wt. empty 1462 lbs.; useful load 738 lbs.; payload with 28 gal. fuel was 376 lbs.; payload included 2 passengers at 170 lbs. each and 36 lbs. baggage; gross wt. 2200 lbs.; max. speed 95; cruising speed 80; landing speed 40; climb 650 ft. first min. at sea level; climb in 10 min. was 4500 ft.; ceiling 8500 ft.; gas cap. 28 gal.; oil cap. 3 gal.; cruising range at 7.5 gal. per hour was 280 miles; price was $7875 at the factory ramp, fully equipped.

The hull framework was built up of spruce and ash members that were fastened together with metal gussets and fittings into a fairly rigid structure, a frame that was covered with sheets of plywood and a final coating of fabric to increase the general strength; the hull bottom was reinforced with double planked cedar. The front cockpit seated two in fair comfort with ample protection and dual joy-stick controls were provided. The rear cockpit seating one, most often

Fig. 137. Landing gear retracted; dolly used for beaching.

Fig. 138. Landing gear extended; simple mechanism operated by hand crank.

turned out to be just a place to put things and could be closed off with a metal panel when not in use. When the rear seat carried no passenger the equivalent weight could be carried in baggage, provided it was lashed down and suitably covered; with three aboard, the baggage allowance for the S-56-B was 37 lbs. and 25 lbs. was allowed for the model S-56-31, which included 19 lbs. for the anchoring gear and 6 lbs. for a tool kit. Under these circumstances we can easily see why the S-56 was usually operated as a 2-place airplane. The wing framework was built up of hollow spruce box-type spar beams with spruce and plywood truss-type wing ribs; the leading edges were covered with plywood sheet to preserve the airfoil form and the completed framework was covered in fabric. Two fuel tanks were mounted in the center

Fig. 139. At days-end of flying fun, S-56-B is taxiied up ramp to be put up in tie-down; versatility of the "amphibian" doubles flying pleasure.

section of the upper wing and a hand-hold for guiding was provided in the tip end of each lower wing. The simple landing gear of 64 in. tread was snubbed by rubber shock-cord and retracted or extended by a hand operated lever in the front cockpit; wheels were 28x4 or 6.50x10 and no brakes were normally provided. The horizontal stabilizer was of a wooden framework covered with plywood sheet, with the vertical fin, the rudder and the elevators of welded steel tubing covered in fabric. Ailerons were in the upper wings only and all movable surfaces were operated by stranded steel cable. A "Paragon" wooden propeller, a Heywood air operated engine starter, navigation lights, fire extinguisher, first-aid kit, an anchor and mooring lines, battery, life preserver cushions, engine cover and cockpit covers were all standard equipment.

Listed below are model S-56-B and model

S-56-31 entries as gleaned from registration records:

NC-324N;	S-56-B	(# 10)	Kinner R-440.
NC-351N;	"	(# 12A)	"
NC-356N;	"	(# 19)	"
NC-386N;	"	(# 29)	"
NC-900V;	"	(# 30)	"
NC-897V;	"	(# 31)	"
NC-906V;	"	(# 36)	"
NC-898W;	S-56-31	(#27)	"
NC-858W;	"	(# 51)	"
NC-67K;	"	(# 52)	"

Serial # 10-19-29-30-36 first as model S-56 before conversion to S-56-B; perhaps due to superstition, serial # 12A was used in place of #13; ser. no. for NC-897V unverified; serial # 27 first as S-56-C special; one S-56 in Brazil registered as PP-TCF; approval for all models in S-56 series expired 9-30-39.

A.T.C. #337
(7-14-30)
WACO, MODEL QSO

Fig. 140. Rare "Waco" QSO mounted 165 h.p. Continental A-70 engine; only one example was built.

Possibly the most exciting thing that one could say about the "Waco" model QSO is that it was extremely rare. It was the only example built in the popular SO-series (straight-wing) to be powered with the newly introduced Continental A-70 engine. Probably best compared with the previous 165 h.p. model BSO (refer to ATC # 168), it might be said that the new QSO compared favorably and was basically typical in most all respects. Considering the fact that "Waco" later used the increasingly popular Continental A-70 engine in several of its up-coming models (such as the QCF and QDC), the lone QSO example might well be likened to a flying test-bed for this particular engine installation. Progressive improvement in the basic airframe of the SO-series occasionally led to slight increases in empty weight, so we find the new model QSO some 56 lbs. heavier than the earlier BSO; however with gross weight increased only by this amount, performance differences in the two airplanes would be hardly noticeable in the course of normal operation. Though the sole example of the model QSO was loosely described as a flying test-bed for the A-70 engine, it was not retained by Waco Aircraft for any extended service and was subsequently sold to a flying-service operator for general-purpose work. Continued production of this version would have been entirely possible in the normal course of things because the only modifications necessary were all ahead of the engine firewall. However, with the new Model F series already in production, "Waco" funneled all of its efforts in that direction and apparently no effort was made to solicit more orders for the model QSO.

The Continental Motors Corp had long been known for its dependable "Red Seal" engines, engines that for many years had powered autos, trucks, boats and even stationary installations where a reliable power source was needed. Drawing from experience gained in the 29 years of building internal combustion engines, Continental announced its aircraft engine early in 1929 and received government approval by Oct. of that year. Even before securing approval for the engine, Continental had received firm orders from several aircraft manufacturers for the new powerplant which was in a popular power range. The American Eagle cabin monoplane and the Rearwin biplane were the first to mount the new 7 cyl. model A-70; Verville, Waco and Swallow soon followed suit. Early experimental testing of the engine far exceeded normal requirements, in an effort to flush out any major faults in the basic design, but all went well and there were none at all. With a rating of 165 h.p. at 2000 r.p.m., the A-70 delivered smooth, responsive and dependable power that was a fitting forerunner to the various other Continental engines that had earned a mark of lasting popularity in the airplane industry.

The "Waco" model QSO was an open cockpit

biplane with seating for three, and was typical of other models in the so-called "straight-wing" series except for its engine installation and other minor details necessary to this combination. Powered with the 7 cyl Continental model A-70 engine of 165 h.p., the QSO was basically arranged for general-purpose work, which came to be a catch-all phrase that covered as many uses as one would care to mention. Performance of the QSO has never been a popular topic of discussion among airmen because of its relative rarity, but it is quite safe to assume that it must have shared all the good qualities that made this particular series (SO-type) of "Waco" biplanes such long-time favorites. Exceptionally smooth running and reasonably quiet, it is most likely that the A-70 engine imparted a feeling to one that the QSO was endowed with an effortless performance. The type certificate number for the "Waco" model QSO was issued 7-14-30 and only one example of this model was built by the Waco Aircraft Co. at Troy, Ohio. Clayton J. Bruckner was president and general manager; Lee N. Brutus was V. P.; R. E. Lees was sales manager, soon to be replaced by Hugh R. Perry; Dick Young was test pilot; and Russell F. Hardy was chief engineer. Russell Hardy left "Waco" about May of 1930 and was succeeded by A. Francis Arcier. Arcier, who had a notable record as an aircraft engineer, both here and abroad, had just left General Airplanes Corp. where he had been V. P. of engineering. Clayton Bruckner was very emphatic in his demands to keep "Waco" biplanes abreast with the best, and Francis Arcier was surely more than equal to the job. This naturally promoted a productive association of many years standing.

Listed below are specifications and performance data for the "Waco" model QSO, as powered with the 165 h.p. Continental A-70 engine: length overall 23'6"; height overall 9'1"; wing span upper 30'7"; wing span lower 29'5"; wing chord both 62.5"; wing area upper 155 sq. ft.; wing area lower 133 sq. ft.; total wing area 288 sq. ft.; airfoil "Aeromarine" 2A modified; wt. empty 1585 lbs.; useful load 1001 lbs.; payload with 63 gal. fuel was 402 lbs.; gross wt. 2586 lbs.; max. speed 120; cruising speed 100; landing speed 45; climb 750 ft. first min. at sea level; ceiling 16,000 ft.; gas cap. 63 gal.; oil cap 7 gal.; cruising range at 10 gal. per hour was 575 miles; price at the factory field was $5575.

The construction details and general arrangement of the model QSO were similar to those of later versions of the model BSO as described in the chapter for ATC #168 of U. S. CIVIL AIRCRAFT, Vol. 2. Wing panels on the QSO were of the latest type, which had been beefed-up considerably to stand the added stresses of greater weight and power, and the leading edges were covered with dural metal sheet to preserve the airfoil form; the wing tips were rounded off a bit more than on previous models, especially at the outboard ends of the ailerons. The familiar gap formed by the trailing edge of the wing at the fuselage junction was faired neatly by a small metal fillet. Fuel was carried in a fuselage tank only, which had 63 gal. capacity, and a direct-reading fuel gauge. Noticed also for the first time on this series was a long head-rest fairing that continued along the turtle-back almost to the vertical fin; this was a familiar characteristic on later "Waco" models of the open cockpit type. The baggage compartment with allowance for up to 60 lbs., was located in the turtle-back section of the fuselage just behind the rear cockpit and was provided with a locking type metal door panel. The landing gear was of the robust out-rigger type with oleo-spring shock absorbing struts; wheels were 30x5 and Bendix brakes were standard equipment. A metal ground-adjustable propeller and navigation lights were also standard equipment. A hand crank inertia-type engine starter was optional. The next development in the standard "Waco" biplane was the model PSO as described in the chapter for ATC #339 in this volume.

Listed below is the only example built of the "Waco" model QSO:
NC-837V; Model QSO (#3133) Continental A-70.

Fig. 141. Eastman model E-2-A was amphibian able to operate off land or water,
note wheels tucked under lower wing.

Very well designed and especially arranged for sport-flying in and around areas abounding in water, the Eastman "Sea Rover" flying boat model E-2 was built in fairly good number. All examples, scattered here and there, ably proved themselves as useful and practical — that is, only for flying and operating off water. As a straight "flying boat" type of airplane the "Sea Rover" was indeed an excellent machine but because of the lack of wheels, it was rather limited in its radius of activities. For some owners and for some operators this was entirely sufficient, but there were those who voiced a preference — and even a need — for a little more utility. For these then, in hopes of offering that extra utility and perhaps extending the useful life of the design, Eastman developed a simple wheeled landing gear that would attach to the hull of the "boat" and thereby transform it into an amphibious aircraft. Considering the down-right simplicity of the installation, it is odd that the "Sea Rover" was not offered as an "amphibian" at the

very beginning, to take full advantage of the versatility available from a craft of this type. With the retracting wheels added, the familiar "Sea Rover" model E-2 became the new "Sea Pirate" model E-2-A, and now was able to take off from land or water. It could take off from some sod-strip close to town, winging its way across lots to land on a distant lake nestled in the woods, or on some busy resort harbor. Quite honestly, it is hard to discount the fact that under certain circumstances, where slightly higher operating costs and a small loss of performance are tolerable, the "amphibian" was the most practical airplane ever built. The E-2-A amphibian was no doubt designed as a last resort to offer the extra utility desired by some, and to extend the waning life of the design a little longer; however the new offering was not altogether enough to stimulate any excitement or any hurried buying, so the model remains as one of the rarer types.

The Eastman "Sea Pirate" model E-2-A was an

open cockpit biplane of the flying boat type, with seating arranged for three, similar to the earlier model E-2 in most respects except for the addition of the retracting wheeled landing gear. The simple landing gear add-on was an assembly, available for attachment at the factory, to any of the existing "Sea Rover" type for about $2000.

Powered with the 6 cyl. Curtiss "Challenger" engine of 185 h.p., there was certainly performance enough to satisfy the typical owner of a sport-craft offering this broad range of utility but because of the additional parasitic drag and the added weight of the wheeled undercarriage installation, the allowable capacity was limited to three occupants (the flying boat seated four), and there was some detriment to the overall performance. Of a biplane configuration that approached that of the sesqui-plane (monoplane and a half) the model E-2-A had adequate visibility but its tractor-mounted engine no doubt caused many anxious moments to occupants entering or leaving the craft with the engine running. Of good aerodynamic proportion the E-2-A, inheriting its basic personality from the E-2, was of sprightly nature in the air, well behaved on land and very maneuverable in the water — just what one might call an ideal craft for the flying sportsman. The prototype for the model E-2-A amphibian was especially built at the factory to test out the retracting landing gear and 2 more "Sea Rover" type were soon after modified into this configuration. The last two airplanes built at the Eastman factory were apparently both model E-2-A amphibians but only one of these, the last one, can be verified as such. The type certificate number for the Eastman "amphibian" model E-2-A was issued 7-17-30 and some 4 or 5 examples of this model were manufactured by the Eastman Aircraft Corp. at Detroit, Mich., a division of the Detroit Aircraft Corp. P. R. Beasley was now the president; James H. Eastman was chief of engineering; and Carl S. Betts was sales manager. Clair Fahy, accomplished aviatrix and wife of top-notch flyer Herb Fahy, was promotional pilot at Eastman for a time, to help demonstrate and sell the advantages of owning a "Sea Rover".

Listed below are specifications and performance data for the Eastman model E-2-A amphibian, as powered with the 185 h.p. Curtiss "Challenger" engine: length overall 26'3''; height on wheels 9'9''; wing span upper 36'0''; wing span lower 20'8''; wing chord upper 68''; wing chord lower 36''; wing area upper 190 sq.

ft.; wing area lower 53 sq. ft.; total wing area 243 sq. ft.; airfoil "Clark Y"; wt. empty 1970 lbs.; useful load 755 lbs.; payload with 40 gal. fuel was 310 lbs.; gross wt. 2725 lbs.; max. speed 102; crusing speed 85; landing speed 50; climb 740 ft. first min. at sea level; ceiling 9500 ft.; gas cap. normal 40 gal.; gas cap. max. 48 gal.; oil cap. 5 gal.; cruising range at 10.6 gal. per hour was 295-360 miles; price at the factory first quoted at $10,985, lowered to $8750 in March of 1931.

The construction details and general arrangement of the model E-2-A amphibian were similar to that of the model E-2 "Sea Rover" flying boat as described in the chapter for ATC # 288 of U. S. CIVIL AIRCRAFT, Vol. 3. The main difference in the model E-2-A was, of course, the incorporation of the wheeled landing gear, a rather simple assembly that was retracted or extended by hand operation. It is quite likely that added members and some strengthening of the hull frame was necessary on the E-2 conversion to E-2-A to take care of the added stresses imposed by the wheeled undercarriage. Because the E-2-A weighed some 225 lbs. more than the E-2 boat when empty, it was generally limited to carrying 40 gal. of fuel in order to allow a 310 lb. payload, which is less than the accepted average for two people; in some cases the E-2-A would even be allowed to carry some baggage. With maximum fuel aboard (48 gal.) the ideal load would be a pilot and passenger at 170 lbs. each and some 87 lbs. of baggage. The landing wheels were 20x6 in. low pressure and brakes were available; shock absorbers were air-oil struts. A metal propeller, Heywood air-operated engine starter, navigation and mooring lights, a battery, fire extinguisher and first aid kit were standard equipment. Dual controls, engine cover and cockpit covers were optional.

Listed below are Eastman model E-2-A entries as gleaned from registration records:

X-592M; Model E-2-A (# 3) Curtiss R-600.

NC-466M;	"	(# 11)	"
NC-470M;	"	(# 15)	"
NC-475M;	"	(# 19)	"
NC-476M;	"	(# 20)	"

Serial # 3 was experimental prototype for amphibian series; serial # 11 and # 15 were first as E-2 flying boat before conversion to amphibian; serial # 19 and # 20 were initially built at the factory as amphibian; registration number for serial # 19 unverified.

Fig. 142. "Waco" model PSO mounted new 140 h.p. Jacobs engine; only one example was built.

The lone "Waco" model PSO under discussion here, has at least the double distinction of being the only example of the popular SO-series (straight-wing) biplane to be powered with the new "Jacobs" engine of 140 h.p., and also of being the very last production version to be developed in this particular series of airplanes except for a few specials built on Group 2 approval. Starting with the "Waco 10" of 1927, this basic design had been variously powered with at least nine different engines to fit a purse or purpose. Checking back into previous discussions we find the "Ten" powered with the Curtiss OX-5 and OXX-6 engines, the Ryan-Siemens (Siemens-Halske) of 97 h.p. and 125 h.p., the Wright "Whirlwind" J5, the "Hisso" (Hispano-Suiza) engines, the 5 cyl. Wright J6, the 7 cyl. Wright J6, the Packard "Diesel" engine, the Continental A-70 and now the Jacobs 140, offering a range of power units spreading from 90 to 225 h.p. Early designations singled out the "Ten" as the OX-5 powered version, the 10-H with the "Hisso" engines, the 10-W with the Wright J5 and so on. In a change from this, the models were linked to their horsepower rating so the basic OX-5 powered "Ten" now became the 90, the Siemens powered version was the 125, the Hispano-Suiza powered versions the 150 and 180, the J5 powered version was the 220 and so on. Later in 1929, designations were linked with a series of code letters that pre-

sented a certain aura of mystery, until it was established that each letter in the designation had a direct bearing on describing the particular version and its general make up. Letters GXE were reserved for the standard OX-5 powered "Model 10" (or Model 90) with the letter G denoting power by Curtiss OX-5, the X denoting the particular wing configuration and its basic construction, the E denoting the standard fuselage configuration with its particular appointments and construction; thus we have GXE to describe this version. With installations of more power, some modifications took place in the fuselage structure to earn the designation O; wing spars, internal drag bracing and interplane bracing were also beefed-up to handle extra power, and were designated S to denote the stronger "straight wing". This was to set them apart from the "tapered wing" panels which were designated T and were also fitted to the basic O fuselage. To complete the identification of designation letters, each engine installation was awarded a code letter; letter A was used for the 220 h.p. Wright J5, D was for "Hisso" engines, and letter B for the 5 cyl. Wright J6 of 165 h.p., thus we have ASO-DSO-BSO. Other engines used in the SO combination were coded C for 7 cyl. Wright J6 of 225 h.p., H for the Packard "Diesel" engine, Q was for the Continental A-70 and P for the Jacobs 140-170; thus we have additional models

Fig. 143. PSO was about the last of the popular "Waco" straight-wing series; shown here on Wings Field in Pennsylvania.

in this series as CSO-HSO-QSO-PSO. Tapered wing versions of this "Waco" biplane were known as ATO-CTO-HTO and could be easily recognized by the T in the code-letter designation. When originally introduced this code-letter system was quickly assessed by many as a bunch of mumbo-jumbo, but it did have a purpose and serves as reliable media in the identification of the numerous models that were manufactured by Waco Aircraft. In view of the fact that "Waco" built so many different models in the next 10 years, a system of this sort was the only practical means of model designation.

The "Jacobs" engine, which became quite a popular powerplant in the late 30's, first came upon the scene in 1928 as the "Jacobs & Fisher" air-cooled engine of 110 h.p. Albert R. Jacobs, designer and co-developer of this new engine, had it mounted in a "Waco 10" biplane (X-6919) and he was off to the National Air Races held at Los Angeles, Calif. where he entered a few events and spent a portion of the time promoting the new engine. The "Jacobs & Fisher" was also reported as an experimental installation in an A-series "Eaglerock" biplane but this could not be completely verified. Al Jacobs was later instrumental in the forming of a company in Camden, N.J., sometime in 1929, which produced the basic "Jacobs & Fisher" in a modified version called the A.C.E. La-1 of 140 h.p. A government approval for this engine was received 9-23-29 and another "Waco 10" (-9546) had already been fitted with this engine for experimental testing. In an amendment issued 2-7-30 the A.C.E. engine was redesignated the "Jacobs" LA-1 with a rating of 140 h.p. at 1800

r.p.m. By 1931 the model LA-1 received a higher horsepower rating for the same basic engine with only a slight increase in compression ratio and higher operating speeds; the rating was then 170 h.p. at 2125 r.p.m. It is worth mentioning that the peculiar valve action on the early Jacobs was reminiscent of that used earlier on the Siemens-Halske (German) engines, and so was the nose-type exhaust collector ring which was also a familiar "Siemens" fixture. The Jacobs 140 and the later 170 came equipped from the factory with the collector ring and a narrow chord "speed-ring" cowling mounted as a standard unit. Al Jacobs, president and chief engineer of the Jacobs Aircraft Engine Co. kept abreast of all developing requirements and held the "Jake" in favor as a dependable power plant for many years to come.

The "Waco" model PSO was an open cockpit biplane of the general-purpose type, with seating for three, and was more or less typical of several other models in this series except for its powerplant installation and some minor details necessary for this particular combination. Powered with the 7 cyl. Jacobs LA-1 engine of 140 h.p. the performance characteristics of this combination was no doubt closely comparable to either the model BSO or QSO, making allowances for the slightly less horsepower available. Otherwise, we can well assume that the PSO shared all other qualities that made the "Waco" SO-series such great favorites the country over. The Jacobs 140 engine cost $2500 at the factory and was considered a good bargain. With this model (PSO), production of the familiar "straight wing" series soon dribbled to an end, but this was only to make way for the many new "stars" that Waco Aircraft had in their line-up for the years to come. The type certificate number for the model PSO was issued 7-17-30 and only one example of this model was manufactured by the Waco Aircraft Co. at Troy, Ohio.

Listed below are specifications and performance data for the "Waco" model PSO, as powered with the 140 h.p. Jacobs engine: length overall 23'6"; height overall 9'1"; wing span upper 30'7"; wing span lower 29'5"; wing chord both 62.5"; wing area upper 155 sq. ft.; wing area lower 133 sq. ft.; total wing area 288 sq. ft.; airfoil "Aeromarine" 2A modified; wt. empty 1508 lbs.; useful load 933 lbs.; payload with 63 gal. fuel was 343 lbs.; gross wt. 2441 lbs.; max. speed 115; cruising speed 97; landing speed 42; climb 700 ft. first min. at sea level; ceiling 12,000 ft.; gas cap. max. 63 gal.; oil cap. 6 gal.; cruising range at 8 gal. per hour was 650 miles; 40 gal. of fuel still offered nearly 5 hours of cruising range and boosted the payload allowance to some 480 lbs.; price at the factory field was $5175; performance with the 170 h.p. Jacobs would be proportionately increased.

The construction details and general arrangement of the model PSO were typical of those of the later versions in the model BSO as described in the chapter for ATC #168 of U. S. CIVIL AIRCRAFT, Vol. 2. Wing panels on the PSO were of the latest type, which had been beef-up considerably to stand added stresses of power and weight, and the leading edges were now covered with "dural" metal sheet to preserve the airfoil form; wing tips were rounded off a bit more than on previous models, especially at the outboard ends of ailerons. All fuel was carried in a tank of 63 gal. capacity that was mounted high in the fuselage just ahead of the front cockpit; a direct-reading fuel gauge projected up through the cowling. Noticed also was the long head-rest fairing that continued along the turtle-back almost to the vertical fin; this feature was probably introduced on the model QSO and was characteristic on all later models of the open cockpit "Waco F" biplanes. The baggage compartment was located in the turtle-back section of the fuselage and was provided with a locking type metal door panel; with a reduction in fuel up to 60 lbs. baggage was allowed but not with maximum fuel load. The landing gear of 78 in. tread was of the out-rigger type using air-oil shock absorbing struts; wheels were 30x5. An adjustable metal propeller and Bendix brakes were standard equipment. Navigation lights, low press semi-airwheels and hand crank inertia-type engine starter were optional. The next development in the "Waco" biplane was the model INF as described in the chapter for ATC #345 in this volume.

Listed below is the only "Waco" model PSO entry as gleaned from registration records: NC-699N; Model PSO (# 3221) Jacobs 140.

This model later eligible with improved Jacobs LA-1 with rating of 170 h.p.; previous experimental installations of the "Jacobs" type engine were X-6843, an A-series "Eaglerock" (ser. #568), X-6919 a "Waco 10" (ser. #1450), and a "Waco 10" (ser. #A-53) with registration number -9546.

Fig. 144. Sikorsky amphibian model S-39-A mounted 300 h.p. "Wasp Junior" engine, seated four.

Feeling there was a necessity for a smaller "amphibion," closely following the pattern set by the very successful S-38 series, the Sikorsky model S-39 was specifically aimed at the sportsman-pilot or the busy business executive. Being able to take off from a stretch of water, then land many miles inland on some grassy strip, was still quite attractive to those who could afford an airplane of this type. It was felt that a good market would yet be assured. Anxious to test its mettle, and knowing the importance of a good shake-down cruise, the first production example of the model S-39-A was flown on the far-flung route of the National Air Tour for 1930 by Geo. Meissner. Finishing in 18th place, it didn't seem to make a very impressive showing in the point standings but the long grind was a good test for the new amphibian and an excellent chance to show it off. Col. E. A. Deeds, a director of United Aircraft, wished to emphasize the unlimited utility of the S-39 type more graphically. He had his 'plane lashed to the deck of his yacht and the S-39 would be lowered into the water to fly back to land as a "yacht tender". Arranged to seat four, the S-39-A showed good performance reserve and an inherent potential to allow extra payload with an increase in the seating capacity. In a modi-

fied form, this improved version came out some months later as the 5 place model S-39-B. Of the 10 examples initially built as the 4-place model S-39-A, most all were later modified to the S-39-B type under the provisions of ATC #375. Of particularly staunch construction and a rugged character that promoted longevity, despite hard service, most of these S-39's were still going strong 10 years later, and some were still flying many years after that.

The very first of the S-39 type as designed by venerable Igor Sikorsky, was basically a scaled-down version of the popular S-38 twin-engined "amphibion." Powered with two 4 cyl. Cirrus "Hermes" (British) engines of 105-115 h.p. each, the first flight for the light "twin" was on Dec. 24 of 1929 with Boris Sergievsky at the controls. Mechanical failure to one engine during take-off on its third flight caused a hurried emergency landing which ended in a complete wash-out; Sergievsky and Mike Gluhareff (project engineer) escaped from the wreckage unhurt. Largely because of the ill-fated flight of the S-39 "twin" and because of other pressures that were brought to bear, it was decided to re-design the light amphibian into a single-engined configuration. The new Pratt & Whitney "Wasp Junior" engine of 300 h.p. was, of course, the

Fig. 145. First production example of S-39-A was flown by George Meissner in 1930 National Air Tour.

logical choice. This second example in the S-39 series was a completely new airplane, although still typical except for its single engine. Because of an accelerated program, it was ready for testing by Feb. of 1930. Flight tests were satisfactory but the overall performance was not up to expectations, so several revisions were hurriedly planned; its twin vertical fins were revised to just one fin and rudder, the complicated landing gear was modified and a general aerodynamic clean-up was deemed necessary. The first production example of the S-39-A had an all-metal hull of slightly modified lines, the long-leg landing gear had been changed to one of much simpler design, and several other changes were made to reduce vibrations and parasitic drag. Production of the S-39-A was launched about mid-1930 with the building of complete assembly sets for 20 airplanes; several airplanes were completed immediately, but most assembly sets were stored — to be put together to make an airplane when an order was received. Heartened by the success of the S-39, despite the rather unfavorable market at this time, another 20 "sets" were nearly finished but there was no justification for so many airplanes (because almost no one was buying) and they were ordered scrapped.

The Sikorsky "Sport Amphibion" model

S-39-A was basically a "parasol" monoplane of the flying boat type with its tractor-mounted engine high in the leading edge of the elevated wing. The fairly large all-metal hull was a seaworthy structure to which was added a wheeled landing gear, a gear that was lowered for operations on land. Equally at home on land or water, the "amphibion" always had a landing place nearby, selecting either land or water with a wheeled undercarriage that could be extended or retracted in about 10 seconds. Comfortable and exceptionally roomy, the cabin had more than ample seating for four, with a wide roof hatch its full length for easy loading and unloading. Powered with the new 9 cyl. Pratt & Whitney "Wasp Jr." engine of 300 h.p., performance was more than adequate for a "boat" of this type and flight characteristics were very pleasant. Whether on land or water, the S-39-A landed gently and take-offs were short and clean. Its behavior on land was admirable despite its somewhat unwieldy appearance, and water handling characteristics were exceptionally good if certain traits were used to advantage. Sikorsky airplanes were never built for competitive pricing and therefore maintained a high standard of quality and capability. The type certificate number for the 4 place model S-39-A was issued 7-22-30 and 10 examples of this model were manufactured by the Sikorsky Aviation Corp. at Bridgeport, Conn., a division of the United Aircraft & Transport Corp. Igor Sikorsky was V.P. and chief engineer; Boris Sergievsky was chief test pilot and Michael Gluhareff was project engineer in charge of S-39 development.

Listed below are specifications and performance data for the 4-place model S-39-A, as powered with the 300 h.p. "Wasp Jr." engine: length overall 31'11"; height overall on wheels 11'8"; wing span 52'0"; wing chord 85"; total wing area 350 sq.ft.; airfoil "Sikorsky" GS-1; wt. empty 2555 lbs.; useful load 1145 lbs.; payload with 65 gal. fuel was 540 lbs.; payload allowed 3

Fig. 146. Original prototype of S-39 series was a small "twin", a scaled-down version of larger S-38.

Fig. 148. S-39-A was sport amphibian seating four; equally at home on land or water.

passengers at 170 lbs. each and 30 lbs. baggage; gross wt. 3700 lbs.; max. speed 115; cruising speed 97; landing speed 50; climb 800 ft. first min. at sea level; climb in 10 min. was 6500 ft.; ceiling 18,000 ft.; gas cap. 65 gal.; oil cap. 6 gal.; cruising range at 15 gal. per hour was 375 miles; price at the factory was $17,500.

The stubby and robust hull framework had internal construction of duralumin bulkheads covered with metal "Alclad" skin, and divided into four water-tight compartments. The two forward seats were individual and the rear seat was a wide bench type with a compartment behind it for stowing baggage, anchor, and mooring gear. Access to the cabin area was through a wide full-length hatch cover in the roof; the cabin was tastefully upholstered and all windows were of shatter-proof glass. Pilot controls were dual wheel type and landing gear extension or retraction was by hydraulic pump with a hand lever. The semi-cantilever wing in 3 sections was built up of duralumin girder-type

spar beams and truss-type duralumin wing ribs; the leading edges were covered with dural metal sheet and the completed framework was covered in fabric. Two fuel tanks of 32.5 gal. capacity each were mounted in the center-section panel of the wing, behind the engine nacelle. The wing was mounted in a "parasol" fashion and braced by a maze of streamlined duralumin struts that also supported the engine nacelle and the two outboard floats; these floats were of construction similar to the hull, of two water-tight compartments that prevented the craft from heeling in the water. The wide tread landing gear, using oil-draulic shock absorbing struts and low-pressure airwheels — complete with brakes, were standard equipment; the landing gear swung out of the way for water landings and extended for alighting on land. Either side of the gear could be lowered individually or together and there was no up-down indicator. Two booms running from the wing center-section and braced to the hull, supported

Fig. 147. Second prototype of S-39 had single engine but retained long-leg landing gear and twin rudders.

Fig. 149. Close-up shows detail of unusual "parasol" arrangement and engine mounting.

the tail-group which was of fabric covered construction similar to that of the wing. A steerable tail-wheel was mounted on the extreme end of the hull and steps were conveniently placed for entry to the cabin. A large exhaust collector ring, a metal propeller, a Heywood air-operated engine starter and fire extinguisher were standard equipment. The next Sikorsky development was the rare "Hornet" powered model S-38-BH as described in the chapter for ATC #356 and the next S-39 development was the improved model S-39-B as described in the chapter for ATC #375, both in this volume.

Listed below are Sikorsky model S-39-A entries as gleaned from registration records:
X-963M; Model S-39 (#1) 2 Cirrus "Hermes."
X-963M; Model S-39 (#2)   Wasp Jr. 300.
NC-42V; Model S-39-A (#900X)   "
NC-887W; Model S-39-A (#901)   "

NC-802Y; Model S-39-A (#903) Wasp Jr. 300.
NC-803W;   "   (#904)   "
NC-804W;   "   (#905)   "
NC-805W;   "   (#907)   "
NC-806W;   "   (#908)   "
NC-807W;   "   (#909)   "
NC-808W;   "   (#910)   "
NC-809W;   "   (#911)   "

X-963M were first two prototype ships having same registration number although they were actually different airplanes; serial #900X was first production ship, later modified to S-39-B; serial #901-903-904-905-907-908-909-910 later modified to S-39-B also; serial #908 exported to England for Cunard Lines as G-ABFN, later returned to U.S.A. and modified to S-39-B; serial #911 later modified to S-39-C on Group 2 approval numbered 2-391.

A.T.C. #341
(7-22-30)
SWALLOW "SPORT," HA

Fig. 150. The beautiful Swallow "Sport" for 1930; model HA shown, powered with 150 h.p. Axelson B engine.

With sales for the classic 3-place "Swallow" biplane all but coming to an end, and sales for the 2-place "Swallow TP" trainer series petering off to barely a trickle, the handwriting on the wall became clearly evident and company executives saw dire need for another good airplane model to take up all the slack. Discussing and analyzing several interesting possibilities, including a 4-place low winged cabin monoplane, they conservatively chose to stick to the tried and true open cockpit biplane as the most likely to succeed at this particular time. Believing that the 3-place all-purpose biplane had yet the best potential to satisfy the needs of the largest percentage of people that flew, Swallow designers worked freely around this concept. Fashioned with a flair by the deft hands of Dan Lake and others, the Swallow "Sport" H-series was enthusiastically planned as versatile sport-type craft with better than average performance, allowing the buyer the choice of several different engines in a comparable power range. Oddly enough the prototype airplane for this new series, labeled as the Model H, was powered with the 100 h.p. Kinner K5 engine; its performance though commendable under the influence of only 100 horses, was hardly spirited enough to justify the name of "Sport." Having a small stock-pile of Axelson engines that were ordered previously for the ill-fated model F-28-AX (refer to ATC 125), Swallow Aircraft

soon prepared another version of the "Sport" (first called "Special") to be known as the Model HA, with an Axelson B-7-R engine of 150 h.p. In this combination the design's capabilities quickly came to the fore. Now aimed specifically at the medium-budget sportsman pilot (who might also be a salesman or businessman), the model HA was not pared down in size in the interests of performance gain and still had room enough for three, with allowance for baggage. Its rather rakish stance and its flashy personality, coupled with a mildly surprising performance, was a combination with sure-fire appeal that should have assured the HA a rousing success. Sadly enough this was not the case at all and the Swallow "Sport" was tragically left begging for customers.

As pictured here, the good-looking Swallow "Sport" was an open cockpit biplane with ample seating for three in an airframe that was thoughtfully arranged to accommodate a useful load in a more practical manner; the "Sport" turned out to be a rather small airplane of light weight without sacrificing any room or any of the normal comforts. Powered with the 7 cyl. Axelson B engine of 150 h.p. the model HA had a good performance reserve and plenty of flash, but certainly not as good as Swallow reports would lead one to believe; pride in their new ship was understandable but the claims were somewhat more fancied than real. Even at that we surely

must say that the "Sport" was a very good airplane that nearly approached the peak of development for a craft of this particular type. The
type certificate number for the model HA as
powered with the 150 h.p. Axelson B-series
engine was issued 7-22-30. It is quite likely
that only one example of this model was manufactured by the Swallow Airplane Co. at Wichita,
Kansas.

Listed below are specifications and performance date for the Swallow "Sport" model
HA: length overall 22'2"; height overall 8'2";
wing span upper 31'0"; wing span lower 23'0";
wing chord upper 66"; wing chord lower 48";
wing area upper 150 sq.ft.; wing area lower 90
sq.ft.; total wing area 240 sq.ft.; airfoil (NACA)
M-12; wt. empty 1416 lbs.; useful load 784 lbs.;
load with 35 gal. fuel was 372 lbs. (baggage 42
lbs.); gross wt. 2200 lbs.; max. speed 130; cruising speed 110; landing speed 42; climb 1100 ft.
first min. at sea level; ceiling 16,000 ft.; gas cap.
35 gal.; oil cap. 4 gal.; cruising range at 8 gal.
per hour was 435 miles; price at the factory
field was $4250., later lowered to $3995.; price
includes Goodyear "airwheels;" oleo shock
absorbing struts and wheel brakes were extra.

The fuselage framework was built up of welded steel tubing into a rigid and robust structure
of Warren truss form, heavily faired with wood-
en formers and fairing strips then fabric covered.
The two open cockpits were deep and well protected with a large door on the left side for entry
to the front pit; the rear cockpit was provided
with a long streamlined head-rest. The wing
panel framework was built up of heavy-sectioned
solid spruce spar beams with truss-type wing
ribs built up of spruce and plywood; the leading
edges were covered to the front spar to preserve
the airfoil form and the completed framework
was covered in fabric. Two gravity-feed fuel
tanks were mounted in the center portion of
the upper wing with a total capacity of 20 gal.
each. The split-axle landing gear was a stiff-
legged structure of welded steel tubing arranged into two rigid pyramids; no shock absorbers were normally used. Large low pressure "airwheels" were calculated to absorb
all the shock, however, oleo-spring shock absorbing struts were available if desired and
wheel brakes were optional. The fabric covered
tail-group was built up of welded steel tubing;
the vertical fin was ground adjustable and the
horizontal stabilizer was adjustable in flight.
Wiring for navigation lights and a metal propeller were standard equipment. A hand crank inertia-type engine starter was listed as optional
equipment. The next version of the Swallow
"Sport" Series was the Wright "Five" powered
model HW, as described in the chapter for ATC
#379 in this volume.

Listed below is the only known example of
the "Sport" (Special) model HA:

NC-109V; Model HA (#102) Axelson B-150
NC-109V, later modified into a model HW by
installation of 165 h.p. Wright J6 engine and
serial number, was changed to #2002; X-108V
with serial #101 was the prototype airplane
for the new H-series.

Fig. 151. Inland "Sport" S-300-E received slight boost in power, engine was 70 h.p. Le Blond 5DE.

There is really not much one can say, for or about, the Inland "Sport" model S-300-E because it was basically so typical of the model S-300 as described in the chapter for ATC #259 of U. S. CIVIL AIRCRAFT, Vol. 3 except for the change in engines. The powerplant for the model S-300-E under discussion here was the 5 cyl. LeBlond 5DE of 70 h.p., an engine basically similar to the LeBlond 5D of 60-65 h.p. except for an increase of some 5-10 h.p. — power which was gained by a slight increase in operating r.p.m., an improvement in the valve operating mechanism, and an improved cylinder head design. Comparatively speaking, the LeBlond 5DE engine was a considerable improvement throughout over the earlier 5D and therefore owners of the Inland "Sport" model S-300, powered by LeBlond 5D, would jump at the chance of the newer installation. There were no available records to show if any of the model S-300-E actually came off the production line as such, but there is record that at least 5 of the existing model S-300, built earlier, were modified to the model S-300-E by installation of the 5DE-70 engine and whatever other modifications were necessary for this particular combination. As a consequence of these changes the immediate gain would be a livelier airplane with a slight improvement in overall performance and surely a considerable gain in engine reliability.

The Inland "Sport" model S-300-E was a pert and proper parasol-type monoplane, with chummy side-by-side seating for two in the sporty atmosphere of an open cockpit. As already well-proven in nearly a year's service, the Inland "Sport" was ruggedly built and well able to withstand the strains and hard knocks of every-day flying by pilot-training schools and private owners. Of good aerodynamic proportion, the flight characteristics and the general behavior of the S-300-E were fairly positive, nimble, and quite friendly; the extra horsepower now offered by the improved LeBlond 5DE engine, however slight, provided a performance that was more than adequate for normal use. Neat in appointment, trim of line and quite appealing, the "Sport" remained popular for a good many years and several were still seen flying 10 years later. The type certificate number for the Inland "Sport" model S-300-E, as powered with the 70 h.p. LeBlond 5DE engine, was issued 7-25-30 and at least 5 examples of this model were modified to the new specifications in the factory of the Inland Aviation Co. on Fairfax Field in Kansas City, Kansas.

Listed below are specifications and performance data for the Inland "Sport" model S-300-E, as powered with the LeBlond 5DE-70 engine; length overall 19'10"; height overall 7'6"; wing span 30'0"; wing chord 60"; total wing area 144 sq. ft.; airfoil (NACA) M-12; wt. empty 786 lbs.; useful load 504 lbs.; payload with 24 gal. fuel was 179 lbs.; gross wt. 1290 lbs.; max. speed 105; cruising speed 90; landing speed 35; climb 720 ft. first min. at sea level; climb in 10 min. 6500 ft.; ceiling 12,000 ft.; gas cap. 24 gal.; oil cap. 2 gal.; cruising range at 5 gal. per hour was 450 miles; price at the factory for modification was $1230. for the new engine plus nominal installation charges. No doubt there was some trade-in value in the older engine that was removed.

The construction details and general arrangement of the "Sport" model S-300-E were typical of the model S-300 as described in the chapter for ATC #259 of U. S. CIVIL AIRCRAFT, Vol. 3. The main changes in the model S-300-E were, of course, centered around the new engine installation which required only a minimum of modification but was of such nature that it could not be readily performed by the owners themselves; without absolute proof to go on, it is therefore assumed that all S-300-E modifications were performed and approved at the factory. It is vaguely remembered, and also has been reported, that several of the S-300-E were later fitted with Townend-type "speed-ring" engine cowling for a gain of several m.p.h. in cruising speeds. The baggage locker was a small bin built up of dural metal sheet with an allowance for 5 lbs., an allowance which con-

sisted mainly of a small tool kit and a few knick-knacks. Dual joy-stick controls were available, with the right hand stick and rudder pedals easily removed when not in use. A wooden propeller was standard equipment. The next development in the Inland "Sport" monoplane series was the model R-400 "Sportster" as described in the chapter for ATC #343 of this volume.

Listed below are Inland model S-300-E entries as gleaned from registration records:
NC-9416; S-300-E (#S-302) LeBlond 5DE-70.
NC-252K;   ”   (#S-308)   ”
NC-254N;   ”   (#S-314)   ”
NC-255N;   ”   (#S-315)   ”
NC-256N;   ”   (#S-316)   ”
All a/c listed were first as model S-300 on ATC #259; serial #S-316 which was first listed as model S-300, was modified to an S-300-E, then later as a model S-300-DF with 85 h.p. LeBlond 5DF engine on Group 2 approval numbered 2-270.

Fig. 152. Several owners of earlier S-300, modified their craft to S-300-E specifications for performance gain.

INLAND "SPORTSTER", R-400

Fig. 153. Inland "Sportster" model R-400 with 90 h.p. Warner "Scarab Jr." engine.

With the Inland "Sport" series monoplane already available with the LeBlond 60 engine as the model S-300, the LeBlond 70 engine as the model S-300-E, and with the Warner "Scarab" 110 h.p. engine as the model W-500 — the new R-400 "Sportster" was carefully groomed to round out and complete the line. This line of airplanes was topped off with an offering that was probably the most attractive of the lot. Of the four differently powered models now available in the Inland sport monoplane series, the "Sportster" model R-400 was accepted well, and was very often described to be the best flying and best behaved airplane of the whole series. Powered with the spunky 5 cyl. Warner "Scarab Jr.," engine of 90 h.p., it was apparently powered to just the correct degree to bring out the best and the most compatible traits in this "parasol" design. It was a happy medium where one might not get caught short for need of extra power in some instances, yet not be excessively powered to cause any bothersome or undesirable characteristics in its normal behavior pattern. Probably best said as once so aptly put, . . . "it's a dog-gone little sweetheart with just the right amount of everything". Arthur Hardgrave, president of the Inland company and an avid sportsman-pilot too, actually preferred the model R-400 "Sportster" for his personal use and, incidently, selected the last of this model to leave the assembly line. Despite the terrific value in this little air-

plane for an out-lay of less than $4000, sales were very light and not very many examples were built; nevertheless, several of these served their respective owners for many years and at least one or two were still flying in the "sixties".

The Inland "Sportster" model R-400, basically typical of all other models in the series, was also a pert little parasol-type monoplane with chummy side-by-side seating for two in the sporty atmosphere of an open cockpit. Leveled especially at the amateur private owner (who tended to appreciate more the added trappings of better appointments and more convenience), the R-400 was bedecked with more complete upholstery of the interior, and the gadgetry was more numerous; even an engine starter was available to eliminate the tedium of prop-swinging by hand. Though finished carefully and in shiny attractive colors, the "Sportster" underneath it all was every bit as rugged as its companion models and could well absorb the ordinary hard knocks of every-day sport flying. Of good aerodynamic proportion, too, the flight characteristics were eager and very nimble, but well behaved at all times. Powered with the Warner "Junior" engine of 90 h.p., its performance was rather high with a comfortable margin of reserve power for an occasional tight spot where that extra burst made the difference. The parasol monoplane certainly had many features in its favor for the average sportsman and the

Fig. 154. Open cockpit and parasol wing of R-400 offered sporty atmosphere for flying fun.

R-400 "Sportster" was one of the best of this particular type. The type certificate number for the Inland "Sportster" model R-400 was issued 7-23-30 and some 9 examples of this model were manufactured by the Inland Aviation Company on Fairfax Field in Kansas City, Kansas. Of the various aircraft companies operating from Fairfax Field, Inland and Rearwin shared the same factory building; Inland occupied the right wing and Rearwin occupied the left wing with a centrally located portion allocated as offices for the two companies. The facility was modest in proportion but well arranged for efficient aircraft manufacture, allowing fly-away deliveries of completed airplanes.

Early in 1930, Inland Aviation was fired with enthusiasm (as were several other manufacturers) at the prospect of this race and announced their intent of entry in the famous All-American "Cirrus Derby" held later that year. The proposed Inland entry was a special racing monoplane, also of the "parasol" type, but quite unlike any of their standard "Sport" series. As a single-seater powered with the 4 cyl. American "Cirrus" (inverted in-line) engine, its wings were small and tapered and if looks were any criteria of speed, it looked fast. Detailed reports of this "Cirrus Race", prior to the actual contest, were very sketchy and good information was hard to come by, so it has been impossible to determine if this Inland "Speedster" started in the race or was even completed. Several pilots who participated in the race could not remember it in the line-up so chances are that it was not completed in time for the race. Besides the four standard "Sport" series monoplanes already covered up to this point, Inland designed one more model to mount the new LeBlond 5DF engine of 85 h.p.; this model was the S-300-DF and only one example was built under a Group 2 approval numbered 2-270. Modified from a model S-300,

with the gross weight held to some 1300 lbs., it is apparent that this version (S-300-DF) must have been quite a performer to say the least. Fighting gamely to beat the odds imposed by a nation-wide business depression, the Inland Aviation Co. held on as best they could, but by late 1931 or 1932 they had to withdraw to the circle of bankrupt airplane companies. They soon became just another memory spawned in that wonderful hey-dey of the "twenties".

Listed below are specifications and performance data for the Inland "Sportster" model R-400, as powered with the 90 h.p. Warner Jr. engine: length overall 19'8"; height overall 7'7"; wing span 30'0"; wing chord 60"; total wing area 144 sq. ft.; airfoil (NACA) M-12; wt. empty 857 [902] lbs.; useful load 564 lbs.; payload with 24 gal. fuel was 234 lbs.; payload includes 2 parachutes at 20 lb. each & 25 lbs. baggage; gross wt. 1421 [1466] lbs.; figures in brackets are revised wts. to allow installation of electric engine starter; max. speed 115; cruising speed 98; landing speed 40; climb 950 ft. first min. at sea level; climb in 10 min. was 7800 ft.; ceiling 16,000 ft.; gas cap. 24 gal.; oil cap. 2 gal.; cruising range at 5.5 gal. per hour was 400 miles; price at the factory field was $3985.

Construction details and general arrangement of the Inland model R-400 was typical of the model W-500 and model S-300-E, as described in the chapters for ATC #315 and #342 of this volume. A metal-lined baggage compartment with a locking door panel was in the turtle-back section behind the cockpit; capacity was 5 cu. ft. with allowance for 25 lbs. Standard wire wheels were 26 x 4 but 6.50 x 10 semi-airwheels with Air Products wheel brakes were optional; the swiveling tail-skid had oleo shock absorbing strut. A metal propeller, wiring for navigation lights and chrome-plated exhaust collector ring were standard equipment. Dual controls

Fig. 155. R-400 often described as best flying airplane of the Inland series.

were optional with the right hand joy-stick easily removable when necessary. An electric engine starter and speed-ring engine cowling (Townend type) were also available for this model. Wing bracing N-type struts were fitted with adjustable ends to allow for rigging, and the ring-type engine mount was easily detachable to allow removal of the engine as a complete unit. The attractive sun-burst design was on main colors of a deep blue and orange-yellow. As the last model built by Inland, the R-400 was the most popular and remains as the only representative example of the Inland series to be actually flying as of this late date.

Listed below are Inland "Sportster" model R-400 entries as gleaned from registration records:

NC-267N;	Model R-400	(# R-401)	Warner 90
NC-266N;	"	(# R-402)	"
NC-264N;	"	(# R-403)	"
NC-453V;	"	(#R-404)	"
;	"	(# R-405)	"
NC-257N;	"	(# R-406)	"
NC-259N;	"	(# R-407)	"
NC-509Y;	"	(# R-408)	"
NC-510V;	"	(# R-409)	"

Registration number for serial #R-405 unknown but it is likely to be 260N or 261N.

A.T.C. #344
(7-25-30)
LINCOLN TRAINER, MODEL PT-T

Fig. 156. Lincoln model PT-T with Brownback "Tiger" engine was rare version of this popular trainer.

The Lincoln model PT was a popular series of two-place biplanes, designed specifically for use by flying schools in either their primary or secondary phases of pilot training. Well balanced and well arranged in the aerodynamic sense with long and easy moment arms, the PT in any version was a stable craft, amiable and gentle and quite forgiving in nature; it was not fussy and didn't get all upset because of a little pilot error now and then. Already available in three different models, the model PT-T as described here, was another version in this interesting series. Otherwise being quite similar to previous models in the "Lincoln PT" series (especially models PT-K and PT-W) there is actually not much of any new consequence to discuss about the model PT-T, except perhaps its engine installation which was the new Brownback C-400 "Tiger" of 90 h.p. The little "Tiger 90" was a 6 cyl. two-row aircooled radial type engine that was actually like two 3 cyl. engines put together, and quite similar in arrangement to the 6 cyl. Curtiss "Challenger" engine that so many will remember. Compared to the two previous models of the Lincoln trainer, the model PT-T with its 90 h.p. "Tiger," was perhaps just a little shy in performance by comparison but it lacked none of the good characteristics inherent in this design. As a pilot-training airplane or a compatible

sport-plane it was every bit as good.

The Brownback "Tiger" was a design similar to the old "Anzani" (French) engine, an engine any pilot is not likely to forget if he has ever flown behind one. H. L. Brownback was the U.S. A. agent for Anzani engines for several years, long enough to become aware of some of their idiocyncrasies. Redesigning the 6 cyl. Anzani to operate closer to American standards, he introduced the model C-400 early in 1929. Several of the model C-400 with 80 h.p. at 1700 r.p.m. were manufactured at the Brownback Laboratories in New York City. Later, in 1930, manufacture was begun at the Light Mfg. & Foundry Co. at Pottstown, Penna., which then built the improved "Tiger 90." This was an engine certainly much more reliable and much easier to put up with than its earlier cousin, the "Anzani." In 1931 the Light Foundry introduced the new "Tiger" of 100 h.p. which was the same engine basically, with several mechanical improvements, and also the "Tiger 125" which used many "Tiger 100" components but was arranged more conventionally as a 7 cyl. single-row "radial." Starting out in 1929 with 80 h.p. the C-400 was listed at $1840.; 1930 improvements jacked the horsepower to 90 and the price was dropped to $1295. Added improvements in 1931 raised horsepower to 100 but Light was finding it hard to flush out enough customers to make the project pay. The C-400 was of 397 cu. in. displacement, was 37 inches in diameter and weighed in at about 270 lbs. dry. These figures remained much the same through all 3 versions of this engine except for compression ratio, rated r.p.m. and other tricks generally used to raise the horsepower output.

The Lincoln trainer model PT-T was a lean and lanky open cockpit biplane, with seating for two in tandem, primarily designed for pilot training. Kept bare of frills and fancy finery but not to the point of gaunt nakedness, the model PT-T did present a neat appearance that would also be of some interest to the economy minded sportsman pilot. Powered with the 6 cyl. "Tiger" of 90 h.p. its performance was ample for average needs, with a smoothness instigated by the effortless power delivered by the slow-turning engine. Attaining its rated horsepower at low r.p.m. turning a fairly large propeller, the "Tiger 90" operated smoothly and efficiently with an economy of operation that was a boon to the average owner or operator. Flight characteristics were of course comparable to other examples in the Lincoln PT line, and despite its lower power it was on its way to becoming quite popular. The type certificate number for the Lincoln model PT-T was issued 7-25-30 and some 5 examples of this model were manufactured by the Lincoln Aircraft Co., Inc. at Lincoln, Nebraska. Victor H. Roos was president and general manager; and Ensil Chambers was chief engineer. Because of the general slump caused by economic conditions the Lincoln working force had been cut to a bare minimum and production had slowed to practically a snail's pace. The 3-place model LP-3 was all but phased out except for a few one-of-a-kind models. The PT trainer was built only on order and experimental work was well along on the new "All-Purpose" monoplane with cross-your-finger hopes attached to it for a somewhat better future for "Lincoln."

Listed below are specifications and performance data for the Lincoln model PT-T as powered with the 90 h.p. Brownback "Tiger" engine: length overall 25'7"; height overall 9'4"; wing span upper 32'3"; wing span lower 27'8" [31'9"]; wing chord both 58"; wing area upper 154 sq. ft.; wing area lower 123 [143] sq. ft.; total wing area 277 [297] sq. ft.; figures in brackets are for earliest versions; airfoil "Goettingen" 436; wt. empty 1164 lbs.; useful load 598 lbs.; crew wt. with 28 gal. fuel was 170 lbs. for each occupant, 20 lbs. for each parachute and 20 lb. for baggage; gross wt. 1762 lbs.; max. speed 102; cruising speed 85; landing speed 38; climb 730 ft. first min. at sea level; climb in 10 mins. was 6400 ft.; ceiling 13,000 ft.; gas cap. 28 gal.; oil cap. 4 gal; cruising range at 5.5 gal. per hour was 380 miles; price at the

factory was $3360. or $2235. less engine and propeller. Earliest listings of PT-T specifications show upper and lower wing span similar to those of PT-K and PT-W but some later listings show about 4 ft. less for lower wing span; further research into this model proved this to be true, for at least the later examples of the PT-T.

The fuselage framework was built up of welded chrome-moly steel tubing in a modified Warren truss, faired to shape with wooden fairing strips and fabric covered. The cockpits were deep and well protected with seats of the bucket type having deep wells for parachute packs; dual controls of the Bloxham safety-stick type were provided. Engine and flight instruments were either provided in both cockpits or placed so that both occupants could see them, while ignition switch, fuel shut-offs, starting choke, cowling shutters and stabilizer adjustment were controllable from either cockpit. Cockpits of 24x42 in. dimension were lightly and neatly upholstered and there was a baggage allowance of 20 lbs. in a small compartment behind the rear seat. The fuel tank was mounted high in the fuselage, just ahead of the front cockpit, with a direct-reading fuel gauge projecting through the cowling. The wing framework was built up of solid spruce spar beams with basswood built-up wing ribs; there was an aileron in each panel and the completed framework was covered in fabric. The split-axle landing gear of 76 in. tread was somewhat taller on the model PT-T, to allow ample ground clearance for the larger diameter propeller; shock absorbers were spools of rubber shock-cord (bungee cord), wheels were 26x4 and no brakes were provided. Low pressure semi airwheels were later optional for $133. extra. The tail-skid was of the spring leaf type with a removable hardened shoe. The fabric covered tail-group was built up of welded steel tubing; the vertical fin was ground adjustable and the horizontal stabilizer was adjustable in flight. A "Fahlin" wooden propeller, a tool kit, Bloxham safety sticks, and a fuel gauge were standard equipment. A metal propeller, navigation lights, engine starter and low pressure semi-airwheels with brakes were optional. The next Lincoln development was the "All-Purpose" monoplane model AP-B5 as described in the chapter for ATC #372 in this volume.

Listed below are Lincoln model PT-T entries as gleaned from registration records:
C-279N; Model PT-T (#951) Brownback C-400.
X-405V;   (#952)
NC-420V;   (#953)
NC-421V;   (#954)
NC-422V;   (#955)

Serial #952 was factory demonstrator for Light Mfg. & Foundry Co.; serial #954 also eligible as model PT-K.

Fig. 157. "Waco" INF mounted 125 h.p. Kinner B5 engine.

Waco Aircraft had hoped to feature the 100 h.p. model KNF as a sort of economy model in their new "F" series line-up, predicting that the lower initial investment and lower operating costs of this model would be more in tune with the leaner times that prevailed. Oddly enough, this was not the case and most prospective buyers were clamoring for a little more power. The model RNF with its 110 h.p. engine was selling in large quantities, but there were also those buyers who had a definite leaning toward the "Kinner" brand of engine, so it was inevitable for "Waco" to mount the new 125 h.p. Kinner B5 in a model called the INF. True, the added 25 h.p. over that of the KNF would seem only a token gesture towards the cause but one must remember that the "Waco F" was an airplane that made every single horsepower count and the boost to 125 h.p. did make a noticeable difference. The boost in price of nearly $300 over that of the cheaper KNF was certainly no deterent to its popularity and the more expensive model INF was accepted with enthusiasm and sold quite well. This was one of the first certificated installations of the newly introduced 125 h.p. Kinner B5 engine and, quite naturally, Kinner Motors bought the first example of the

INF to come off the line. Proud of this versatile combination, Kinner Motors marshalled Leslie Bowman, their capable test pilot, off to the 1930 National Air Races held at Chicago where he entered several events with the new "Waco" INF. Les Bowman finished second in one event and fourth in another; these were speed events amongst some mighty fast company too. A week or so later, Bowman and the same INF were entries in the National Air Tour for 1930, and captured 10th place in the long grind in the company of a very determined field. The "Model F" was certainly not blessed with a great amount of speed in any of its versions but what it lacked in that respect it made up for in a nimble performance and a willing response.

The "Waco" model INF was an open cockpit biplane seating three, and was basically typical of the RNF and the KNF except for its engine installation and some minor modifications necessary to this combination. Powered with the 5 cyl. Kinner B5 (R-440) engine of 125 h.p. its short-field performance was greatly improved (take-off run 125 ft. & landing roll about 100 ft.) and though the top speed was boosted but little, the extra maneuverability and the flashing response was enough to warm the heart of any good man. With this spirited performance and sharp response, the model INF was popular for teaching aerobatics in the secondary phases of pilot training courses, even popular in the CPTP program of the late "thirties." It was popular, too, with the private-owner who aspired to play sportsman-pilot but had to do it on a lower budget. In direct contrast to what was just said, Thomas Fortune Ryan, wealthy enough to afford something more expensive, bought a "Waco" model INF, then learned to fly in it for the sport. When used solely for sport flying the INF was most often decked out with a low drag "speed ring" engine cowling and streamlined "pants" over the wheels; when thus fitted the INF was good for at least 125 m.p.h. top speed. The type certificate number for the "Waco" model INF was issued 8-2-30 and 50 or more examples of this model were manufactured by the Waco Aircraft Co. at Troy, Ohio.

Listed below are specifications and performance data for the "Waco" model INF, as powered with the 125 h.p. Kinner B5 engine: length overall 20'8"; height overall 8'4"; wing span upper 29'6"; wing span lower 27'5"; wing chord both 57"; wing area upper 130.5 sq. ft.; wing area lower 111 sq. ft.; total wing area 241.5 sq. ft.; airfoil (NACA) M-18; wt. empty 1156 [1171] lbs.; useful load 740 lbs.; payload with 32 gal. fuel was 360 lbs.; gross wt. 1896 [1911] lbs.; figures in brackets show revised allowable wts.; max. speed 118; cruising speed 100; landing speed 35; climb 800 ft.

Fig. 158. INF high over California.

first min. at sea level; ceiling 16,000 ft.; gas cap. 32 gal.; oil cap. 3 gal.; cruising range at 8 gal. per hour was 370 miles; price at the factory

Fig. 159A. Ideal for all-purpose service, INF was also favored for sport.

field was originally announced as $4285 but held at $4365 through 1930 with a hike to $4565 in April of 1931.

The construction details and general arrangement for the model INF were similar to the RNF and KNF, as described in the chapters for ATC #311 and #313 in this volume. Although a speed-ring engine cowling and streamlined wheel pants were optional equipment they were quite often used on this model; the standard propeller installation was a wooden "Hartzell" but a metal propeller was very often used too. Other optional equipment was a Heywood engine starter, navigation lights and a battery. For special purpose, a 52 gal. fuel cap. was allowed and this extra weight was naturally deducted from the payload allowance. The individual wheel brakes were operated by the rudder pedals in conjunction with a hand-lever to transmit the braking tension desired, and a locking arrangement was used for parking. The baggage allowance was 15 lbs. and a later amendment boosted allowable gross weight to 1938 lbs. The next development in the "Waco" biplane was the model CRG as described in the chapter for ATC #362 of this volume.

Listed below is a partial listing of model INF entries as gleaned from registration records:

X-864V; Model INF (#3265) Kinner B5.
NC-133Y;   ''   (#3320)   ''

NC-601Y; Model INF (#3331) Kinner B5.
NC-615Y;   ''   (#3351)   ''
NC-618Y;   ''   (#3352)   ''
NC-619Y;   ''   (#3364)   ''
NC-620Y;   ''   (#3365)   ''
NC-627Y;   ''   (#3366)   ''
NC-622Y;   ''   (#3367)   ''
NC-625Y;   ''   (#3368)   ''
NC-640Y;   ''   (#3369)   ''
NC-634Y;   ''   (#3370)   ''
NC-635Y;   ''   (#3371)   ''
NC-637Y;   ''   (#3372)   ''
NC-638Y;   ''   (#3373)   ''
NC-643Y;   ''   (#3381)   ''
NC-644Y;   ''   (#3382)   ''
NC-645Y;   ''   (#3383)   ''
NC-648Y;   ''   (#3384)   ''
NC-649Y;   ''   (#3385)   ''
NC-670Y;   ''   (#3387)   ''
NC-11200;   ''   (#3397)   ''
NC-11205;   ''   (#3398)   ''
NC-11203;   ''   (#3399)   ''
NC-11209;   ''   (#3400)   ''
NC-11207;   ''   (#3401)   ''
NC-11214;   ''   (#3402)   ''
NC-11226;   ''   (#3403)   ''
NC-11236;   ''   (#3404)   ''
NC-11245;   ''   (#3405)   ''

Serial #3390 in Brazil as PP-TDK believed to be a model INF.

Fig. 159B. Ryan "Foursome" model C-1 offered as family-type airplane or for men of business.

Patterned very much after its sister-ship the famous "Brougham," the new smaller Ryan "Foursome" was tailored more to the needs and tastes of the flying businessman or the flying family man. To be more attractive and thereby more appealing to this rather selective buyer, several changes were made in arrangement and appointment — to offer more practical coach-work and the little conveniences so appreciated by the average private owner pilot. Showing influence coming out of Detroit, the four-place "Foursome" was outfitted more like a custom motor-car; the interior lacked the usual "air-plane clutter," the seats were deep, roomy, and richly upholstered; one rear seat had a reclining back for snoozing or just lounging. Spacious inside, there was plenty of leg room to let one stretch and relax, and a wide door eliminated the usual gymnastics of entry or exit. Giving some thought to the pilot also, there was now more convenience, still plenty of visibility and a fairly quiet interior. Flight characteristics were more in tune for the accomplished amateur pilot, who excelled more in enthusiasm than he did in piloting technique. Typical of several other craft on the market, the Ryan "Foursome" no doubt would have stood up well in the thick of competition, but it came rather late and at a very bad time, so it was left begging for customers. The new "Foursome" was developed early in 1930 at St. Louis; it was hope-

fully displayed at the New York Aircraft Salon in May, but the Ryan plant was sold in Oct. of 1930 and its manufacture was transferred to the plant in Detroit. Here another version of the "Foursome" was being tested with the new Packard "Diesel" engine as the model C-2, but it went begging too, and before very long the parent Detroit Aircraft Corp. had only the Lockheed Company to look to for any semblance of normal production amongst their "family" of aircraft manufacturers. Of the three Ryan "Foursomes" that were built, all led varying useful lives in business and sport for many years. One example was acquired from the Horlick Malted Milk Co. in mid-1938 by Alexander Loeb and fitted with extra fuel capacity for long distance flying. Taking off from a strip in Nova Scotia in Aug. of 1939 for a rumored flight to Palestine, he was apparently lost at sea and not heard from again.

The Ryan "Foursome" model C-1 was a sporty high-winged cabin monoplane with more than ample seating for four in an arrangement of comfortable richness and subdued good taste. Aimed more closely at the wealthier private owner, it was a blend of features more attractive to the wants and needs of business and sport — a ship one might be easily proud to own and operate. Powered with the 7 cyl. Wright J6 series (R-760) engine of 225 h.p. in its prototype version and later with the improved J6-7 of 240 h.p., its performance was calculated to offer a match with the best in this class. Flight characteristics were good and described as traditionally "Ryan," with perhaps a little more stability and just a little more sharpness, in its maneuvering response. Take offs were short and clean, the landings were true and final with ground maneuvering that was greatly enhanced by easily operated individual wheel brakes and a swiveling tail wheel. Rugged of character and robust of frame, the "Foursome" wore well in all kinds of service and operated for many years. Stemming from a design that led to the famous ocean-crossing "Spirit of St. Louis," the "Foursome" was fortunate to inherit this lineage, but it was left with the unwanted distinction of being the last airplane of the "Mahoney-Ryan" series. The type certificate number for the Ryan "Foursome" model C-1 was issued 8-11-30 and three examples of this model were manufactured by the Ryan Aircraft Corp. at St. Louis, Mo. and at Detroit, Mich., a division of the Detroit Aircraft Corp. P. R. Beasley was president; James Work was V.P.; John C. Nulsen was general manager; W.A. Mankey was chief engineer and Carl (Karl) S. Betts was sales manager. The Detroit Aircraft Corp. had twelve companies in its family of subsidiaries and the operation of all but two was now centered in and around the motor-city area of Detroit, Mich.

Listed below are specifications and performance data for the Ryan "Foursome" model C-1, as powered with the 225 h.p. Wright J6 (R-760) engine; length overall 27'7"; height overall 8'6"; wing span 39'3"; wing chord 75"; total wing area 230 sq.ft.; airfoil "Clark Y"; wt. empty 2133 lbs.; useful load 1217 lbs.; payload with 70 gal. fuel was 568 lbs. (this included 3 passengers at 170 lbs. each and 58 lbs. baggage); gross wt. 3350 lbs.; max. speed 128; cruising speed 108; landing speed 55; climb 800 ft. first min. at sea level; ceiling 14,000 ft.; gas cap. 71 gal.; oil cap. 7 gal.; cruising range at 12.5 gal. per hour was 600 miles; price at the factory originally quoted at $10,900., lowered to $9985. in May of 1931.

The fuselage framework was built up of welded chrome-moly steel tubing in a rigid truss form, liberally faired with formers and fairing strips then fabric covered. All cabin seats were of the individual car-type and richly upholstered to match the cabin roof and sidewalls; a baggage compartment with allowance for 58 lbs. was behind the rear seats, and pilot's controls were of the joy-stick type with "duals" optional. There was a large entry door on the right side, forward windows were of shatterproof glass and cabin heat was provided; to minimize fouling, all control tubes and cables were under the floor. The wing framework in two halves, was built up of laminated spruce spar beams with stamped-out duralumin wing ribs; the leading edges were covered with dural metal sheet, the ailerons were of a welded steel tube frame and the completed framework was covered in fabric. A fuel tank was mounted in the root end of each wing half, and wing bracing struts were of heavy-sectioned streamlined steel tubing with adjustable end fittings for accurate rigging. The wide tread landing gear was of the out-rigger type assembled into the wing bracing struts to make a rigid truss; shock absorbing struts were of oleo-spring units, wheels were 30x5 and Bendix brakes were standard equipment. The fabric covered tailgroup was built up of welded steel tubing and the horizontal stabilizer had an improved adjusting mechanism to allow fine settings for flying trim. A metal propeller, navigation lights, cabin heater and a tail wheel were standard equipment. Dual controls, low pressure semi-airwheels, a speed-ring engine cowling (narrow chord Townend-type) and an inertia-type engine starter were optional.

Listed below are Ryan "Foursome" model C-1 entries as gleaned from registration records:

X-556N; Model C-1 (#400) Wright R-760.
NC-557N; Model C-1 (#401) Wright R-760.
NC-558N; Model C-1 (#402) Wright R-760.
Serial #401 first as model C-2 with Packard "Diesel" DR-980 engine as approved on a Group 2 certificate numbered 2-263, issued 8-20-30, later modified to C-1; all examples also eligible with Wright R-760 engine of 240-250 h.p.

Fig. 160. Curtiss "Kingbird" model D-1 mounted two Wright J6-7 engines.

The twin-engined Curtiss "Kingbird" was a medium-sized transport designed especially to operate safely in the case of failure to any one engine, the remaining engine being capable of flying the airplane safely to a haven of refuge. In tests proving that aerodynamic arrangement was all-important in the multi-engined design, Curtiss engineers placed the engines on the "Kingbird" so far forward and so close together that the center-line of engine thrust moved laterally only to a small degree in the case of failure to one engine. This fact, coupled with responsive rudder control, practically eliminated the tendency of the airplane to lurch to one side when power was lost on one engine or the other, therefore it was reported to handle almost as well as a single-engined airplane. First introduced at the St. Louis Air Exposition in Feb. of 1930, the twin-motored "Kingbird" soon embarked on a nation-wide tour where its single-engined performance was one of the cardinal features demonstrated time and time again. Taking off with eight people aboard and a full load of fuel, the "Kingbird" would "lose an engine" when barely 50 ft. off the ground and continue its climb-out quite steadily to a safer altitude. Normal turns in either direction were quite safe and procedure to a landing was more or less normal, but we must say in all fairness that it did take a certain amount of technique, and was not a chore that could be lightly turned over to just any pilot.

Featuring good comfort and economical air-travel, the operating cost per mile of the "Kingbird" twin was just about one-half that of the average 12 to 14 place tri-motor, so theoretically two "Kingbirds" could be operated at the same cost to offer more seats, more frequent schedules and many other extra advantages. Eastern Air Transport (now Eastern Air Lines) studied the new design with interest and decided it fit their needs very well. They placed an order for 12 ships to add to their mail and passenger carrying fleet; the "Kingbirds" hauled passengers and small amounts of mail on certain segments of the E.A.T. route from New York to Miami. Eventually, E.A.T. ordered two more of the "Kingbird," making a total of 14 in operation, and together they piled up many miles of dependable service. Pan American-Grace Lines, operating routes in So. America, ordered one of the earliest "Kingbird" for operation in the mountainous Latin-American countries; so did Consolidated Air Lines, who operated shuttle service in California.

Designed and developed at Garden City, Long Island by the Curtiss corps of engineers under the direction of T. P. Wright with F. Al Wedberg in charge of the development project, the "Kingbird" was more or less a larger "Thrush" with

Fig. 161. Twin-engined "Kingbird" offered reliability and power reserve; several features were innovation in multi-motored design.

two engines. The prototype was first powered with 2 Curtiss "Challenger" engines as the model C, but this combination sorely lacked any power reserve so these engines were hastily replaced with 225 h.p. Wright J6 engines. Built in 3 examples and thoroughly tested, the "Kingbird" project was then turned over to the Curtiss-Robertson Div. at St. Louis to ready for production. The engineering and flight testing were under the direction of H. Lloyd Child with E. M. Flesh and Knute Hendrichson as project engineers who were also responsible for the development of various interior arrangements. Back in Garden City, Ken Perkins was project engineer on the development of a "Kingbird" as a troop-transport and air-borne ambulance for the U. S. Marines. Things at the St. Louis division were beginning to taper off, with "Robin" and "Thrush" production slowly coming to an end, so the plant force had been noticeably thinned out to about 100 key men. With an order for twelve of the "Kingbird" practically in their hip-pocket, and the outlook very bright on the new Curtiss "Junior" program, some 200 men who had been previously laid off, due to lack of work, were rehired to bring the working force back to 300 or more. Tooling-up had already been completed and some production on the "Kingbird" was getting underway by July of 1930.

The Curtiss "Kingbird" model D-1 was a twin-engined, high wing cabin monoplane with ample seating for eight; and its basic arrangement was attuned to utility and economy of operation. Offered in several arrangements, the interior could be fitted with eight seats as a coach-style transport, or in an arrangement cutting the total seating capacity to 6 or 7 to make allowance for 200 or 300 lbs. of mail. The model D-1 in any combination was capable of fairly good performance

and efficient operation despite its stress on economy, but regardless of claims to the contrary, the power reserve available was not entirely sufficient to insure peace of mind in case of failure to one engine. The model D-1 was powered with two 7 cyl. Wright J6 (R-760) engines of 225-240 h.p. each, but prospective customers seemed to prefer the higher powered model D-2 with two 300 h.p. engine and as a consequence no examples of the D-1 version had been sold as such. It is certainly likely that several examples of the "Kingbird" must have been fitted with 7 cyl. Wright J6 engines, at least to pass certification tests, but even these must have been later modified to the D-2 version because no model D-1 "Kingbird" appears in any of the civil aircraft registers. The type certificate number for the Curtiss "Kingbird" model D-1, as powered with two Wright J6 (R-760) engines of 240 h.p. each, was issued 8-16-30 to the Curtiss-Robertson Div. which, by late 1930, was already reorganized into the Curtiss-Wright Airplane Co. of St. Louis, Mo.

Listed below are specifications and performance data for the "Kingbird" model D-1, as powered with two Wright J6 engines of 240 h.p. each: length overall 34'10"; height overall 10' 0"; wing span 54'6"; wing chord 100"; total wing area 405 sq. ft.; airfoil Curtiss C-72; wt. empty 3754 lbs.; useful load 2361 lbs.; payload with 120 gal. fuel was 1361 lbs.; gross wt. 6115 lbs.; max. speed 130; cruising speed 112; stalling speed 65; landing speed 62; climb 820 ft. first min. at sea level; climb in 10 min. was 6430 ft.; ceiling 14,700 ft.; gas cap. 120 gal.; oil cap. 14 gal.; cruising range at 25.4 gal. per hour was 455 miles; price at the factory was quoted at $23,333.

The fuselage framework was built up mainly of seamless duralumin tubing that was riveted into dural and chrome-moly steel wrap-around

Fig. 162. "Kingbird" engines mounted close to fuselage center-line to eliminate excessive thrust-line changes in operation on one engine.

fittings to make the proper joints; welded chrome-moly steel tubing was used at all highly stressed points. The framework was faired to shape with formers and fairing strips, then fabric covered except for the door panels and the forward nose-section cover. Eight seats were arranged 4 on each side, with an aisle down the middle; a baggage compartment was to the rear with allowance for 170 lbs. The wing framework in two panels, was built up of spar beams that were welded steel tube trusses with wing ribs of riveted aluminum alloy tubing; a center-section panel extending out to engine attach points was built up of duralumin tubular spar beams with riveted aluminum alloy wing ribs, and the completed framework was covered with fabric. Fuel tanks were mounted in the center-section panel of the wing flanking the fuselage, in close prox-

imity for gravity feed to the engines. The landing gear of wide tread (13'2") was fastened to a point just under each engine nacelle using oleo-spring shock absorbing struts; wheels were 36x8 and Bendix brakes were standard equipment. A swiveling tail-wheel was provided for better ground maneuvering and ground handling. The fabric covered tail-group was built up with spar beams and former ribs of stamped-out "Alclad" metal sheet that was riveted together into various assemblies; the tail-group was of the biplane type with twin rudders. Curtiss-Reed metal propellers, inertia-type engine starters and navigation lights were standard equipment. The next "Kingbird" development was the higher-powered model D-2, as described in the chapter for ATC # 348 in this volume.

Fig. 163. Curtiss "Kingbird" model D-2 mounted two 300 h.p. Wright J6 engines.

Practically the only difference in the Curtiss "Kingbird" model D-2 over the model D-1, as described just previous, was an increase of some 120 h.p. to the total available power but this was apparently enough reason to warrant its preference over the lower powered D-1 version. Of typical aerodynamic arrangement, the model D-2 also had its engines placed far forward and relatively close together, to practically eliminate severe thrust line changes in the case of failure to any one engine; in fact, the propeller blades were beyond the front end of the fuselage nose and less than one foot separated the tips as they swung through their arc. With production slowly getting underway by July of 1930, on a large order for Eastern Air Transport (E.A.T.), Walter Beech watched the progress with great interest and selected the fourth ship off the line for entry in the National Air Tour for 1930. As usual, Walter Beech demonstrated his prowess, that was always fired by dogged determination, and flew the new "Kingbird" model D-2 to 6th place amongst some very fierce competition. Owen G. Harned, sales manager, went along as the co-pilot for Beech. Previously, at the National Air Races held in Chicago, Frank Kern had flown a "Kingbird" in the ten-lap multi-motored race but dropped out at the fifth lap, after being "dusted off" by a fast Ford and a Bach tri-motor. The E.A.T. was soon accepting delivery on their new fleet of "Kingbirds" and by year's end of 1930 they were linking New York with Miami,

by air, in passenger service at a rate of 7 cents per seat-mile; some of the "Kingbird" were later fitted with a mail compartment to carry mail and cargo also. Flying shoulder to shoulder with veteran "Mailwings," some huge Curtiss "Condors" and a Ford "Tri-Motor," the "Kingbird" gave very dependable and popular service, and most saw active duty with the line into 1935. Back in the Garden City parent plant, project engineer Ken Perkins had already developed plans to modify a model D-2 into a troop-transport and air-borne ambulance for the U. S. Marines, which was called the RC-1. When the "Kingbird" were phased out of service by E.A.T. it has been reported that some were shipped to Turkey for air-line use.

The Curtiss "Kingbird" model D-2 was a medium sized, twin-engined, high winged cabin monoplane also seating eight, and was more or less typical of the model D-1 except for its engine installations which were now the 9 cyl. Wright J6 (R-975) engines of 300 h.p. each. Offered in several interior arrangements for versatility in airline service, the "Kingbird" was basically arranged as a coach-style transport for a pilot and seven passengers, with 210 lbs. of their baggage. Other available arrangements were the mail-passenger type that carried a pilot and six passengers with 155 lbs. of their baggage and some 200 lbs. of mail; and the so-called cargo type that carried pilot, co-pilot, two passengers and 794 lbs. of cargo, some of which was passenger baggage. Of course any

Fig. 164. "Kingbird" model D-2 was flown by Walter Beech in 1930 National Air Tour; shown waiting to be waved off from starting line.

version, as equipped for service on E.A.T. routes, was later fitted with radio and night flying equipment. Performance of the model D-2, with a total of 600 h.p., was naturally much better than that of the model D-1 that only had a total of 480 h.p. Take-off time was lowered considerably, climb-out was increased by some 180 ft. in the first minute and cruising speed was some 10 m.p.h. better. Flight characteristics were compatible and general behavior was quite lively for a ship of this size and type. The type certificate number for the "Kingbird" model D-2, as powered with 2 Wright R-975 engines of 300 h.p. each, was issued 8-16-30 and 14 examples of this model were manufactured by the Curtiss-Wright Airplane Co. (formerly the Curtiss-Robertson Div.) on Lambert Field in St. Louis, Mo. Ralph S. Damon was the general manager and literally the acting president until Walter Beech was sent over from Wichita to act as the president; E. M. Flesh and Knute Hendrichson were project engineers and H. Lloyd Child, who came from the Garden City parent plant with the "Thrush" project, was the chief engineer and chief test pilot. Karl H. White was added to the engineering staff when the "Moth" division was moved to St. Louis and both Ted Wells and Herb Rawdon came over from the Travel Air Co. Some time later both Al Wedberg and Ken Perkins were temporarily assigned to the St. Louis Div. but were recalled to the Curtiss division in Buffalo, N. Y., for other assignments.

Listed below are specifications and performance data for the Curtiss "Kingbird" model D-2, as powered with 2 Wright J6 engines of 300 h.p. each: length overall 34'9"; height over-

all 10'0"; wing span 54'6"; wing chord 100"; total wing area 405 sq. ft.; airfoil Curtiss C-72; wt. empty 3877 lbs.; useful load 2238 lbs.; payload with 120 gal. fuel was 1238 lbs.; gross wt. 6115 lbs.; max. speed 142; cruising speed 122; stall speed 65; landing speed 62; climb 1000 ft. first min. at sea level; climb in 10 min. was 9000 ft.; ceiling 16,000 ft.; gas cap. 120 gal.; oil cap. 14 gal.; cruising range at 36 gal. per hour was 415 miles; price at factory was $25,555.; the model D-2 was later eligible with 330 h.p. Wright R-975 engines for a slight increase in performance. The cargo-type model D-2 weighed in at a standard gross of 6055 lbs.; the mail-passenger type 4280 lbs. empty, useful load was 2080 lbs., payload with 120 gal. fuel was 1080 lbs. and gross wt. was 6360 lbs.; performance was only slightly affected.

In a special test flight of the "Kingbird" Mail-plane (D-2 type) the ship was loaded to a 5050 lb. gross which included 2 pilots, 675 lbs. of fuel and oil, 200 lbs. of mail, radio equipment, a parachute for each pilot and oxygen apparatus. Thus loaded the "Kingbird" was flown to an official reading of 25,525 ft.; at the 20,000 foot level the ship was still climbing at 330 ft. per min. Climb in the first minute at sea level was a rousing 1700 ft.; at 5000 ft. the climb was still 1280 ft. per min. and 10,000 ft. was reached in 8.3 minutes. With a normal top speed of 140 m.p.h., this speed had just fallen to 125 m.p.h. at 18,000 ft. and range at this altitude was at least 225 miles; the absolute ceiling with only one engine was 10,000 ft. These tests intended to demonstrate the capability of service in So. America over the high peaks and passes of the Andes mountains; as a consequence, one "King-

Fig. 165. "Kingbird" served seaboard routes of Eastern Air Transport into 1935.

bird" Mailplane was built for service with Pan American-Grace Airways.

The construction details and general arrangement of the model D-2 was typical of that of the model D-1, as described in the previous chapter. The nose-section of the fuselage was covered with a bulbous metal panel that was easily removed for the servicing of instruments and the control mechanism; dual controls were provided by a central column with a swing-over control wheel. Entry to the cabin was by a large door and a convenient step on either side; a baggage compartment for up to 210 lbs. and a metal-lined mail compartment for 200 lbs. were to the rear of the cabin area, and some ships were fitted with the convenience of a small lavatory. Due to the forward position, and a sky-light overhead, the visibility from the pilot's station was excellent; all windows were of shatter-proof glass and slid open for ventilation. Reclining seats formed of aluminum were upholstered in broadcloth and the interior was finished in harmonious trim; cabin heaters and dome lights were added for passenger comfort and convenience. The wings were braced to the fuselage by parallel struts of streamlined aluminum alloy tubing, and combined with various trusses of streamlined tubes that supported the engine nacelles and provided attachment for the landing gear. Balanced-hinge ailerons of the Friese type were built up of duralumin beams and stamped-out "Alclad" metal ribs. The large tail-group was of the biplane type and braced to the fuselage by a complicated number of strut trusses; the top plane of the horizontal stabilizer was adjustable in flight. The engines were well muffled by large volume exhaust collector rings and were shrouded with "Townend" type speed-rings for a bit of extra speed; these "speed rings" were de-

veloped by Curtiss, using the shape of the Curtiss C-72 airfoil section. Wheel tread was 13 ft. & 2 in., wheels were 36x8 and Bendix brakes were standard equipment. Metal propellers, electric inertia-type engine starters, navigation lights, battery, fire extinguisher and a full panel of the latest instruments were also standard equipment. A complete set of night flying equipment and wheel pants were optional. The next "Kingbird" development was the rare model D-3 as described in the chapter for ATC #440. The next Curtiss-Wright development was the "Travel Air" monoplane model 6-B, as described in the chapter for ATC #352 in this volume.

Listed below are Curtiss "Kingbird" model D-2 entries as gleaned from registration records:

NC-585N; Model D-2 (#2001) 2 Wright R-975		
NC-586N;	″	(#2002)	″
NC-588N;	″	(#2003)	″
NC-589N;	″	(#2004)	″
NC-599N;	″	(#2005)	″
NC-600V;	″	(#2006)	″
NC-601V;	″	(#2007)	″
NC-602V;	″	(#2008)	″
NC-620V;	″	(#2009)	″
NC-621V;	″	(#2010)	″
NC-622V;	″	(#2011)	″
NC-626V;	″	(#2012)	″
NC-628V;	″	(#2014)	″
NC-629V;	″	(#2015)	″

Serial #2001-2002-2003-2004 passenger type alternately modified to mail-passenger type; serial #2004 also modified to cargo type; all the rest were mail-passenger type; NC-3133 was model J prototype (#G-1) on Group 2 approval 2-122; NC-31ON was model J-3 (#G-2) on Group 2 approval 2-196; NC-374N was model J-2 (#G-3) as prototype for model D-1; serial #2013 may have been U. S. Marines ambulance.

Fig. 166. "States" model B-3; note novel cowl around Kinner K5 engine.

During the closing months of 1929, many manufacturers began concentrating on the two place sport-trainer type that also harbored features attractive to the average sport flyer. Several outstanding airplanes built to these requirements had already been introduced and several more were in the offing. Of course not many 'plane builders could agree as to what the ideal form should be for a craft of this type because there were several open-cockpit biplanes in this sport-trainer class, a few cabin monoplanes, open monoplanes with the wing on the bottom or in the middle and even supported high on top in a "parasol" fashion. Each arrangement had its avid following, and each had its particular advantages, so perhaps it would be unfair to single out any particular configuration as the ideal or a better one than any of the others. However, the parasol monoplane had many features in its favor for this type of use and was becoming increasingly popular. An addition to this fast-growing family of airplanes was the "States" sport-trainer monoplane, a craft that first impressed one as being rather stolid and plain, but a slow measured look soon revealed feature after feature that was specifically planned for a practical purpose and nothing was added just to make it look cute. Perhaps somewhat unexciting, because it had not the flashing personality nor even the gay stance of some, the "States" monoplane was nevertheless an honest air-

plane with devotion to duty and would have eventually enticed a considerable following had there been enough time under more normal conditions.

In designing the first example of the "States" monoplane, Frederick H. Jolly was largely concerned in fashioning a craft that would be docile and easily manageable at the lower air speeds, still retaining a good measure of control into the stall, with good recovery when lost in a spin. There were still a large number of accidents during this period (1929-30) resulting from the "stall," and fledgling pilots especially were experiencing some difficulty in mastering the "spin recovery." On the average, instruction in stalls and spins was not a phase looked forward to by most student pilots so Fred Jolly made it a particular point to take on the development of a trainer inheriting a nature that would help instill confidence in the pilot and help him to learn the proper techniques, without exposing his neck to needless dangers. Selecting the parasol monoplane as the most likely configuration to start with, because of its inherent stability, ample wing area was provided to slow down accelerating forces. Aerodynamic geometry was calculated so that good control was provided throughout the low speed range and a novel differential action kept the large control surfaces from getting too sensitive at the higher air speeds. The prototype version was powered

Fig. 167. Prototype "States" with Rover engine; Frederick Jolly in foreground.

with the 4 cyl. Michigan "Rover" engine of 75 h.p. and characteristics lived up to most expectations, except for the delay in waiting for the "Rover" engine to be approved for manufacture. Two additional examples of this version were built, being awarded a Group 2 approval numbered 2-254; a slight redesign had already been planned for the new Kinner-powered model B-3. Built in quite modest surroundings by only a handful of craftsmen and rarely advertised nationally, the "States" monoplane was fairly unknown to most people except for occasional mention in some of the periodicals and stories brought back by pilots who visited the Chicago area during the National Air Races held there in 1930.

The "States" model B-3 (sometimes as S-E-5F) was an open cockpit monoplane of the parasol type with seating for two in tandem; rugged of frame, the fuselage was unusually deep for weather protection and there was stretch room for even the biggest pilot. With the wing perched atop a cabane of struts, visibility was good in most all directions and the landing gear was clearly visible during take-off and landings. The spraddle-legged landing gear was fitted with roly-poly "airwheels" to iron out even the worst landings and the wide wheel tread tended to keep the ship on track when taking off or landing on rough ground. Because of its comparative scarcity, lore on the "States" model B-3 has been buried deep with the passage of

Fig. 168. Parasol-wing "States" suitable for training or sport.

Fig. 169. Beefy construction of "States" insured more hours on the flight line.

time, but occasional stories credit it with a very compatible nature and a splendid perform-ance. Not otherwise particularly outstanding, the "States" had a feature quickly noticed and that was the unusual form of engine cowling. The engine fairing was more like a funneled N.A.C.A. cowling with the cylinder heads pro-truding; air entered through the front aperture and was vented through a circumferential slot in the rear. There is no doubt this afforded ef-ficient cooling to the engine, but whether it improved the airflow to lessen the parasitic drag is open to some question. The type certificate number for the "States" monoplane model B-3 was issued 8-11-30 and some 6 examples of this model were manufactured by the States Aircraft Corp. in Chicago Heights, Ill., through 1931; it is probable that at least 5 more examples were built in the next few years and the type certifi-cate finally expired on 9-30-39. There has never been a listing of company officers but we do know that Frederick H. Jolly was the chief en-gineer, at least at the beginning. The States Aircraft Corp. was planning a move to Center, Texas about July-Aug. of 1933 on the city's new municipal airport; a factory building was being erected and States was planning to go into pro-duction, but whether they did or not we cannot verify.

Listed below are specifications and perform-ance data for the States model B-3, as powered with the 100 h.p. Kinner K5 engine: length overall 22'0"; height overall 7'10"; wing span 32'0"; wing chord 72"; total wing area 180 sq. ft.; airfoil Clark Y; wt. empty 1083 lbs.; useful load 552 lbs.; payload with 24 gal. fuel was 220 lbs.; 2 parachutes at 20 lbs. each were part of the payload; gross wt. 1635 lbs.; max. speed 108; cruising speed 90; landing speed 35; climb 750 ft. first min. at sea level; climb in 10 min. was 6000 ft.; ceiling 14,000 ft.; gas cap. 24 gal.; oil cap. 2.5 gal.; cruising range at 6 gal. per hour was 325 miles; price at the factory was $3645.

The fuselage framework was built up of weld-ed 1025 steel tubing, heavily faired to shape with wooden formers and fairing strips, then fabric covered. The cockpits were deep and well pro-

tected, having removable seat cushions to allow wearing of a parachute pack; there was a con-venient step for entry to the rear cockpit and a large door on left side for entry to the front cockpit. The interior was fully upholstered in leather and a small baggage locker, with al-lowance for 10 lbs., was in the dash panel of the front cockpit. The semi-cantilever wing framework, in two halves, was built up of solid spruce spar beams with spruce and plywood truss-type wing ribs; the leading edges were covered with duralumin sheet and the completed framework was covered in fabric. The wing halves were fastened to an inverted-vee cabane over the fuselage and braced to the lower long-eron by vee-struts of faired steel tubing; a 12 gal. fuel tank was mounted in the root end of each wing panel. A large cut-out in trailing edge of the wing allowed overhead vision from the rear cockpit; narrow-chord ailerons of the Friese type, with differential action, spanned nearly full length of the wings. The long legged land-ing gear of 84 in. tread used oleo shock absorb-ing struts; wheels were 22x10-4 and brakes were available. The fabric covered tail-group was built up of welded 1025 steel tubing; the fin was ground adjustable and the horizontal stabili-zer was adjustable in flight. A Hartzell wooden propeller, Goodyear airwheels and dual joy-stick controls were standard equipment. A metal propeller, engine starter, navigation lights and wheel brakes were optional. Custom colors were available on prior order. The next develop-ment in the "States" monoplane was the model B-4 as described in the chapter for ATC #477.

Listed below are "States" model B-3 entries as gleaned from registration records:
NC-10369; Model B-3 (#103) Kinner K5
NC-10556;   ″   (#104)   ″
NC-943N;   ″   (#105)   ″
NC-10719;   ″   (#106)   ″
NC-10723;   ″   (#107)   ″
NC-12043;   ″   (#108)   ″
This approval for serial #103 and up; there was evidence of a serial #113 but details on it are unknown.

Fig. 170. St. Louis "Cardinal" model C2-85 mounted 85 h.p. Le Blond 5DF engine, only one built.

When LeBlond Motors introduced their improved model 5DF engine of 85 h.p. (4-8-30), several manufacturers had already laid plans to use it in forthcoming models. Davis Aircraft was the first out with their model D-1-66 (D-1-85) parasol monoplane and others were to follow suit. The St. Louis Aircraft Corp. was producing the "Cardinal" monoplane series in three different versions and decided to bring forth a new model mounting the new 5 cyl. LeBlond 5DF. The "Cardinal" was already available in a C2-90 version, which mounted the 7 cyl. Le Blond 7D engine of 90 h.p. (ATC #264), and in production for some nine months now but the new 5DF engine was smaller, lighter, more efficient mechanically, with less fuel consumption, so it was certainly well worth the try. Using a "Cardinal" monoplane that already mounted the 7 cyl. engine as a model C2-90 (ser. #113), company engineers replaced this 7 cyl. engine with the 5 cyl. 5DF and only slight changes were necessary for this new combination. The switch in powerplants resulted in a ship that was 59 lbs. lighter when empty, it carried 22 lbs. more in useful load and was still 37 lbs. lighter when fully loaded. Performance between the two versions was still more or less comparable but the advantages were in the new 5DF engine which was a considerable improvement over the older model 7D-90. The new 5 cyl. LeBlond 5DF developed 85 h.p. at 2125 r.pm.

with 266 cu.in. displacement, it weighed in at 219 lbs. dry, and improvements in lubrication and cylinder head design enabled it to operate with more reliability for longer periods of time. St. Louis Aircraft felt that its introduction of the model C2-85 would replace the earlier C2-90 in its line-up, and perhaps even incite modification of all the older models into the new C2-85, that is, by owners preferring to invest in the new LeBlond 5DF engine.

The St. Louis "Cardinal" model C2-85 was a light high winged cabin monoplane with side-by-side seating for two; basically typical of the earlier model C2-90, the only difference would be slight advantages gained by installation of the new 5DF engine. Specifically leveled at the sport-type flyer who was forced to operate on a limited budget, the model C2-85 was tastefully remodeled, both in and out, to stand it slightly apart from the average ship of this type and it delivered quite a sporty performance at a minimum of expense. Flight characteristics and general behavior have never been clearly defined for the "Cardinal" but have always been described as a good average for a ship of this type. Due to the basic similarity between the models C2-90 and C2-85, we suggest reference to the chapter for ATC #264 of U.S. CIVIL AIRCRAFT, Vol. 3 for various applicable details. The type certificate number for the "Cardinal" model C2-85 was issued 8-11-30 and it is quite likely that no more than one example of this model was manufactured by the St. Louis Aircraft Corp., a subsidiary of the St. Louis Car Co. in St. Louis, Mo.

Listed below are specifications and performance data for the St. Louis "Cardinal" model C2-85, as powered with the 85 h.p. Le Blond 5DF engine: length overall 21'0"; height overall 7'0"; wing span 32'4"; wing chord 60"; total wing area 162 sq.ft.; airfoil Clark Y; wt. empty 940 lbs.; useful load 580 lbs.; payload with 30 gal. fuel was 211 lbs.; gross wt. 1520 lbs.; max. speed 116; cruising speed 98; landing speed 40; climb 880 ft. first min. at sea level; ceiling 10,500 ft.; gas cap. 30 gal.; oil cap. 3 gal.; cruising range at 5.5 gal. per hour was 490 miles; price at the factory was $3750.

The construction details and general arrangement of the "Cardinal" C2-85 were, of course, similar to that of the model C2-90, as described in the chapter for ATC #264 of U.S. CIVIL AIRCRAFT, Vol. 3. The fuselage framework, as on "Cardinals," was built up mostly of welded 1025 steel tubing with chrome-moly steel tubing and gussets in all points of stress. The cabin area measured 35x44x39 in. high with a 4 cu.ft. baggage shelf behind the seat; baggage allowance was normally 41 lbs. but not allowed when carrying 2 parachutes. Cabin windows were of "Charmoid" plastic, as was the sky-light in the cabin roof; controls were dual with a novel swing-over stick. The out-rigger landing gear had tread of 75 in. with oleo-spring shock absorbing struts; wheels were 26x4 equipped with brakes. The spring-leaf tail skid was fitted with a novel "Micarta" ball to decrease friction on hard-surface runways. The model C2-85 was also available in custom colors on prior order. A Hartzell or Supreme wooden propeller & navigation lights were standard equipment. A Townend type speed-ring engine fairing was optional. For a complete resume of all "Cardinal" models to this date, we suggest reference to the chapters for ATC #264-273-277 of U.S. CIVIL AIRCRAFT, Vol. 3. A 1931 model of the "Cardinal" was built as the C2-100, with the 110 h.p. Warner "Scarab" engine, and this appears to be the very last effort for the St. Louis Aircraft Corp. until sometime later in 1939 when they developed a primary trainer biplane for an Army Air Corps competition.

Listed below is the only known example of the "Cardinal" model C2-85 as gleaned from registration records:
NC-559N; Model C2-85 (#113) LeBlond 5DF-85
Serial #113 originally built as a model C2-90 with LeBlond 7D-90 engine.

Fig. 171. "Aeronca" model C-2 with 26 h.p. Aeronca engine; craft shown was of later type.

Introduced in Feb. 1930 at the St. Louis Air Show, as the first truly light-weight airplane to be certificated by the government for the civilian market, the "Aeronca" finally had appeared at a very opportune time. Coming out to show itself when economic distress in this country was affecting everyone in aviation, and many of the new converts to the airplane had not the means to keep flying as often as they would like to. The low first-cost and the extremely low upkeep of the new "Aeronca" C-2 brought regular flying well within the reach again to many that were faced with giving up flying altogether, or possibly indulge in a half hour or so on some other airplane after scrimping and saving for weeks. Under the circumstances the "Aeronca" C-2 was tailor-made to improve this situation, so to lightly say that this airplane emerged as a boon to private flying is indeed an understatement. It could be better said that the "Aeronca" started a new era in aviation. Brought forth into flying circles at a time when the open biplane was still considered the mainstay of private flying, and the small cabin monoplane was rapidly emerging to the fore as a useful machine for the average private flyer's use, the distinctive "Aeronca" was greeted first as some sort of aerodynamic novelty that seemed like a caricature of a real airplane. With its bulging fuselage shaped somewhat like the bill of a pelican and its broad wing braced only by flimsy-looking wires, it was subject to the poking of fun and every flight was sure to bring gales of laughter

from the assembled on-lookers. This skepticism and near-ridicule did not need to be endured for long because the laughter soon turned to friendly smiles of indulgence; the phenomenal performance of this putt-putt airplane began drawing queries of interest and it flew its way straight into the heart of flying America.

Going back into its lineage, the "Aeronca" C-2 as shown here, stems from a nearly-identical prototype that was introduced as the "Roche-Dohse" flivver-plane back in 1925. Designed by Jean A. Roche, a senior aeronautical engineer with the Air Service, this first model powered with a 2 cyl. Wright-Morehouse engine, actually had not caused much of a stir in flying circles because of the popularity of the war-surplus "Jenny" and "Standard" biplanes. In 1927, Jean Roche revived the early design with a few slight changes but the time for this type of airplane was not yet right; the pace of aviation in this historic year was certainly not conducive to a market for the feather-weight airplane. With a firm and determined belief in their convictions, though often discouraged by the pace of others, a small band of hardy pioneers formed the Aeronautical Corporation of America late in 1928 to continue development on the "Aeronca" concept, and to develop a small and reliable 2 cyl. engine to power it. Ironically enough, the "depression" that was triggered off in late 1929, bringing on circumstances that caused so many aircraft manufacturers a great concern, was the turning point that assured the

Fig. 172. Early version of C-2 showing buggy-wheel landing gear.

economical "Aeronca" a good future. Before long, several other aircraft manufacturers, inspired by the acceptance of the "Aeronca," quickly realized the possibilities lying ahead for an airplane of this type and began feverish development of low-powered light airplanes of their own. By the end of 1931 the feather-weight airplane movement was gaining great momentum and was providing just the stimulus needed for an increase in private flying. Solo flying time was now available for as little as $5.00 an hour in the "Aeronca" and nearly everybody could manage to scrape up a few bucks now and then for a few turns around the air-field.

Despite its very nominal power and its rather comical proportions the "Aeronca" C-2 was really quite an airplane, a capable airplane of a good many accomplishments. Its comparatively high performance spurred many of the young-bloods on to a rash of record attempts and soon the diminutive "Aeronca" C-2 held title to many official and unofficial record flights. Equipped with twin "floats" as a seaplane, a C-2 was gamely pushed to 15,082 ft. for a light-plane record; the C-2 landplane also recorded several flights to 20,000 ft. and beyond during 1930. A non-refueled endurance record for light airplanes was set by staying aloft for some 26 hours and several long-distance flights of 800 miles or more were satisfactorily performed. With its gliding ratio of more than 11 to 1 the "Aeronca" could easily glide from 16,000 ft. to any point within a 64-mile diameter circle and many pilots would thus fly their gas out and then "sailplane" for a half hour to a landing.

The loveable "Aeronca" C-2 was a high wing wire-braced light monoplane with what might be called an open cockpit for the seating of one person, with the seat placed down so low in the fuselage that one could actually touch the ground while seated in the airplane. Devoid of all mechanical frills and finery, the C-2 was equipped

with a joy-stick, a rudder bar and very little else; it was a ship in which you sat back relaxed and flew happily just by the feel and sound. Powered with the 2 cyl. "Aeronca" E-107A engine of 26-30 h.p., performance was adequate, and quite surprising, for this nominal power and much better than one would really expect. With its success well assured after only a few months out in active service, the "Aeronca" production schedule was geared to a furious pace to meet the country-wide demand. By the time 50 airplanes had been built, a seaplane version was also available and the C-2 was offered in a deluxe version with many improvements. First certificated on a Group 2 approval numbered 2-216 (issued 5-19-30) this approval was superseded by an approved type certificate (ATC) issued 8-13-30 for both the landplane and seaplane. The model C-2 in some three versions was in active production during the last half of 1930 and most of 1931, by which time well over 100 examples were built. With the introduction of the 2-seated model C-3 in early 1931, production of the C-2 series fell off to some degree and was finally discontinued in favor of the more practical "Duplex". However, improved and more powerful versions of the C-2, called the "Scout" and the "Cadet", were still available on order as late as 1933. The "Scout" was a C-2 deluxe with 36 h.p. and the "Cadet" was a 36 h.p. sport model (model C-1) arranged with less wing area and a more robust framework for the sportsman-pilot whose repertoire at some time would include "stunt flying". Some accounts say all models of the one-seater were available into 1933, but the C-3 series had firmly taken over by then and the standard C-2 was finally discontinued from production entirely. The "Aeronca" model C-2 was manufactured by the Aeronautical Corp. of America at Lunken Airport in Cincinnati, Ohio. Taylor Stanley was the president; Conrad G. Dietz was V.P. and

Fig. 173. Aeronca C-2 on APC floats was excellent seaplane.

general manager and Robert B. Galloway, formerly with Curtiss, was the chief engineer but Jean A. Roche, originator of the series, was still consulting engineer to the firm. Maj. Brower of the U.S. Army Air Corps performed all tests on the "Aeronca" prototype as developed in 1929. Conrad Dietz, himself an accomplished pilot, also did some of the test-flying on craft coming off the line.

Listed below are specifications and performance data for the "Aeronca" model C-2, C-2 Deluxe and PC-2 seaplane as powered with the 26 h.p. Aeronca E-107A engine; length overall 20'0"; height overall 7'6"; wing span 36'0"; wing chord 50"; total wing area 142 sq. ft.; airfoil "Clark Y"; wt. empty 398 [426] lbs.; useful load 274 [274] lbs.; crew wt. 170 [170] lbs.; allowable baggage 50 [50] lbs.; gross wt. 672 [700] lbs.; max. speed 80; cruising speed 65; stall speed 38; landing speed 32; climb 550 ft. first min. at sea level; climb in 10 min. was 5000 ft.; climb to 10,000 ft. was 30 mins.; normal ceiling 16,000 ft.; gas cap. 8 gal.; oil cap. 3 qts.; cruising range at 2 gal. per hour was 240 miles; price at the factory for C-2 Standard was $1555., lowered to $1545. and reduced to $1245. in late 1931; 1931 model C-2 Deluxe was $1695. at factory; figures in brackets are for C-2 Deluxe; permissable gross wt. later amended to 675 and 703 lbs. respectively. The following figures are for the PC-2 seaplane mounted on APC model A-1680 twin-float gear; wt. empty 470 [498] lbs.; useful load 224 [224] lbs.; crew wt. 170 [170] lbs.; no baggage allowed; gross wt. 694 [722] lbs.; figures in brackets are for PC-2 Deluxe seaplane; performance figures would be slightly lower than for landplane; price for

standard seaplane at factory was $1990.; pontoons installed on landplane at factory for $825. Later revised wts. for seaplane as follows; wt. empty 482; useful load 274; crew wt. 170; baggage and mooring gear 50 lbs.; gross wt. 756 lbs.; 30 h.p. engine offset losses by the extra weight.

The fuselage framework was built up of welded 1025 and chrome-moly steel tubing, in an odd form that was more or less square in the forward portion and triangular in the aft portion, with the top longeron forming the apex of the triangle and continuing across the cockpit area to form a mount for the wing attach; the framework was lightly faired with wooden fairing strips and fabric covered. The seat was placed down low in the portion just under the wing and a small baggage hamper was behind the seat back; a pyralin sheet was fastened across the gap between the wing root anchors to form a narrow sky-light for vision overhead. The wing framework in two halves, was of solid spruce spar beams with spruce truss-type wing ribs; the leading and trailing edges were covered with dural metal sheet and the completed framework was covered in fabric. Ailerons were a duralumin framework covered in dural sheet, being interchangeable from either left to right. The wing was stoutly braced with streamlined steel wires both above and below; the upper landing wires were fastened from the wing to an inverted-vee cabane struts and the flying wires were fastened from the wing to the lower longeron of the fuselage. The landing gear on the earlier models was a straight-axle type with 2 large buggy-type wheels; the axle and the rubber shock-cord were housed within the fuse-

Fig. 174. Comical lines of the C-2 fostered many smiles but no one objected to its economy.

lage to minimize drag. Later versions of the C-2 used low-pressure airwheels on this same type of gear and still later the landing gear was changed to a wider tread tripod type; the tripod gear was stiff-legged with no shock absorption other than the cushioning effect of the "airwheels". No brakes were provided. A fuel tank of 8 gal. capacity was mounted just behind the firewall and an extra fuel tank of 5 gal. capacity could be mounted in the baggage area, reducing baggage allowance to 10 lbs. or a 6 gal. tank could be mounted with no allowance for baggage. The windshield on the 1930 models was painfully small but 1931 models had a rather large windshield forming a sort of semi-cabin. The fabric covered tail-group was built up of welded 1025 steel tubing; the vertical fin was built integral to the fuselage and horizontal stabilizer was adjustable only on the ground. 1930 models C-2 had what was known as the narrow-type fuselage and 1931 models C-2 Deluxe had a wide-type fuselage; all types could be fitted with the new wider-tread "tripod" landing gear with Goodyear airwheels and all types could also be fitted with APC model A-1680 floats. The next "Aeronca" single-seater developments were the model C-1 "Cadet" and the C-2N "Scout", as described in the chapters for ATC # 447 and # 448. The next "Aeronca" development immediately following the model C-2 as discussed here, was the 2-seated model described in the chapter for ATC # 396 in this volume.

Listed below are various "Aeronca" model C-2 entries as gleaned from registration records:

-3774;	Aeronca C-2 (#1)	Aeronca E-107.
X-626N;	” (#2)	Aeronca E-107A.
X-627N;	” (#3)	”
NC-289V;	” (#4)	”
X-288V;	” (#5)	”
-522V;	” (#6)	”
-523V;	” (#7)	”
-524V;	” (#8)	”
NC-525V;	” (#9)	”
NC-560V;	” (#10)	”
NC-561V;	” (#11)	”
-562V;	” (#12)	”
NC-563V;	” (#13)	”
-564-V;	” (#14)	”
NC-565V;	” (#15)	”
-566V;	” (#16)	”
-567V;	” (#17)	”
NC-568V;	” (#18)	”
-569V;	” (#19)	”
NC-640W;	” (#20)	”
NC-11275;	” (#A-105)	”
NC-11276;	” (#A-106)	”
NC-11278;	” (#A-108)	”
NC-11279;	” (#A-109)	”
NC-11280;	” (#A-110)	”
NC-11281;	” (#A-111)	”
NC-11283;	” (#A-115)	”

Registration number for serial #8 unverified; serial #12-14-15-16-17 listed both as 1 pl. and 2 pl; NC-641W was serial #21 and numbers ran consecutively to NC-659W which was serial

#39; NC-10164 was serial #301-40 and numbers ran consecutively to NC-10174 which was serial #301-50; X-10175 was serial #A-51 and numbers ran consecutively to NC-10189 which was serial #A-65; serial #A-51 was the prototype seaplane; NC-10300 was serial #A-66 and numbers ran consecutively to NC-10325 which was serial #A-91; NC-650Y was serial #A-92 and numbers ran consecutively to NC-656Y which was serial #A-98; serial #A-99 and #A-102 were model C-3 prototypes; no listing for serial #A-100, A-101, A-103; NC-301Y was serial #A-104; listing for serial #A-105 and up in table above; serial # A-107 converted to model C-3; no listing for serial #A-112, A-113, A-114; serial #A-115 appears to be last entry for model C-2; serial #A-51, A-68, A-74, A-77, A-91, A-98 as PC-2 seaplane on floats.

A.T.C. #352
(8-13-30)
CURTISS-WRIGHT "TRAVEL AIR SEDAN", 6-B

Fig. 175. Model 6-B was a "Travel Air" favorite in new dress.

A good many will easily recognize the Curtiss-Wright "Sedan" model 6-B as a slight modification of the earlier "Travel Air" models 6000-B and S-6000-B, a popular 6-place cabin monoplane that was built in fairly good number and used practically everywhere for all manner of services. Coming nearly a year and a half later, the model 6-B was still quite typical except for some apparent face-lifting that changed its form somewhat, and a generous allowance added for more useful load. The so-called improvements no doubt added to the ship's utility but the super-imposed bulge in front tended to distract from the flowing lines of its former beauty. Introduced about mid-year, the new 6-B was carefully groomed for entry in the 1930 National Air Tour in Sept., which would serve as a good shake-down cruise for the new ship and also a good chance to show it off. Flown by Truman T. Wadlow, expert "Travel Air" pilot, who was naturally expected to bring out the best there was in it, together they finished the grueling grind to 9th place amongst a particularly determined field. Especially leveled at the needs of big-business, one of the first examples was used by a pipe-line company to survey their 1450 mile pipe-line route from the Oklahoma refineries to various junctions along the way, and finally clear to Chicago and Milwaukee terminals; the airplane was the only sensible way to lay out such a job. The pipe-line company used its 6-B through all phases of construction and servicing too; pleased company officials judged it one of its most valuable pieces of equipment. A meat-packing company and an oil company were other satisfied users of the 6-B

and years later, when most of them were replaced in service by newer equipment, the intrepid flyers of the "bush country" jumped at the chance to put them in their own service. The load-toting ability and the rugged character of the 6-B found its place in the various parts of the country that were of like nature.

The Curtiss-Wright "Travel Air Sedan" model 6-B was a big high-winged cabin monoplane with spacious seating for 6. It was quite typical to the earlier S-6000-B except for numerous modifications calculated to improve its general utility and to help extend the life of the basic design just a little longer. In view of the small number that were actually built it seemed a rather futile effort but these were fast becoming futile times in the aircraft industry. The well apparent bulged-out cabin in front might have been some improvement, but there seemed to be no urgent reason for it unless it was to afford a bit more visibility and as an extension of head-room for the pilot. With all the improvements in the pilot's cockpit, the main cabin interior, the ventilating and heating systems, provisions for a chemical toilet and some changes in the basic airframe, the 6-B weighed in about 100 lbs. heavier when empty. Over 90 lbs. was added to the useful load and this all added up to nearly 200 lbs. heavier at gross weight. As a consequence, this was some detriment to the overall performance but not enough to make a great deal of difference; a normal take-off run was still reported to be some 600 ft. at gross load. Of a gentle and obedient nature, the buxom model 6-B wore quite well in hard service and stretched out its usefulness

Fig. 176. 6-B "Sedan" spruced up with speed-ring and wheel pants; speed was boosted to 144 mph.

over a period of many years. The type certificate number for the Curtiss-Wright model 6-B was issued 8-13-30 and at least 4 examples of this model were manufactured by the Travel Air Div. of the Curtiss-Wright Airplane Co. of St. Louis, Mo. By Nov. of 1930, "Travel Air" was a company no more but was now the Travel Air Div. of the parent company in St. Louis. Walter Beech had sold out his interests and so his company and its designs were merged into the workings of the Curtiss-Wright organization; as further

er compensation he became president of the St. Louis Div.

Listed below are specifications and performance data for the model 6-B as powered with the 300 h.p. Wright J6 (R-975) engine; length overall 31'5"; height overall 9'3"; wing span 48'7"; wing chord 78"; total wing area 282 sq. ft.; airfoil Clark Y-15; wt. empty 2707 lbs.; useful load 1713 lbs.; payload with 90 gal. fuel was 950 lbs. (5 passengers at 170 lb. each & 100 lb. baggage); gross wt. 4420 lbs.; max. speed 135 (144

Fig. 177. 6-B as seaplane on Edo floats.

with speed-ring & wheel pants); cruising speed 115; stall speed 66; landing speed 62; climb 750 ft. first min. at sea level; ceiling 15,500 ft.; gas cap. 90 gal.; oil cap. 7 gal.; cruising range at 16 gal. per hour was 575 miles; price at factory $12,435.; wts. later amended to 2739 empty, useful load 1713 and 4452 as gross wt.; also eligible with Wright R-975 of 330 h.p. at 4500 lbs. gross wt.

The fuselage framework was built up of welded chrome-moly steel tubing, the portion from the firewall to end of cabin area was braced in a Pratt truss form with steel tubing, and from there to tail-post it was braced with round steel tie-rods in an X-truss form; the framework was faired with wooden formers and fairing strips, then fabric covered. From the front wing strut forward, the fuselage was covered with metal panels. The cabin walls were sound-proofed and insulated and all windows were of shatter-proof glass; the windows were split panels that could be slid open for extra ventilation. The seats were comfortable, well upholstered in fine fabrics and could be easily removed for hauling bulky cargo in the cabin of 112"x52"x36" dimension. Custom interiors sometimes included a divan or an extra passenger seat; cabin heat was available and also a chemical toilet. Baggage allowance was 100 lbs.; 25 lbs. of it, usually a tool-kit and other airplane gear, was carried in a bin under the co-pilot's seat and 75 lbs. of it was stowed in a compartment to the rear. Controls were dual "Dep" wheels and the ship was normally flown with 2 pilots as the crew. The wing framework of semi-cantilever design, was built up of spruce and plywood box-type spar beams with spruce and plywood truss-type wing ribs; the leading edges were covered in dural metal sheet and the completed framework was covered in fabric. Two gravity-feed fuel tanks were mounted in the wing, flanking each side of the fuselage; long narrow-chord ailerons of the Friese type were operated by a differential action. The wing bracing struts and the landing gear struts were of heavy gauge chrome-moly steel tubing and encased in balsawood fairings covered with fabric. The outrigger landing gear of 108 in. tread used "Aerol" (air-oil) shock absorbing struts; wheels were 32x 6 and Bendix brakes were standard equipment. A large steerable tail wheel was provided for ease in ground handling or maneuvering. The fabric covered tail-group was built up of welded chrome-moly steel tubing; the rudder had aerodynamic balance and the horizontal stabilizer was adjustable in flight. A metal propeller, Townend type speed-ring engine cowl, navigation lights, dual control wheels, electric inertia-type engine starter and mountings for radio equipment were standard. Optional equipment included parachute flares, battery, generator, radio gear and wheel streamlines; also included was an engine cover for $10, landing lights for $200 and two-tone color schemes for $100 extra. The next Curtiss-Wright "Travel Air" development was the model 12-Q sport biplane, as described in the chapter for ATC #401.

Listed below are Travel Air 6-B entries as gleaned from registration records:

NC-452N; Model 6-B (#6B-2037) Wright R-975
NC-431W;   "   (#6B-2038)   "
NC-432W;   "   (#6B-2039)   "
NC-447W;   "   (#6B-2040)   "

Some serial numbers prior to #6B-2037 were sometimes listed as model 6-B but were not eligible for this approval; serial #6B-2041, 2042, 2043, 2044 were assembled by the Airtech Flying Service of San Diego, Calif., according to factory drawings.

A.T.C. #353
(8-20-30)
LAIRD "COMMERCIAL," LC-B300

Fig. 178. Laird "Commercial" model LC-B300 mounted 300 h.p. Wright J6-9 engine.

Since the introduction of the LC (Laird-Commercial) type back in 1926, E. M. "Matty" Laird had been building slowly and carefully, producing the type of aircraft that possessed a personality of such subtle flavor that most everybody, if fortunate to get close enough, was compelled to walk to and around for a better look, and perhaps even gingerly stroke its glistening finish. Not especially outstanding in shape or form, this aircraft nevertheless was a frequent topic of conversation when flying folk got together because it possessed that measure of dignified beauty and good breeding that was somehow hard to ignore. Likened to a fine hand-made piece of machinery, the "Laird Commercial" was always built unhurriedly with great care and then was proudly delivered to a customer for his complete satisfaction. Since introduction of the LC-B type (refer to ATC #86 of U. S. Civil Aircraft, Vol. 1) in 1928, quite a few of them were already in service by 1930 and all were operated by discriminate sportsman-pilots and men of business. Designed specifically as a high-performance utility craft and generally used for what is loosely called all-purpose service, the LC-B was a more or less typical 3-place open cockpit biplane that was fashioned in planned harmony, both aerodynamically and structurally. Suf-

ficiently powered to deliver a performance that was always a mite better than the average, the LC-B biplane was kept abreast of progressive requirements by periodic refinement. The latest version in this series, brought out early in 1930, was now known as the model LC-B300, an improved example of a fine design that was aided by the extra stomp and drive of a 300 h.p. engine. The Wright J6 "Nine" was a perfect mate for the personality of this airplane and together they performed well in a spirited harmony. Most always in limited production because they were more or less a custom-built airplane, the model LC-B300 was built individually to suit the customer's particular fancies or requirements. An impressive airplane without doubt, its owners were also an impressive roster of business houses and sportsmen. Refined continuously in the following years by addition of horsepower, more and better accessories, improvements for better comfort, refinements for better performance and more speed, the LC-B300 and also a deluxe version called the LC-1B300, were built through 1934 and were available even into the latter "thirties." Proudly called "The Thorobreds of the Air" they had justified that claim time and again and have never lost their deserving reputation for mechanical excellence.

Fig. 179. Leveled at the sportsman, LC-B300 featured high performance and many extras.

The Laird model LC-B300 was an open cockpit biplane with seating for 3 and was typical to earlier models in the series except for its engine installation and other minor modifications necessary to this new combination. Aimed specifically at the flying sportsman or men of business, the LC-B300 was not built at a competitive price so it was therefore well equipped with various niceties and accessories for comfort and operating convenience. Powered now with the 9 cyl. Wright J6 (R-975) engine of 300 h.p. the performance over earlier 220 h.p. models was considerably improved and its short-field performance was such that it could operate from just about anywhere; normal loaded take-offs averaged about 300 ft., climb-out was a rousing 1600 ft. in the first minute and landing rolls were easily held to 250 ft. with wheel brakes. Flight characteristics offered sure-footed response and a compatible behavior offered pleasure-filled hours of flying; well endorsed by those who flew it, the LC-B300 was perhaps one of the finest airplanes of its type. As progressive demands increased and as progressive knowledge in aerodynamics was learned, the LC-B300 was refined into a deluxe version offering still better performance with comparable utility, incorporating many features heretofore offered only as options. The engine on this new LC-1B300 had power increased by 30 h.p., was shrouded with a low-drag NACA-type engine cowling and consequently the fuselage was widened in front to fair out the large diameter; the lower wing roots were faired in at the fuselage for better airflow and the general aerodynamic clean-up netted up to 165 m.p.h. Here now was a real dandy but because of the depressed times it was left begging for customers. The type certificate number for the Laird model LC-B300 was issued 8-20-30 and this was later amended to include the high-performance model LC-1B300. Scant records show that perhaps no more than 4 or 5 examples of the LC-B300 were built by the E. M. Laird Airplane Co. at Chicago, Ill. It is remarkable testimony of reliability that of the 25 or so LC-B type that were built, at least 20 were still flying in active service by 1939.

Listed below are specifications and performance data for the Laird model LC-B300 as powered with the 300 h.p. Wright J6 engine; length overall 23'9"; height overall 9'3"; wing span upper 34'0"; wing span lower 30'0"; wing chord upper 65'5"; wing chord lower 57'5"; wing area upper 170 sq.ft.; wing area lower 124 sq.ft.; total wing area 294 sq.ft.; airfoil "Laird" # 2; wt. empty 1958 lbs.; useful load 1064 lbs.; payload with 74 gal. fuel was 390 lbs. (2 passengers at 170 lbs. each & 50 lbs. baggage); gross wt. 3022 lbs.; max. speed 145 (155 with wheel pants & speed-ring cowling); cruising speed 115; landing speed 50; climb 1600 ft. first min. at sea level; ceiling 21,000 ft.; gas cap. 74 gal.; oil cap. 8 gal.; cruising range at 16 gal. per hour was 500 miles; price at factory first quoted at $13,500. and lowered to $10,850. in June of 1931. This model also eligible with Wright R-975-E of 330 h.p.

The fuselage framework was built up of dur-alumin tubing fastened together by steel clamp-joints and trussed together by heavy steel tie-rods; points of greater stress were fortified with welded chrome-moly steel tubing. The framework was amply faired to shape with wooden formers and fairing strips, then fabric covered. The wing framework was built up of laminated box-type spruce spar beams with spruce and plywood truss-type wing ribs; the leading edges were covered with dural metal sheet and the completed framework was covered in fabric. The main fuel tank was mounted high in the fuselage ahead of the front cockpit and extra

Fig. 180. Built with care, LC-B300 was rugged and handsome.

fuel was in a tank mounted in the center-section panel of the upper wing. The fabric covered tail-group was built up of welded steel tubing, except the horizontal stabilizer which was built up of spruce spar beams and former ribs; the fin was ground adjustable and the horizontal stabilizer was adjustable in flight. The split-axle landing gear of 90 in. tread used two spools of rubber shock-cord as shock absorbers; wheels were 30x5 and Bendix brakes were standard equipment. The tail skid was of the steel spring-leaf type with a removeable hardened shoe. A metal propeller, navigation lights, Eclipse hand crank inertia-type engine starter and 8-day cockpit clock were standard equipment. A speed-ring engine cowling (Townend type), dual controls, wheel pants and front cockpit cover panel were optional. The following are various improvements added to the model LC-1B300; the engine was shrouded with an NACA-type low drag cowling and the fuselage was wider in front to fair out the large diameter. The fuselage was covered in metal panels to the rear edge of pilot's cockpit and the lower wing root was faired into the fuselage junction. There were foot-steps to aid entry into the cockpits and the rear windshield was of shatter-proof glass; the front windshield folded down and the gaping hole of the front cockpit could be covered with a metal panel. Boots were attached to base of

controls in pilot's cockpit to keep draft from coming through floor-boards, there were several handy pockets for maps, gloves, etc.; cockpit heaters and ventilators were also provided. The pilot's seat, with deep well for a parachute pack, was adjustable and dual joy-stick controls were optional. Shock absorber spools were covered with streamlined metal cuffs and landing gear was fitted with 27 in. "General" streamlined tires; a swiveling tail wheel was optional. A metal propeller, navigation lights, electric engine starter, cockpit clock and fire extinguisher were standard equipment. Extra instruments, wheel fairings, landing lights, parachute flares, oil-cooling radiator and radio set were optional. The next Laird development was the swift "Speedwing" model LC-RW300 described in the chapter for ATC # 377 of this volume.

Listed below are model LC-B300 entries as gleaned from registration records:
NC-632; Model LC-B300 (#176) Wright R-975.
NC-867M;   "   (# 184)   "
NC-170N;   "   (# 186)   "
NC-10402;   "   (# 188)   "
Serial # 176 first as "Speedwing" model LC-R300, converted into LC-B300 under Group 2 approval numbered 2-189; serial # 184 and # 186 also built under Group 2 approval 2-189; ATC # 353 for serial # 188 and up; approval expired on 9-30-39.

GREAT LAKES "SPORT," 2-T-1E

Fig. 181. Great Lakes "Sport" model 2-T-1E featured 90 h.p. inverted "Cirrus" Hi-Drive engine.

The flashing form of the swept-wing "Great Lakes" biplane had graced a lot of sky since its introduction in 1928. Small, wiry and quite sassy, it was almost an immediate hit with the flying-public and was soon seen in just about all parts of the country. Dubbed as a "trainer" but more of a "sport" at heart, the "Great Lakes" biplane offered many the opportunity to play at sportsman-pilot on limited funds and offered performance enough to satisfy even the most adventurous. Judged "a little too hot" for pilot-training at first, it was not long before flying schools appreciated the fact of offering flight training in the spirited "Great Lakes" for both primary and secondary phases. T. Claude Ryan, famous airman who operated one of the largest flying-schools in the country, used the 2-T-1A almost exclusively and even had one ship equipped for instrument flying "under the hood." Tex Rankin up in the north country of Oregon, taught winter-flying in a "Great Lakes" equipped with skis.

Blessed with a back-log of orders they could hardly keep up with, the Great Lakes Aircraft Corp. (G. L. A. C.) was dealt a staggering blow late in 1929 when this country fell into economic distress, forcing the cancellation of hundreds of orders. Dazed but not beaten, the G.L.A.C. redesigned the model 2-T-1A for 1930 and also introduced the model 2-T-1E which was power-

ed with the new "Inverted Cirrus" engine. This engine received its baptism under fire in the grueling All-American "Cirrus Derby" of 1930. Flying the new model 2-T-1E with a "Cirrus Hi-Drive" engine, Chas. W. Meyers flew to 16th place in the National Air Tour for 1930. Due to rather extensive promotion and the extra expense of various new aircraft developments, G.L.A.C. found themselves rather short financially so a refinancing for $300,000 was arranged in June of 1930. This caused some re-shuffling and a thinning-out of company officers and plant personnel. Prices were also lowered on the two new sport-trainers as equipped with either the upright or inverted A.C.E. "American Cirrus" engines. During the National Detroit Air Show held in April of 1931, Great Lakes displayed both versions in "trainer" and "deluxe" models offered in a variety of color schemes with prices pared down to an unbelievable low.

Following the show, sales took a slight jump, T. C. Ryan in San Diego ordered a car-load of both types, but the lush and happy days of 1929 were gone. In the National Air Tour for 1931, Joe Meehan flew a "Great Lakes" trainer to 9th place in the company of impressive competition. Several of these swept-wing biplanes were flown in various air-shows about the country and slowly but surely these spirited airplanes were earning a nation-wide reputation as an

Fig. 182. Great Lakes "Special" flown to 3rd place in famous "Cirrus Derby;" Chas. W. Meyers in foreground.

ideal machine for the intricate art of "acrobatics." Upholding this reputation well throughout the "thirties," the "Great Lakes" biplane had been progressively modified with engines of higher horsepower; this extra punch of power increased their capabilities in "aerobatics" to the point that even as late as 1960, the "Lakes" was unanimously judged as the best airplane available for general air-show work. Engines of up to 220 h.p. have been installed in the frame of this biplane in recent years as compared to its original 90 h.p. It is still one of the most sought-after airplanes for restoration, with restorable airframes bringing a premium price.

The "Great Lakes" sport-trainer model 2-T-1E was a small swept-wing open cockpit biplane with seating for two in tandem and in general was typical to the newer version of the 2-T-1A except for its inverted engine installation. With the fuselage padded with extra fairing and engine cowling rounded out to a greater fullness, the 2-T-1E was more buxom in appearance and looked like a much bigger airplane. Judged slightly more stable and relieved of some of its earlier sass, the "Great Lakes" biplane for 1931 was perhaps a little more amiable, still quite handsome and a terrific bargain for the money. Powered with the 4 cyl. in-line inverted A.C.E. "Cirrus Hi-Drive" engine of 95 h.p. the model 2-T-1E had a performance that compared favorably with the model 2-T-1A at its best. Though seemingly not as popular as the 2-T-1A with the "upright" Cirrus engine, we cannot say that this indicates anything in particular because no airplane was selling in any great number at this time. The development of the model 2-T-1E stems directly from the sport-racer that was flown to 3rd place by Charlie Meyers in the torturous "Cirrus Derby" of 1930. This particular

ship was later modified to seat 2 and was certificated as the 2-T-1E under a Group 2 approval which in turn was superseded by ATC #354. A model 2-T-1A (ser. #124) had its upright engine removed and an inverted "Cirrus" was installed for further testing of this new combination. The production version of the 2-T-1E, as shown in the heading, began rolling off the line in early 1931; although the "Great Lakes" sport-biplane was available into 1933, not many were built in the next two years. The type certificate number for the "Great Lakes" model 2-T-1E was issued 8-20-30 and about 12 examples of this model were manufactured by the Great Lakes Aircraft Corp. at Cleveland, O. In the re-shuffle of company officers, Chas. F. Barndt became president and general manager and P. B. Rogers, formerly a project engineer, replaced Capt. Holden C. Richardson as chief of engineering. Chas. W. Meyers, test-pilot in charge of preliminary design and development, left G.L.A.C. employ in the winter of 1931 to help continue flight-tests on "de-icer boots" for Goodrich Rubber Co. and was replaced by Cal Johnston. Cal Johnston later took on extra duties as sales manager. The G.L.A.C. was also contractor to the Army Air Corps and the Navy, building all-purpose torpedo-bombers and scouting airplanes.

Listed below are specifications and performance data for the "Great Lakes" model 2-T-1E as powered with the 95 h.p. Cirrus Hi-Drive "Ensign" engine; length overall 21'0"; height overall 7'10"; wing span upper & lower 26'8"; wing chord both 46"; wing area upper 97 sq. ft.; wing area lower 90.6 sq. ft.; total wing area 187.6 sq. ft.; airfoil (NACA) M-12; wt. empty 1012 lbs.; useful load 568 lbs.; payload with 26 gal. fuel was 218 lbs.; gross wt. 1580 lbs.; max.

Fig. 183A. "Cirrus Derby" racer was basis for development of 2-T-1E.

speed 110; cruising speed 95; landing speed
45; climb 780 ft. first min. at sea level; climb to
5000 ft. was 8 min.; ceiling 12,000 ft.; gas cap.
26 gal.; oil cap. 3 gal.; cruising range at 6.3
gal. per hour was 375 miles; price at factory
first quoted at $4990., lowered to $3985. and
reduced to $3730. for the deluxe model and
$3155. for the trainer (stripped down) in April
of 1931.

The fuselage framework was built up of
welded chrome-moly steel tubing in a Warren
truss form, heavily faired to shape with metal
formers and fairing strips, then fabric covered.
A hinged panel in the rear portion of the cockpit cowling was split to swing up and make room
for easy access to the front cockpit; bucket seats
were provided with wells for a parachute pack.
As a single-seater, the 2-T-1E had a hinged
panel covering the front cockpit opening which
was then used for 200 lbs. of extra baggage or
the installation of an extra fuel tank. Factory
literature frequently mentioned that the sport-
trainer was also available as a closed-in "coupe"
but outside of tests on 2 factory airplanes, no
other examples were equipped in this manner.
The robust wing framework was built up of
solid spruce spars that were routed to an I-
beam section, with wing ribs of stamped-out
17ST dural metal sheet; the leading edges were
covered with dural metal sheet and the completed framework was covered in fabric. The

upper wing panels were swept back to an angle
of 9 deg. 13 min. each, the lower wing was set
back to 25 in. of positive stagger and dihedral
was 3 deg. on the upper wing and 2 deg. on the
lower wing. A gravity-feed fuel tank of 26 gal.
capacity was mounted in the center-section
panel of the upper wing with a fuel gauge suspended on the underside. The Friese type
offset-hinge ailerons were mounted in the
lower panels and operated directly by metal
push-pull tubes and bellcranks. The rudder
and elevators were actuated by stranded steel
cable and adjustable limit-stops were used on
all controls. All of the interplane struts were of
heavy gauge chrome-moly steel tubing of a
streamlined section and interplane bracing was
of heavy gauge streamlined steel wire. The
out-rigger type landing gear of 70 in. tread
used "Aerol" shock absorbing struts; wheels
were 24x4 or 22x10 and brakes were available.
The fabric covered tail group was built up of
riveted dural tubes, formers and ribs; the
vertical fin was fixed and the horizontal stabilizer was adjustable in flight. The 2-T-1E and
later versions of the model 2-T-1A were called
"the big tail jobs," having increased area in the
fin and rudder and a slightly altered shape. The
standard sport-trainer version of the 2-T-1E
(Type A) had bare cockpit interior, high-pressure 24x4 wheels with no brakes, a Hartzell
wooden propeller and dual controls. The

deluxe model (Type B) had neatly upholstered cockpit interiors, low-pressure 22x10 airwheels with brakes, a metal propeller, wiring for navigation lights, optional color combinations, a fire extinguisher and first-aid kit. Navigation lights, battery and Eclipse electric engine starter were extra.

Listed below are "Great Lakes" model 2-T-1E entries as gleaned from registration records:

NC-700K;	2-T-1E	(# 54)	Cirrus Hi-Drive.
X-863K;	"	(# 124)	"
NC-302Y;	"	(#189)	"
NC-304Y;	"	(#191)	"
NC-306Y;	"	(#193)	"
NC-314Y;	"	(#201)	"
NC-11320;	"	(#237)	"
NC-11321;	"	(#238)	"
NC-11322;	"	(#239)	"
NC-11332;	"	(#245)	"
NC-11333;	"	(#246)	"
NC-11335;	"	(#248)	"

Serial #54 as 2-T-2 sport-racer in 1929, modified to "Cirrus Derby" sport-racer in 1930, then modified as 2-T-1E on Group 2 approval numbered 2-261; serial #124 was G.L.A.C. test airplane as 2-T-1E; serial #189 first as 2-T-1A; serial #201 first as 2-T-1A; approval expired 11-19-36; for resume of earlier Great Lakes models refer to chapter for ATC #167 of U.S. CIVIL AIRCRAFT, Vol. 2 and ATC #228 of Vol. 3.

Fig. 183B. Rare "Monocoupe" 90-J mounted 90 h.p. Warner "Scarab Jr." engine.

Under the circumstances there is very little one can say about the "Monocoupe" model 90-J because for all intents it was practically identical to the standard "Model 90," except for its engine installation. As a companion model to the more powerful Warner-powered "Model 110," the "Ninety-J" mounted the smaller 5 cyl. Warner "Scarab Jr." engine of 90 h.p. This smaller engine was basically identical to the familiar and popular "Scarab" except for having 2 less cylinders. Because of similarity a big portion of the parts were interchangeable between either the 5 cyl. or 7 cyl. engine. Its use in the "Monocoupe 90-J" was the second certificated installation of this new engine and before long this engine was to enjoy popularity among several manufacturers of light airplanes. Equipped with a narrow-chord "Townend type" speed-ring engine fairing, the model 90-J could easily maintain top speeds up to 120 m.p.h. and was a fitting companion to the standard "Model 90" with 90 h.p. Lambert engine. More or less custom-built to order, it appears likely that no more than two examples of the Model 90-J were built, at least not through 1931, and it is doubtful if the 90-J could have successfully competed with the bargain-priced standard "Ninety" during the coming days of scant airplane buying.

The "Monocoupe 90-J" was also a high-winged cabin monoplane with side-by-side seat-ing for two and was identical to the standard 90-series (as described in the chapter for ATC #306) of this volume, except for certain modifications necessary to this combination and the addition of some deluxe equipment. Performance and flight characteristics were judged practically identical, so selection of this particular model over the Model 90 would be wholly dominated by engine preference. With a price tag boosted to $575 over that of the standard Model 90 as powered with the Lambert engine, this model was certainly not handed any favors or a bright and popular future. The first few examples of the model 90-J were reportedly manufactured by the Mono Aircraft Corp. at Moline, Ill., and it was kept on the production roster as an available model even after the move to Robertson, Mo., but any production totals for this model could not be determined from records available. The type certificate number for the model "Ninety-Jay" as powered with the 90 h.p. "Scarab Junior" engine was issued 8-20-30. It is pertinent to mention at this time that Vern Roberts who had been test-pilot for Mono Aircraft since 1927, graduated progressively into a valuable member of the "Monocoupe" staff; he also became project engineer, and finally factory manager, before leaving in 1934 because of ill health.

Listed below are specifications and performance data for the "Monocoupe" model 90-J as powered with the 90 h.p. Warner engine; length overall 20'10"; height overall 6'11"; wing span 32'0"; wing chord 60"; total wing area 132.3 sq. ft.; airfoil Clark Y; wt. empty 902 lbs.; useful load 609 lbs.; payload with 30 gal. fuel was 244 lbs. (payload included 2 parachutes at 20 lb. each & 30 lb. baggage); gross wt. 1511 lbs.; max. speed 118; cruising speed 102; landing speed 40; climb 850 ft. first min. at sea level; ceiling 15,000 ft.; gas cap. 30 gal.; oil cap. 2.5 gal.; cruising range at 5.5 gal. per hour was 500 miles; price at factory field with standard equipment was $3,950.

The construction details and general arrangement of the "Model 90-J" were typical to those as described in the chapter for ATC #306, including the following: a metal ground-adjustable propeller, speed-ring engine cowling, Heywood air-operated engine starter, dual controls, 6.50x10 semi-airwheels, wheel brakes and a cabin heater were standard equipment. Available as optional equipment were navigation lights, tail wheel, and wheel streamlines (wheel pants). A 28 gal. tank was available in place of the 30 gal. fuel tank to boost payload to 15 lbs. extra. Custom "Berryloid" color combinations were also available at extra charge. The next "Monocoupe" development was the Model 125 described in the chapter for ATC #359 of this volume.

Listed below are the only known entries for the Model 90-J as gleaned from registration records:

NC-528W; Model 90-J (#5J82) Warner 90
NC-530W;   "   (#5J93)   "

Serial #5J82 modified as 90-J from a model 90; both serial #5J82 and #5J93 later modified to model 90 with Lambert 90 engine; this approval for serial #5J82 and up; approval expired 9-1-33. In the chapter for the "Monocoupe 90" (ATC #306) we mentioned that by 1935, Tom Towle was the chief engineer; actually, Tom Towle's tenure as chief engineer was from 1932 to 1936. Such models as the 90-A, the D-145 and a special "Coupe" for Chas. Lindbergh were developed during this time.

*Fig. 184. Sikorsky model S-38-BH being readied for flight; high performance
prompted setting of several records.*

Typical to the many other Sikorsky S-38-B twin-motored "Amphibions", the model S-38-BH was unusual mostly in the fact that it harbored the extra grit and muscle provided by the addition of 300 h.p. Instead of the two 425 h.p. Pratt & Whitney "Wasp" engines that were normally mounted in this type, this thundering "boat" mounted two big "Hornet B" engines of 575 h.p. each. From gaining early experience and learning a few lessons with a "Hornet A" powered S-38-AH built in the latter part of 1928, Sikorsky used this fund of knowledge to advantage in preparing the model S-38-BH. With all the extra performance uncovered in the more powerful version, Sikorsky Aircraft happily eyed the probability of setting a few national and international records; as it came to be, 1930 was certainly a good year for aviation records. Preparing the S-38-BH version early in 1930 by converting a standard S-38-B, it was hastily approved in March and Boris Sergievsky, Sikorsky's talented test-pilot, was off to set some records. On March 4 of 1930 he flew the S-38-BH as a seaplane, climbing to an altitude record of 19,065 ft. with a 2000 kilogram (4409 lbs.) payload. Just over a week later

Sergievsky flew the same craft with a 1000 kilogram (2204 lbs.) payload to the altitude record of 23,222 ft. On the same day, with the same payload, flying a 100 kilometer course, he established a speed record for seaplanes of this category at 165.73 m.p.h. Also on the same day with ballast now increased to a 2000 kilogram payload and flying a 100 kilometer course, he established a seaplane speed record of 143.77 m.p.h. All trials were held at North Beach, N.Y. and to qualify as a "seaplane" the S-38-BH amphibian had its wheeled landing gear removed. Other than international recognition for this outstanding performance, Sikorsky gained proof also that the basic S-38 design, with a little muscle added to its frame, was capable of transporting worthwhile increases in payload and at better speeds. The idea for a slightly larger craft of this same basic type led to the successful development of the 16 passenger model S-41 for the Pan American Airways System.

The Sikorsky "Amphibion" model S-38-BH was a large sesqui-plane basically of the flying boat type with seating for ten. Typical to other Sikorsky amphibians of this type it was equipped

Fig. 185. 575 h.p. "Hornet" B engines unleashed extra performance; Boris Sergievsky, testpilot, in cockpit.

with a wheeled landing gear that extended or retracted for operation off land or water. Actually, the model S-38-BH was but a conversion of an S-38-B, by replacing the "Wasp" engines with two of the more powerful "Hornet" engines, and the increase in power available was of course translated into more speed and a healthy improvement in take-offs and the climb-out. Because of the larger engines which brought on the penalty of added weight, the S-38-BH was 350 lbs. heavier when empty than the S-38-B so consequently there was a comparable loss in the useful load with the gross weight held to certain limits. The penalty of weight was also influencing the payload so as a 10-place airplane, the S-38-BH was limited to a 275 gal. fuel load; with its maximum fuel load of 330 gals. its payload was held to a pilot, co-pilot and 6 passengers. However, for the bonus of performance promised by this airplane, it was no inconvenience to adjust payloads according to specific needs of load or range. That the S-38-BH was capable of sustaining much more load than the limits allowed is well reflected in the fact that twice it carried a 2000 kilogram load to achieve international speed and altitude records. Flight characteristics and general behavior of the standard S-38-B type were excellent and maneuverability was surprising for a

ship of this size. So, for the S-38-BH we can assume more of the same with perhaps a little more verve, requiring stricter command in piloting. Take-offs from land were reported as averaging 300 ft. and landings required less than 300 ft. so with all this inherent short-field performance, the S-38-BH was certainly not limited to its landing places and could drop in just about anywhere. It seems likely from data available that the first example of the S-38-BH was certificated on a Group 2 approval numbered 2-190, issued sometime in early March of 1930. Another example was prepared under the conversion specifications allowed by ATC # 356 and was eventually sold to Pan American Airways for use by an affiliate line in China. The type certificate number for the Sikorsky model S-38-BH was issued 8-23-30 and both examples of this model were probably converted at the new plant of Sikorsky Aviation Corp. at Bridgeport, Conn. For a complete resume of the S-38 type in all its normal variants we suggest reference to the chapters for ATC # 60 of U.S. Civil Aircraft Vol. 1 for the model S-38-A, ATC # 126 for the S-38-B and ATC # 158 for the S-38-C, both in Vol. 2. Other variants of this design, mostly in the special category, were built under various Group 2 approvals.

Listed below are specifications and performance data for the Sikorsky model S-38-BH as powered with two "Hornet B" engines of 575 h.p. each; length overall 40'5"; height overall on wheels 13'6"; height overall on water 10'2"; wing span upper 71'8"; wing span lower 36'0"; wing chord upper 100"; wing chord lower 59"; wing area upper 574 sq.ft.; wing area lower 146 sq.ft.; total wing area 720 sq.ft.; airfoil "Sikorsky" GS-1; wt. empty 6900 lbs.; useful load 3580 lbs.; payload with 330 gal. fuel was 1165 lbs.; gross wt. 10,480 lbs.; max. speed 143; cruising speed 120; landing speed 55; climb 1100 ft. first min. at sea level; ceiling 20,000 ft.; gas cap. max. 330 gal.; oil cap. 34 gal.; cruising range at 56 gal. per hour was better than 600 miles; price at the factory was $53,000.

The construction details and general arrangement of the model S-38-BH were typical to that of the model S-38-B as described in the chapter of ATC #126 of U. S. Civil Aircraft, Vol. 2, including the following: a mooring cockpit in the very nose of the hull provided for stowing of anchor, ropes, mooring gear, etc.; further back in the bow was an 82 cu. ft. baggage compartment with allowance for up to 200 lbs. The pilot's cockpit was equipped with a swing-over control wheel to provide dual controls; side windows of the pilot's cockpit lowered for entry or exit. Access to the main cabin was by way of a ladder and large hatchway in aft end of the hull. The hydraulically operated retractable landing gear was easily removable for operation strictly as a flying-boat seaplane; this removed some 500 lbs. from the empty weight and allowed a comparable increase in the payload. Custom-interior arrangements and toilet equipment were available on order. The big "Hornet" engines were shrouded with Townend-type "speed ring" cowlings for better airflow around the engine cylinders and a consequent increase in speed; the big engines were equipped with short exhaust stacks in the record trials for better performance but a large-volume collector ring was normally provided. The S-38-BH was also equipped with oil-cooling radiators. Hamilton-Standard metal propellers, hand crank inertia type engine starters and wheel brakes were standard equipment. The next Sikorsky development was the improved single-engined model S-39-B described in the chapter for ATC #375 in this volume.

Listed below are the only known examples of the Sikorsky model S-38-BH as gleaned from various records:

NC-25V; S-38-BH (#414-16) 2 Hornet B-575.
NC-40V;   "   (#514- 4)   "
Both examples shown were converted from the model S-38-B by installation of the larger engines; a Group 2 approval numbered 2-190 was first issued to the S-38-BH but this was later superseded by ATC #356.

A.T.C. #357
(8-25-30)
FAIRCHILD, MODEL 51

Fig. 186. Fairchild model 51; 300 h.p. Wright J6-9 engine just faintly disguises airframe of early FC-2.

Stemming from a popular design first introduced back in 1926, the Fairchild FC-2 cabin monoplane soon became known and respected far and wide as a winged work-horse on the various frontiers just opening up in the sphere of commercial aviation. Purposely lean, sinewy and not particularly pretty, the FC-2 proved itself especially valuable in handling the tough and peculiar assignments found in the "bush country" of the U. S. and in Canada; it also helped pioneer some of the earliest air-line routes here in this country. Because of its rugged frame, its hardy character, its practical, economical utility and its unflinching attitude towards any duty, it is not remarkable then that its useful life-span was to be prolonged far beyond the average norm. It is admirable that it was yet deemed worthwhile to add a bit more muscle to the old frame of this craft and let it continue on. Although hiding slightly behind a new designation, it is still very plain to see that the "Fairchild 51" is the same old FC-2 of several years back, that had been rejuvenated somewhat, and now sported the verve and extra ability made possible by the installation of a 9 cyl. Wright J6 (R-975) engine of 300 h.p. Yes, the old FC-2 was still hard to beat for the type of work it was usually engaged in and no one need be skittish that the upholstery would get scuffed or the floor get dirty. In all fairness, it seems rather odd to refer back to this venerable old craft of 1927-28, since the art of airplane design had come such a long way since then, but the Fairchild FC-2 was an airplane that simply mellowed with the oncoming years and it would

certainly be some time yet before it even approached being obsolete.

The Fairchild model 51 was a high-winged cabin monoplane, converted from the earlier model FC-2 of 1927-28, that was progressively developed from the earliest razor-back (or pinch-back) type to the turtle-back type that had a 4-cornered fuselage all the way back to the tail-post. The model 51 were probably all of this latter type with ample space for seating five. Because of its use in places where it was almost imperative to haul things, all passenger seating was easily removed to provide clear floor space for the cartage of miscellaneous cargo. The folding-wing feature, found so useful in the north country, was still retained and the ship could be easily converted to operate as a float-seaplane or on skis. The old FC-2 was known to lift tremendous loads, sometimes as much as 1000 lbs. over gross and anything could be carried that could be gotten in or strapped on. Although the model 51 would surely be capable of doing the same it is doubtful that they were subjected to such devil-may-care treatment at this stage of the game. Aviation by now was becoming a little more sophisticated and some of the spontaneous actions of the pioneer flyer, usually prompted by an urgent need, were now being frowned upon as hazardous. Flight characteristics and general behavior of the new "51" were still gentle and quite obedient, but the added horsepower surged through its old frame like a tonic for a substantial increase in all-round performance. On 6-28-29 a Group 2 approval numbered 2-86 was issued for all FC-2 that were to

Fig. 187. Beautiful setting of 51 as seaplane; pontoons were also built by Fairchild.

be reworked at the Fairchild factory into the Model 51 by installation of the 300 h.p. Wright J6-9-300 engine, plus any modifications necessary for this new combination. On 8-14-30 a Group 2 approval numbered 2-259 was issued for serial #102 as a seaplane on Fairchild model P-4 floats. The type certificate number for the model "Fifty-One" was then issued 8-25-30 for both the landplane and seaplanes with possibly some 7 or more examples of the FC-2 converted to the "51" at the factory of the Fairchild Airplane Mfg. Corp. at Farmingdale, Long Island, N. Y.

Listed below are specifications and performance data for the Fairchild model 51 as powered with the 300 h.p. Wright R-975 engine; length overall 31'0"; height overall 9'0"; wing span 44'0"; wing chord 84"; total wing area 290 sq. ft.; airfoil Goettingen 387 modified; wt. empty (landplane) 2375 lbs.; useful load 1500 lbs.; payload with 85 gal. fuel was 745 lbs. (4 passengers at 170 lb. each & 85 lbs. baggage); payload with 105 gal. fuel was 625 lbs.; gross wt. 3875 lbs.; max. speed 130; cruising speed 110; landing speed 50; climb 900 ft. first min. at sea level; ceiling 16,000 ft.; gas cap. normal 85 gal.; gas cap. max. 105 gal.; oil cap. 10 gal.; cruising range at 16 gal. per hour was 525-670 miles. The following figures are for the 51 fitted with Fairchild model P-4 twin-float gear; wt. empty 2702 lb.; useful load 1298 lbs.; payload with 85 gal. fuel was 543 lbs. (3 passengers at 170 lb. each & 48 lb. baggage); gross wt. 4000 lbs.; max. speed 122; cruising speed 105; landing speed 55; climb 750 ft. first min. at sea level; ceiling 14,000 ft.; cruising range at 16 gal. per hour was 500 miles; a later amendment allowed gross weight increase to 4000 lbs. for landplane also, boosting payload allowance by some 125 lbs.;

performance would be proportionately decreased.

The construction details and general arrangement for the Model 51 are typical to that as described for the Model 51-A on ATC #358 of this volume, including the following. Normal fuel capacity was 85 gal. with a max. fuel capacity at 105 gal.; the payload allowance was cut according to extra amount of fuel carried. All model 51 were also eligible with the improved Wright R-975 engine of 330 h.p. for slight increases in performance. Most examples had an engine exhaust system with long tail-pipes that allowed the fitting of two heater-muffs for ample cabin heat. Some examples were fitted with a swiveling tail wheel for better ground handling. A metal propeller, hand crank inertia-type engine starter, Bendix wheel brakes and navigation lights were standard equipment. The next development in the 51-series was the model 51-A described in the chapter for ATC #358 of this volume.

Listed below are Fairchild model 51 entries as gleaned from registration records:

NS-8;	Fairchild 51	(# 10)	Wright R-975.
NC-3637;	"	(# 40)	"
NC-4755;	"	(# 87)	"
C-5364;	"	(#102)	"
C-5574;	"	(#108)	"
NC-8037;	"	(#169)	"
NC-9120;	"	(#171)	"

NS-8 was with Bureau of Standards; serial #40 was with Rogers Air Lines operating out of New York and Miami; serial #87 with Curtiss Flying Service and first mounted a Curtiss C-6-A engine; serial #102 operated in U. S. as FC-2 then operated in Alaska with Pacific-International Airways as model 51, out of service and dismantled Sept. 1932; serial #108 with Fairchild Aerial Surveys.

Fig. 188. Fairchild model 51-A with 300 h.p. "Wasp Jr." engine; ship shown, used by RCAF.

The Fairchild cabin monoplane model 51-A was a companion offering to the "Fifty-One", as discussed here just previous, and most likely its development was brought about by those that would prefer the 9 cyl. Pratt & Whitney "Wasp Jr." (R-985) engine instead of the 300 h.p. Wright J6. There was actually very little difference between these two powerplants, they were both very fine engines, so it was just a matter of picking the "brand" of engine that was more favored. The Pratt & Whitney engine must have been the preferred choice in Canada because most of the converted FC-2 were of the model 51-A type and nearly all were in service with the Royal Canadian Air Force. A thorough search through all pertinent records disclosed only one example of the Model 51-A here in the U.S.A. and almost as force of habit, it was used as a photographic-ship. The Southwestern Aerial Surveys, Inc. with headquarters in Austin, Texas were engaged in air-mapping and aerial surveys since 1928 on various projects involving highway lay-out and dam-site mapping. In the course of this specialized work they had used several airplanes of various makes, both open, closed, biplanes and monoplanes but each had some unsatisfactory traits of one kind or another. Late in 1930 they purchased a Fairchild model 51-A, an airplane that had served many hours already as an FC-2, on the strength of the Fairchild's well-known reputation; on their order it was fitted to handle various equipment for aerial photography. Fitted

also with extra fuel tanks to allow a capacity of up to 140 gals., the 51-A could stay aloft for nearly 10 hours at slow-cruise r.p.m. (105 m.p.h.) and its stable unerring flight path allowed the holding of a predetermined altitude close enough to produce sharp photographs of a uniform scale. From the actual words as quoted by one of the survey-company officials — "it was the best ship for aerial photography that we have ever found". We have not bothered to check into the usage of the model 51-A in Canada but from past performance we can guess that it was kept very busy doing all-purpose work, a blanket term that just about covers anything and everything.

The Fairchild model 51-A was also a high-winged cabin monoplane arranged with seating for five and was of course typical to the model 51 in that both models were conversions of the early Fairchild FC-2. Conversion at the factory consisted mainly of removing the Wright "Whirlwind" J5 engine of 220 h.p. and replacing it on a new mount with the 9 cyl. "Wasp Jr." engine of 300 h.p. plus other minor modifications necessary to this new combination. Although the switching of engines only amounted to a boost of some 80 h.p., the gain in all-around performance was considerable, and the use of a more modern engine also paid off in less of the tedious maintenance and a bit more reliability in the long run. A comparison of performance data between the early FC-2 and of the new 51 or 51-A will point out that gains were made where the pilot needed it the most. Take-off runs were shortened considerably, climb-out put you about 300 feet higher every minute and top speed was some 8 m.p.h. better, at least. Cruising speed was also better; the altitude ceiling was extended by a good margin, and all the good characteristics of the old FC-2 seemed to be noticeably improved. It seems that only one example of the model 51-A was operating here in the U.S.A. but the praise voiced by its owner for a job well done was ample testimony of its good character. The model 51-A was first awarded a Group 2 approval numbered 2-241 some time in July of 1930, but this was superseded by a type certificate issued 8-25-30, and all conversion work was done at the factory of the Fairchild Airplane Mfg. Corp. at Farmingdale, Long Island, New York. We can assume that conversion of the FC-2 into the 51-A for the RCAF was most likely made at Fairchild's factory-branch in Longueuil, P.Q. in Canada.

Listed below are specifications and performance data for the Fairchild model 51-A as powered with the 300 h.p. "Wasp Jr." engine; length overall 31'0"; height overall 9'0"; wing span 44'0"; wing chord 84"; total wing area 290 sq.ft.; airfoil Goettingen 387 modified; wt. empty 2440 lbs.; useful load 1480 lbs.; payload with 85 gal. fuel was 725 lbs. (4 passengers at 170 lb. each & 65 lb. baggage); payload with 140 gal. fuel was 375 lbs. (2 passengers & 45 lb. baggage); gross wt. 3920 lbs.; max. speed 130; cruising speed 110; landing speed 50; climb 900 ft. first min. at sea level; ceiling 16,000 ft.; gas cap. normal 85 gal.; gas cap. max. 140 gal.; oil cap. 10-12 gal.; cruising range at 16 gal. per hour was 525-945 miles.

The fuselage framework was built up of welded chrome-moly and 1025 steel tubing, lightly faired to shape with wooden formers and fairing strips, then fabric covered. The pilot was seated in front portion of the cabin which narrowed towards the nose, and behind the pilot were 4 metal-framed passenger seats, any of which was removable to make room for cargo or camera equipment. The cabin was upholstered in serviceable fabrics and all windows were of shatter-proof glass. Two cabin entry doors were on the right side, one in front for the pilot and one in the rear for passenger entry. Ventilation and cabin heat was provided. The baggage compartment was to the rear of the main cabin with a door for access from the outside. The semi-cantilever wing framework in two halves, was built up of spruce and plywood box-type spar beams with spruce and mahogany plywood gusseted girder type wing ribs; the leading edges were covered with mahogany plywood sheet and the completed framework was covered in fabric. The wing halves were hinged to a center-section panel built into the cabin roof and braced to the bottom longeron by 2 vee-type struts of streamlined steel tubing; a novel mechanism allowed folding of the wings without any disconnection of fuel lines or controls. One fuel tank was mounted in the root end of each wing half and one fuel tank was mounted in the cabin roof. The split-axle landing gear was of the normal tripod type with oleo-spring shock absorbing struts; wheels were 32x6 and Bendix brakes were standard equipment. The fabric covered tail-group was built up of welded steel tubing; the fin was ground adjustable and the horizontal stabilizer was adjustable in flight. All movable control surfaces were aerodynamically balanced. A metal propeller, hand crank inertia-type engine starter, Bendix wheels and brakes and navigation lights were standard equipment. The next Fairchild development was the KR-21-B sport biplane described in the chapter for ATC # 363 of this volume.

Listed below is the only known entry of the model 51-A in the U.S.A. as gleaned from various records:

NC-6800; Fairchild 51-A (# 141) Wasp Jr. 300. Serial # 141 was of course an FC-2 first and it was even listed as an FC-2W for a time, before its conversion to 51-A; the number of 51-A examples in Canada has been undetermined; another interesting 51-type was the model 51-R registered as X-8017 (ser. # 157) that was used for early tests of the V-12 "Ranger" engine.

As the earlier "Monosport 1" had a stable-mate in the Kinner-powered "Monosport 2", likewise the "Monocoupe 110" had a running-mate in the Kinner-powered "Monocoupe 125". The Model 125 for all intents was practically identical to the Model 110 in all basic respects except for its engine installation. Powered with the new 5 cyl. Kinner B5 (R-440) engine of 125 h.p. the "One Twenty-Five" was principally developed to cater to the buyers who would prefer the "Monocoupe" with this engine; Kinner Motor Co. naturally acquired the first ex-ample of this model to promote the sales of their new engine. This was the third certificated installation of this powerplant and soon many other manufacturers would be using this engine to power their new models. The "Monocoupe 125" received its baptism under fire in the National Air Races for 1930 held at Chicago; Les Bowman, chief pilot for Kinner Motors, flew the "125" daringly and stayed on the tail of Vern Roberts in a "Monocoupe 110" to place second in a 25 mile speed event around the pylons. Additional shake-down for the Kinner B5 engine took place in the grueling National Air Tour for 1930; Les Bowman, this time in a Waco INF, finished the hard-fought tour in 10th place. Offered as an ideal high-performance airplane for the sportsman, the "Monocoupe 125" was made-to-order for rough and tumble use but was never accorded the chance to achieve any great popularity. As a consequence it was built only in very small number. The Aeronautics Branch of the Dept. of Commerce procured two of the "Model 125" for use by their inspectors in the field and these appeared to be the only other sales through 1931.

The "Monocoupe 125" was a small high-winged cabin monoplane with side-by-side seating for two and it was nearly identical to the standard "Model 110" except for the in-stallation of the Kinner B5 engine and some minor modifications necessary to this combina-tion. The performance of this model, although listed as identical in various reports, was be-lieved to be just a shade better than the "One-Ten" and its flight characteristics were more or less the same, so the selection of this model would be dominated entirely by a preference for the B5 engine. In an effort to describe its nature, some have said of the "One Twenty-Five" that it fairly throbbed with eagerness to go but others jokingly diagnosed its throbbing eagerness as just engine vibration. To sum it all briefly, there is no doubt that the "125" was a fitting companion model to the highly-touted

"One-Ten" and had a deserving place in the expanded "Monocoupe" line-up but it had no support to achieve popularity and was built in scant number. The type certificate number for the "Monocoupe 125" was issued 8-23-30 and at least 4 or more examples of this model were built.

Listed below are specifications and per-formance data for the "Monocoupe 125" as powered with the 125 h.p. Kinner B5 engine; length overall 20'8"; height overall 6'10"; wing span 32'0"; wing chord 60"; total wing area 132.3 sq.ft.; airfoil Clark Y; wt. empty 1007 lbs.; useful load 583 lbs.; payload with 30 gal. fuel was 203 lbs. (1 passenger at 170 lb. & 33 lb. baggage); gross wt. 1590 lbs.; max. speed 133 [144]; cruising speed 113 [124]; landing speed 45; climb 1100 ft. first min. at sea level; ceiling 18,000 ft.; figures in brackets are with speed-ring engine cowling and wheel pants; gas cap. 30 gal.; oil cap. 4.5 gal; cruising range at 7.5 gal. per hour was 440-480 miles; price at factory field was $4500., later raised to $4750. A "Mono-coupe 125" with good-running Kinner B5 en-gine was offered for sale in July of 1931 for $2350. in like-new condition; an indication that the "depression" was beginning to take its toll.

The construction details and general ar-rangement of the Monocoupe 125 were typical to that as described in the chapters for ATC # 306 and # 327 of this volume, including the following. The baggage allowance was held to 33 lbs. and 2 parachutes at 20 lbs. each were part of the payload. A custom-made "speed ring" engine cowling with streamlined "humps" over the cylinder heads had to be built up be-cause increased diameter of the B5 engine had a tendency to hamper visibility when regular Townend-type cowling was used. A metal ground-adjustable propeller, a Heywood air-operated engine starter, navigation lights and first-aid kit were standard equipment. Semi-airwheels with wheel brakes, a cabin heater, speed-ring engine cowling and wheel stream-lines were optional equipment. The next de-velopment in the "Monocoupe" line was the rare model 70-V described in the chapter for ATC # 492.

Listed below are the only known "Monocoupe 125" entries as gleaned from registration rec-ords: NC-529W; Model 125 (# 5K84) Kinner B5.

NS-10;	" (# 6K03)	"
NS-21;	" (# 6K04)	"
NC-797H;	" (# 6K19)	"

Serial # 6K03 later as NC-3771; serial # 6K04 later as NC-13914.

A.T.C. #360
(8-27-30)
BELLANCA "AIRBUS," P-100

Fig. 189. Bellanca "Airbus" model P-100 with 600 h.p. Curtiss "Conqueror" engine;
view shows unusual beauty.

Many years noted for its aircraft of high efficiency, that is, large load-carrying ability without detriment to high cruising speeds, Bellanca Aircraft now proudly introduced the big, unusual "Airbus." Arranged in a very useful form best described as a sesqui-plane, with nearly every square foot of airframe devoted to some useful purpose, the "Airbus" was about the pinnacle of achievement in airplane efficiency for these times. Designed primarily for hauling passengers and air-mail in scheduled airline service or hauling cargo-freight on more or less unscheduled flights, there never was until now an airplane that could match the "Airbus" in lifting capacity on a horsepower for horsepower basis. Nearing its completion at the Bellanca factory in April of 1930 and flight-tested some time in May, the name "Airbus" although somewhat prosaic, was selected for this many-seated craft to convey the impression of a big, reliable, every-day air transport. The "Airbus" model P-100 as shown, was not exactly a new design because its basic configuration stemmed directly from the 1928 model K, a large load-toting airplane designed specifically for long-distance flying. The illustrious "Roma" was an example of this early model K and shows the similarity between this type and the improved "Airbus" model P-100. The new "Airbus" actually carried a gross loaded weight that would be more of a job normally for a tri-motored airplane but Bellanca chose wisely to do the job with just one engine of 600 h.p. and did it remarkably well too. Guiseppe M. Bellanca never was an advocate of the multi-engined airplane, believing that the adding on of engines only multiplied the chances for some sort of trouble. The comparatively

low safety record of the multi-motored craft in general, tended to prove this more or less true. One well maintained engine of proper power output with simple fuel and oil systems still had the better chance of recording the best performance and the best reliability.

The 12 cyl. vee-type liquid cooled Curtiss "Conqueror" engine, chosen by G.M.B. to power the first example in the new "Airbus" series, delivered 600 h.p. at 2400 r.p.m. with the propeller-shaft geared down 2 to 1 to allow the use of a large slow-turning propeller for better efficiency. The liquid-cooled engine required all sorts of plumbing and a large radiator, which incidentally was mounted right on the nose-end; not entirely suitable in a transport of this type for many mechanical reasons, this particular installation was used on only the one example of the P-100. Actually, the "Conqueror" installation seemed rather archaic even for these times and only Guiseppe Bellanca could get away with it. In extended service the nose type radiator was also cited as a poor installation because centrifugal force at the propeller hub had a tendency to cause a wall of airflow not penetrated by cooling air, thus cutting down efficiency of the radiator and a need for larger coolant capacity. The single-engined installation for a transport of this type and size, however, did have its merits as well proven time and again by Geo. W. Haldeman, chief pilot for Bellanca; Haldeman flew the new P-100 "Airbus" on a country-wide tour of over 4400 miles early in 1931 to promote interest among airlines and other carriers. Covering the tour in some 35 hours of actual flying time, the 600 h.p. Curtiss

Fig. 190. Unusual sesqui-plane arrangement practical and very efficient.

"Conqueror" consumed 33 gal. of fuel per hour at an average speed of 125.4 m.p.h. for the entire jaunt. Figuring the cost of fuel, oil and some 100 hours of maintenance work, the trip was made for slightly more than $.08 per mile. This pointed to the economy gained by a single-engine transport with one large engine over that of a tri-motored craft of similar gross weight and similar total horsepower. A tally of the facts literally promised a great future for the new "Airbus," but this particular version carried a rather high price-tag and the "Conqueror" engine, perhaps the finest powerplant in its class, was not particularly an enticing choice for this kind of work and consequently the model P-100 didn't sell. The model P-100 "Airbus" was available on order through 1932 but no more examples were built.

The Bellanca "Airbus" model P-100 was a large high-winged monoplane with an unusual configuration closely approaching that of a sesqui-plane; with ample room in the long spacious cabin, there was seating for 12 or 14 with a generous allowance for baggage, depending on the particular interior arrangement. Probably the most significant features on the "Airbus" were the large lower stub-wings and the large airfoiled wing bracing struts, a practical combination that contributed heavily to airframe strength and the overall lifting area. Typical of "Bellanca" design, most everything within reason had to double in duty, so we find that the short well-faired landing gear, equipped with large roly-poly "airwheels" was attached to the inverted apex where the wing struts and the stub wings came together. Of ample cross-section, the wing stubs were fitted with compartments for passenger baggage, mail, or miscellaneous cargo. By removal of all passenger seating there were many cubic feet of clear floor area for up to 3000 lbs. of paying load. Powered with the Curtiss "Conqueror" (GV-1570) engine of 600 h.p. there was ample power reserve at full gross weight for a good performance throughout the whole range. Just newly approved for its ATC number, the model P-100 "Airbus" was flown to the 1930 National Air Races held at Chicago by chief pilot George Haldeman. In the Air Transport Efficiency Race, Haldeman came in 3rd for speed (123.8 m.p.h.) in 10 laps over a 5 mile course and then placed 2nd in

the race for efficiency, carrying a 2570 lb. payload, beaten only by a Bellanca "Pacemaker" of course. The short-field performance of the big "Airbus" was outstanding, proving it quite suitable for shuttle-lines operating on frequent schedule out of the smaller airports or for mixed cargo hauling in the depths of the rugged "bush country." The stately "Airbus" with its serene behavior and its majestic air attracted interested crowds wherever it went and "Bellanca" reputation, of course, instilled confidence in pilots and passengers alike. It is fair to speculate that had the "Airbus" been favored with better times, with people enjoying a more normal economy, it is quite likely that it would have been seen far and wide in fairly substantial number. The type certificate number for the Bellanca "Airbus" model P-100 was issued 8-27-30 and only one example of this model was manufactured by the Bellanca Aircraft Corp. at New Castle, Delaware.

Listed below are specifications and performance data for the "Airbus" model P-100 as powered with the 600 h.p. Curtiss GV-1570 engine; length overall 40'8"; height overall 10'4"; wing span 65'0"; wing chord 94"; total wing area 652 sq.ft.; airfoil "Bellanca"; wt. empty 5490 lbs.; useful load 4010 lbs.; payload with 200 gal. fuel was 2580 lbs. (11 passengers at 165 lb. each & 765 lb. baggage & cargo) or (13 passengers & 435 lb. baggage-cargo); payload with 130 gal. fuel was 3000 lbs.; gross wt. 9500 lbs.; max. speed 145; cruising speed 122; landing speed 55; climb 750 ft. first min. at sea level; ceiling 16,000 ft.; gas cap. max. 200 gal.; oil cap. 8 gal.; cruising range at 33 gal. per hour was 700 miles; price at factory first quoted at $38,500. lowered to $37,500. later in 1930.

The fuselage framework was built up of welded chrome-moly steel tubing, faired to shape with formers and fairing strips, then fabric covered. Two rows of steel framed seats for passengers were arranged on a raised dais some 12 in. high, which provided convenient space for the personal baggage of each passenger; outlets for cabin heat and ventilation were provided at each seat. Main cabin entry was through two large doors in the aft cabin with a large, convenient folding step; a center-aisle ran full length of the cabin providing entry or exit to pilot's compartment also. The cabin walls were sound-proofed and insulated against extremes

Fig. 191. Coach-style interior of "Airbus", other arrangements optional.

in noise and temperature and all windows were of shatter-proof glass. A small door on each side up front provided separate entry or exit for the pilots; a crew of 2 was normally used. The large semi-cantilever upper wing was built up of heavy sectioned solid spruce spar beams with spruce and plywood truss-type wing ribs; with each wing-half built up in two sections, the leading edges were covered with dural metal sheet and the completed framework was covered in fabric. The lower stub wing slanted downward into anhedral and the wing bracing struts slanted upward into an extreme dihedral; the intersection of these two assemblies was stiffened by streamlined steel tube struts and steel wire interplane bracing as would normally be used in a biplane cellule. The inverted apex of these two units, where they joined together, also provided fork-type mounts for attachment of the large wheels; the landing gear of some 18 foot tread was equipped with 2 oleo shock absorbing struts on each side, large low-pressure "airwheels" and wheel brakes were standard

equipment. The 12 cyl. "Conqueror" engine, using a nose-type coolant radiator, was provided with adequate ventilation; adjustable scoops and shutters were actuated for regulating the engine temperature. The fabric covered tail group was built up of welded chrome-moly steel tubing; the large rudder was aerodynamically balanced and the horizontal stabilizer was adjustable in flight. A wooden propeller, navigation lights, wheel brakes, steerable tail wheel, electric engine starter, battery, fire extinguishers, cabin lights and clock were standard equipment. A large 3-bladed metal propeller was optional. The next development in the "Airbus" series was the model P-200 described in the chapter for ATC # 391 of this volume.

Listed below is the only entry for the Bellanca "Airbus" model P-100 as gleaned from registration records:

NC-684W; Airbus P-100 (# 701) Curtiss GV-1570.

Serial # 701 also eligible for conversion to model P-200 with Wright "Cyclone" engine.

Fig. 192. "Airbus" on tour to promote interest in new series transport.

A.T.C. #361
(8-30-30)
VIKING "FLYING BOAT", V-2

Fig. 193. Viking flying-boat model V-2 mounted 225 h.p. Wright J6-7; several used in Coast Guard Service.

The handsome Viking "flying boat", clearly showing a graceful "continental" flavor in its general form and mechanical make-up, was actually a well-proven and popular French design manufactured by F.B.A.-Schreck. For operation here in the U.S.A., the "Viking" version was modified slightly in lifting areas and form to better fit accepted American standards. The vee-type "Hisso" (Hispano-Suiza) water-cooled engine as normally used in the European model was replaced with the air-cooled 7 cyl. Wright J6 (R-760) engine of 225 h.p. Seating 4, all out in the airy open, the Viking V-2 was leveled more towards the needs and pleasures of the flying sportsman; however records hint that a few were being used in Coast Guard service somewhere on the Atlantic shores. Rugged of frame and quite seaworthy, we can assume that the Viking was well suited and stout enough to handle rescue in the open seas within a reasonable distance from the shores and the other varied duties usually performed on water by this branch of the service. Arranged and equipped to operate off water only, the Viking "Flying Boat" would have been more versatile as an amphibious airplane and perhaps more attractive to potential customers here in this country; but as it was not, its utility for pleasure or commercial purposes was narrowly limited and only a few examples were built. Although a classic example of a fine flying-boat design, its production here in the U.S.A. was stifled by the practically non-existent market for this type

of craft and it was left sadly begging for customers.

Designed by Louis Schreck, a naval architect and yacht designer who turned aeronautical engineer, the graceful "Schreck" flying boats were manufactured by the Hydravions Schreck-F.B.A. plant in France, whose activity in the manufacture of flying boat types dates back to the early part of the first World War. Developed for some 15 years by this time, through models of various sizes for various purposes, the popular 17-HT-4 was the model chosen by "Viking" to become their V-2. Early in 1930, two of the 17-HT-4 type were approved here in this country on a Group 2 certificate numbered 2-113; these were also 4-place open flying boats and were powered with the 8 cyl. vee-type Hispano-Suiza engine of 180 h.p. Meanwhile another version of this "flying boat" was nearing completion by early 1930 for tests in a modified configuration that harbored more power and several changes in the basic design, a change for a better all-round performance. This prototype was to become the Viking model V-2.

The Viking "Flying Boat" model V-2 was an open cockpit biplane of the flying "boat" type with spacious seating for four in the forward portion of a seaworthy wooden hull. Arranged in a position that was actually best suitable for this type of craft, the radial-type engine was mounted high up between the wing panels in a "pusher" fashion to be well clear of the water and to minimize danger to the occupants. Level-

Fig. 194. Graceful form of the "Shreck" was typical of the classic flying boat design.

ed especially at the flying sportsman, the Viking V-2 was rugged of frame and hardy of character, as a fitting aerial conveyance for sporting or every-day commercial operations off water. With all occupants seated in the sporty atmosphere of open cockpits up in the forward bow, we can well imagine that each flight essentially combined some of the thrills of speed-boating with the varied pleasures of flying for a double dose of enjoyment. Assuming that factory data and descriptions bear a good percentage of truth, the flight characteristics of the Viking were pleasant and graceful with a maneuverability on water or in the air that was exceptional for a "boat" of this type. Powered with the 7 cyl. Wright R-760 engine of 225 h.p., the model V-2 offered good honest performance in a range quite suitable for its type of operation, and its stamina under severe types of usage was perhaps its cardinal feature. On speculation, it appears that the model V-2 compared closely and favorably with the Rogers "Sea Eagle" as described in chapter for ATC #274. The type certificate number for the Viking "Flying Boat" model V-2 was issued 8-30-30 and perhaps some 6 or more examples were manufactured by the Viking Flying Boat Co. on Causeway Island in New Haven, Conn.

Listed below are specifications and performance data for the Viking model V-2 as powered with the 225 h.p. Wright "Seven" (R-760)

engine; length overall 29'4"; height overall 10'4"; wing span upper 42'3"; wing span lower 38'7"; wing chord upper 65"; wing chord lower 65"; total wing area 401 sq. ft.; airfoil Eiffel 359 modified; wt. empty 2300 lbs.; useful load 1150 lbs.; payload with 60 gal. fuel was 580 lbs. (3 passengers at 165 lb. each & 85 lb. baggage); gross wt. 3450 lbs.; max. speed 105; cruising speed 90; landing speed 46; climb 750 ft. first min. at sea level; service ceiling 14,000 ft.; gas cap. 60 gal.; oil cap. 5 gal.; cruising range at 13 gal. per hour was 380 miles; price at the factory was approx. $15,000 plus extra equipment.

The all-wood hull framework was built up of an ash keel, ash chines with some 34 poplar frames all glued and bolted together; every fourth frame was a solid bulkhead, two heavy reinforced bulkheads provided mounting for the lower wings and the completed framework was covered with plywood sheet. Two large open cockpits, each seating 2 side-by-side, were in the forward hull and well protected by large metal-framed windshields. The wing framework was built up of heavy-sectioned spruce spars that were routed out and laminated into box-type beams with spruce and plywood web-type wing ribs; the leading edges were covered with plywood sheet and the completed framework was covered in fabric. Plywood covered ailerons, actuated by stranded steel cable, were in the upper wings only. A system of N-type struts supported the center-section panel of the upper wing and also provided mounting for the "pusher" type engine; interplane bracing struts were of spruce shaped to a streamlined form and a cut-out in the trailing edge of the upper wing provided clearance for the propeller. One 48 gal. fuel tank was mounted in the hull behind the rear cockpit and a 12 gal. fuel tank was mounted under the rear seat; we must assume that some sort of fuel pump must have been used to feed fuel to the high-mounted engine. The fabric covered tail-group was an all-wood structure; lower part of the vertical fin was built up as an integral part of the aft hull. A baggage compartment of 3 cu. ft. capacity with allowance for up to 85 lbs. was located somewhere in the hull; the anchor, mooring gear and life preserver jackets were part of the baggage allowance. Wood-framed and plywood covered tip floats were mounted on underside of each lower wing half to stabilize the craft during water operations. A wooden propeller, inertia-type engine starter, navigation lights, anchor and mooring gear were standard equipment. A metal propeller and dual controls in front cockpit were optional. The next Viking development was the 3-place "Kitty Hawk" model B-8 described in the chapter for ATC #392 of this volume.

Listed below are the only known Schreck-Viking "flying boats" as gleaned from registration records:
NC-37V;   Viking V-2 (#1) Wright R-760
NC-519M;   17-HT-4 (#63) Hispano-Suiza 180
NC-136N;   17-HMT-2 (#64)   "
NC-792K;   17-HT-4 (#133)   "

Serial #133 as 4-place on Group 2 approval numbered 2-113; serial #64 as 2-place on Group 2 approval numbered 2-168; serial #162 eligible on Group 2 approval 2-113 also, registration number unknown; no information on examples used by the U. S. Coast Guard service.

Fig. 195. "Waco" model CRG, shown here during 1930 National Air Tour.

One can hardly say that the "Waco" model CRG was a standard production type airplane because, in a sense, it was specifically designed and developed to incorporate performance features that were carefully calculated to win the 1930 National Air Tour. That it did not win the so-called "tour" is but a fickle quirk of fate and the nature of a scoring formula. From the experience gained in winning the "reliability tour" for both 1928 and 1929, Waco Aircraft knew just about what it would take to gather most points in this type of contest, so they felt reasonably confident that they could incorporate these requirements into a new design to come up with a potential winner. That they did, almost, is a deserving compliment to "Waco" design and engineering. Top speed, for the first time in these contests, was a primary factor in the point scoring formula for the 1930 tour, but final scoring also produced a direct power factor advantage to 3-engined aircraft; so it appeared almost as if a handicap had been planned beforehand. While the "tour formula" was so scored that no single-engined airplane could hardly hope to win, the significant improvement of "Waco" performance in 1930 over that of the 1929 "Waco" entries showed proof of the advances that were possible in a year's time. As it finally turned out, the winning airplane in the 1930 National Air Tour was a Ford "Tri-Motor" model 7-AT

and the two CRG entries flown by Johnnie Livingston and Art Davis were a very close second and third. The obvious arguments that developed over this contest and its revised scoring formula were actually never settled to anyone's satisfaction and many of the contestants felt they had a legitimate "beef." Waco Aircraft of course hollered "we wuz robbed" but "Ford" was determined to win this important event and to finally gain possession of the coveted "Edsel B. Ford Trophy"; so, many had found out there was really not much to gain in the arguing. Unpopular reaction to the scoring method and its imposed penalties caused many to forestall entry in the next annual tour; for 1931 the first and second place winners were Ford "Tri-Motors," the "trophy" was awarded permanently to the Ford Motor Co. and this popular annual tour was ended. What had started out in 1925 as an interesting event, destined to show and prove the reliability of the average airplane to the public at large, graduated into proportions of a personal feud between manufacturers and it all ended upon a rather sour note.

The flashy "Waco" model CRG was an open cockpit biplane seating 1 or 3 and was basically typical to the normal designs, as patterned by this company for several years now, but its haughty stance and its determined behavior while in action soon conveyed that it was really something quite special. With an open cockpit for the pilot, the space up front normally used for a two-place cockpit was arranged more like a cargo-hold for freight and had the capacity for some 540 lbs. of payload; total useful load was actually 47.7 per cent of the gross weight and this was an increase of some 269 lbs. over "Waco" model CSO entries of the year previous. Powered with the improved 7 cyl. Wright "Whirlwind" R-760 engine of 240 h.p. the model CRG could carry more load at a substantially higher speed and still take off or land within the space of some 50 yards; here indeed was a performance calculated to win. With throttles pushed nearly wide-open for the entire 4800 mile grind, it is quite remarkable that the "Waco CRG" was able to maintain a 148.4 m.p.h. average for the distance. Besides maintaining the highest average speeds for the tour, "Waco's" two remarkable entries also demonstrated shortest take-offs and shortest landings of any airplane in the contest. Bearing the burden of what amounted to nearly a 40 per cent handicap, the CRG entries lost first place by only a margin of 5 per cent, which indicates again that a potential winning performance was surely present in these airplanes. The peculiar stance of this airplane, caused by the unusually long landing gear legs, was dictated by the selection of an "airfoil" that could "hold on to its lift" at higher angles of attack, so a 3-point landing attitude was at a much greater angle to the ground than normally used. This translated into "braking area" for lower landing

speeds and consequently a shorter ground roll. This particular "wing section" was also capable of much higher speeds than the highly-cambered "Aeromarine" section that was normally used on "Waco" straight-wings so it is easy to see why the model CRG was so much faster than the earlier CSO. Flight characteristics and general behavior of the model CRG were quite good but the small ailerons on the lower wing panel were somewhat inadequate for good sharp control, especially at higher speeds. The tall landing gear was probably a source of some discomfort, because of the rather large angle it created in a 3-point position, but this was only a matter of getting used to it. Fully loaded take-off time for this airplane was as little as 6.83 seconds and landing roll to a stop was 5.03 seconds; this was short-field performance at its very best. Timed maximum speed was 157 m.p.h. with a provision added later for placarded speeds as follows: 154 m.p.h. in level flight or climb and 185 m.p.h. in glide or dive. The model CRG was certificated just a scant week before its entry in the 1930 National Air Tour. As it first rolled out of the factory for tests, the entry flown by Johnnie Livingston was equipped with the Wright R-760 "geared engine," but approval for this combination was not in the offing just yet, so the standard engine was hurriedly installed just a day before the start of the tour, thus both entries were then alike. The type certificate number was issued on 9-2-30 and the two "tour" entries were the only examples of this model that

Fig. 196. Slightly modified from original lines, CRG later used for sky-writing.

were manufactured by the Waco Aircraft Co. at Troy, Ohio. Francis Arcier, Waco's new chief engineer, came to Waco Aircraft about May of 1930 so he probably didn't get in on much design work for the CRG; the CRG most likely was designed by Clayton Bruckner and Russell Hardy. Andy Stinis, a pilot of great talent, flew one of the surviving CRG (NC-600Y) in 1938 for "skywriting"; by now the ship had been modified with a full NACA-type engine cowling, wheel pants, a high-profile turtle-deck and smoke writing gear. It has been reported that this ship is still in existence and awaits only to be rebuilt by someone who'd take a fancy to her special nature.

Listed below are the only available specifications and performance data for the "Waco" model CRG as powered with the 240 h.p. Wright R-760 engine; wt. empty 1359 lbs.; useful load 1241 lbs.; payload with 80 gal. fuel was 541 lbs.; gross wt. 2600 lbs.; max. speed 154; cruising speed 130; landing speed 45; climb 1200 ft. first min. at sea level; ceiling 17,000 ft.; gas cap. 80 gal.; oil cap. 7 gal.; cruising range at 14 gal. per hour was nearly 700 miles; price at factory was not announced.

The construction details and general arrangement of the "Waco CRG" were quite typical to the earlier model CSO, as described in the chapter for ATC # 240 in U.S. Civil Aircraft, Vol. 3, including the following. The fuselage framework was more or less of typical construction, but faired out to a somewhat larger diameter in front, to blend in with the engine cowling arrangement; the front cockpit, with an actual capacity for two passengers, was faired over with a metal panel cover, the pilot's windshield was of a special streamlined shape and a head-rest fairing extended nearly to the base of the vertical fin. A 64 gal. fuel tank was mounted in the fuselage ahead of the front cockpit, and a 16 gal. fuel tank was mounted in the center-section panel of the upper wing; baggage allowance was 60 lbs. in a lined compartment behind the rear cockpit. The wing panels were basically of normal "Waco" construction in a slightly thicker section with the lower panel neatly faired into the fuselage junction; ailerons of the balanced-hinge type were covered with a finely corrugated aluminum alloy sheet. The interplane struts and the interplane bracing of somewhat heavier cross-section were of normal "Waco" arrangement. The outrigger type landing gear was unusual only in its length; long telescopic legs used oleo-spring shock absorbing struts, center vees were partially faired and the low-pressure "airwheels" were streamlined by fairings of a rather unusual type. The engine fairing was unusual too in that it had a venturi-type opening in the nose with air-venting slits at the firewall and the outer diameter of the engine was shrouded with a speed-ring cowling; the Wright R-760 engine normally had a nose-type exhaust collector ring but the CRG used only short curved "stacks". The fabric covered tail-group was of typical "Waco" construction and configuration and the horizontal stabilizer was adjustable in flight. A ground adjustable metal propeller, airwheels with wheel brakes, hand crank inertia-type engine starter, wheel fairings and speed-ring engine cowling were standard equipment. The next development in the "Waco" biplane was the Menasco-powered model MNF as described in the chapter for ATC # 393 of this volume.

Listed below are the two "Waco" model CRG entries as gleaned from various records:
NC-600Y; Model CRG (# 3349) Wright R-760.
NC-660Y; " (# 3350) "
Model CRG also eligible with 250 h.p.; approval expired 7-1-32.

Fig. 197. Fairchild model KR-21-B with 125 h.p. Kinner B5 engine.

The Fairchild KR-21 series (Kreider-Reisner C-6) was a handsome sport-biplane of robust proportion, with an easy stance that was somewhat apart from the other sport-trainer craft of this type. Already designed with robust and rugged frame to stand the gaff of pilot-training and every-day flying by private-owner pilots, the KR-21 biplane was now fitted with slightly more power to boost performance, and other niceties were added to attract more of the sport-flyer customers. Doubly braced where it counts and amply stressed for all manner of air-borne gyrations, the KR-21 could be "wrung out" with enthusiasm, whether by fledgling or expert, without any signs of weakness in the structure or its character. Possessing a flexible nature that ranged from docile to exhuberant, the KR-21 biplane adjusted itself very well according to the experience wielded at the "control stick". With the adding of 25 more horsepower in the new model KR-21-B, all good traits were noticeably amplified, it handled somewhat better and the new craft showed promise at least of becoming very popular. Introduced rather late in 1930, the KR-21-B was greeted warmly and with interest, but sales were hard to come by so it remained scarce and rather rare. Because of its inherent strength and good performance, the KR-21-B was eligible for aerobatic stages of the Civilian Pilot Training Program (CPTP) as instituted by the government a few years later.

The Fairchild model KR-21-B was a light open cockpit biplane with seating for two in tandem, and in this more deluxe version, was appointed better and equipped with more pilot conveniences and aids to enhance the enjoyment of its operation. With tapered wing panels set in a rather large interplane gap, plus a cut-out in the trailing edge of the upper wing, visibility was excellent in all directions; the KR-21 was often used for "stunt flying" so the heavy wing bracing trusses offered freedom from worry when snap-rolling the wings around. Powered with the 5 cyl. Kinner B5 (R-440) engine of 125 h.p., the all-round performance of the KR-21-B was slightly improved over that of the earlier KR-21 with 100 h.p. but somewhat penalized by the increased gross weight allowed. Stout of frame and of heart, the improved KR-21-B was eager and sensitive, but not "tricky", and quite a satisfying pleasure to fly. Its performance and behavior in all other aspects has always been commended by those who flew it and of course generally compared to the very best. The type certificate number for the model KR-21-B was issued 9-10-30 and perhaps only 3 examples of this model were built; two additional examples were modified from the earlier KR-21. Manufactured by the Kreider-Reisner Aircraft Co., Inc. at Hagerstown, Md., a subsidiary of the Fairchild Airplane Mfg. Corp. John S. Squires was president and Louis E. Reisner was V.P.

Listed below are specifications and performance data for the Fairchild model KR-21-B as powered with the 125 h.p. Kinner B5 engine; length overall 21'7"; height overall 8'6"; wing span upper 27'0"; wing span lower 24'6"; wing chord both 57" at root and 41" at tip; total wing area 193 sq.ft.; airfoil USA-45 modified; wt. empty 1120 lbs.; useful load 610 lbs.; payload with 30 gal. fuel was 237 lbs.; (payload includes 1 passenger at 170 lbs., 2 parachutes at 20 lb. each & 27 lb. baggage); gross wt. 1730 lbs.; max. speed 115; cruising speed 98; landing speed 48; climb 800 ft. first min. at sea level; climb in 10 min. 6500 ft.; ceiling 12,500 ft.; gas cap. 30 gal.; oil cap. 3 gal.; cruising range at 8 gal. per hour was 370 miles; price at factory $4525.

The fuselage framework was built up of welded chrome-moly steel tubing into a light but rigid truss, faired to shape with wooden formers and fairing strips, then fabric covered. The cockpits were deep and well protected and bucket-type seats had wells for parachute pack; engine instruments were provided in both cockpits. A baggage bin of 1.5 cu. ft. capacity had allowance for 27 lbs. The wing panels which were tapered both in plan-form and in section, were built up of heavy spruce spar beams that were routed out for lightness and girder-type

wing ribs were built up of spruce members and plywood gussets; the completed framework was covered in fabric. The 4 narrow-chord ailerons were connected together in pairs by a streamlined push-pull strut; a gravity-feed fuel tank of 30 gal. capacity was mounted in the center-section panel of the upper wing. The split-axle landing gear of 66 in. tread used oleo-spring shock absorbing struts; low-pressure airwheels (22x10) were equipped with brakes. Semi-airwheels with brakes were also available and usually encased with streamlined wheel pants. The fabric covered tail-group was built up of welded chrome-moly steel tubing; the fin was ground adjustable and the horizontal stabilizer was adjustable in flight. Standard equipment for the KR-21-B included, dual controls, tool kit, log books, first-aid kit, fire extinguisher, booster magneto and custom colors. Engine starter and wheel pants were optional. The next Kreider-Reisner development was the model KR-125 biplane described in the chapter for ATC #'368 of this volume.

Listed below are Fairchild model KR-21-B entries as gleaned from registration records: NC-367N; Model KR-21-B (# 1022) Kinner B5.

NC-962V;	"	(# 1053)	"
NC-247V;	"	(# 1501)	"
NC-954V;	"	(# 1502)	"
NC-955V;	"	(# 1503)	"

Serial # 1022 and # 1053 modified into KR-21-B from earlier KR-21.

Fig. 198. KR-21-B typical of earlier trainer but leveled more at the sportsman.

A.T.C. #364
(9-13-30)
PITCAIRN
"SUPER MAILWING", PA-8M

Fig. 199. The last of the "Mailwing" series, PA-8 mounted 300 h.p. Wright J6-9 engine for boost in payload.

The jaunty single-seated Pitcairn "Mailwing" biplane had an aura of romance and excitement around it that was brought about by its colorful service record on various mail-routes around the country; an aura built up too by the almost legendary stories that came from "Mailwing" pilots in the course of 3 years since its introduction to carrying the mail. A trim and beautiful airplane with a sure-footed, flashing performance, one feels the "Mailwing" was certainly deserving of all the praise and the love that was heaped upon it. Cocky, with a handsome stance and with plenty of heart, this series of mail-carrying biplanes was progressively improved through 3 different models up to this time. Pioneering in the hauling of mail by private contractor, the earliest "Mailwing" was small and quick to get about, sort of a "Pony Express" on short-haul routes that soon crisscrossed southern and eastern portions of the country. In the development of the new PA-8, the biggest and perhaps the finest of the "Mailwing", Pitcairn drew heavily on a background of over 5 million miles of successful operation in airmail service by more than a dozen different operators. Inheriting all the proven basic qualities that were essential to efficient and profitable transporting of mail, plus new requirements that were dictated by an increase

in annual tonnage, the Pitcairn model PA-8 was able to carry a far greater payload in its cavernous hold at increased speeds with longer range. Designed specifically for the job it had to do and not an adaption of a high-performance sportplane, the PA-8 was arranged in minute detail for efficient operation and maintenance; its "buddy-buddy" nature was also cooperative at all times in making a trying chore a more pleasant one. With more varied equipment now available to the different air-lines, the big "Mailwing" (PA-8) found its circle of customers somewhat limited so, as it turned out, it was built only for Eastern Air Transport (EAT). Pressed into service on the new seaboard route from New York to Florida, the PA-8 was not long in setting an impressive service record. As the last version in the "Mailwing" series, the model PA-8 was the most unobtrusive and therefore the least remembered but, with legend and sentiment put aside, it may rank as the finest. By Dec. of 1930. Pitcairn Aircraft had all but divorced itself from the manufacture of regular airplanes and was already planning production of the new and revolutionary "Autogiro".

The Pitcairn "Super Mailwing" model PA-8 was a single-place open cockpit biplane designed especially for the transport of mail and cargo. Basically typical to earlier "Mailwing,"

Fig. 200. Model PA-8 served eastern seaboard routes for Eastern Air Transport.

the PA-8 was however extensively redesigned to meet extra requirements imposed by progressive demands in the air-mail system. Carrying a payload nearly twice that of its earlier sister-ships, the PA-8 was essentially a larger airplane and what it lacked in dash and flash, it made up for in many other ways. Because of its buxom proportion it could hardly be called jaunty but the PA-8, in spite of its increased bulk, turned in a very good performance in a surprisingly nimble attitude. Pilots who flew it claim it was a nice airplane to fly, having no annoying traits, and though it never displayed playful exhuberance, it was extremely deft and capable. Pilots were also grateful for its dependability and the line-mechanics enjoyed working on it. Powered with a 9 cyl. Wright J6 (R-975) engine of 300 h.p., the PA-8 used every horsepower wisely and came up with quite a lot of performance on 300 horses. The type certificate number for the model PA-8 was issued 9-13-30 and some 6 or more examples of this model were built by Pitcairn Aircraft, Inc. on Pitcairn Field in Willow Grove, Penna. Harold F. Pitcairn was president; veteran pilot James G. Ray was V.P. and general manager; H. H. Hoeberly, Jr. was sales manager; Agnew Larsen was chief engineer for aircraft and Walter C. Clayton was chief engineer for the "autogiro" projects. Moving into new quarters at Willow Grove, Pitcairn now had 80,000 sq. ft. in a modern plant adjacent to its own airport, for fly-away deliveries.

Listed below are specifications and performance data for the Pitcairn "Super Mailwing" model PA-8 as powered with the 300 h.p. Wright J6 engine; length overall 24'10"; height

overall 9'9"; wing span upper 35'0"; wing span lower 32'1"; wing chord upper 58"; wing chord lower 52"; wing area upper 161 sq.ft.; wing area lower 117 sq.ft.; total wing area 278 sq.ft.; airfoil "Pitcairn" # 2; wt. empty 2294 lbs.; useful load 1706 lbs.; payload with 78 gal. fuel was 1008 lbs. (a parachute at 20 lbs. & night-flying equipment deducted from payload); gross wt. 4000 lbs.; max. speed 145; cruising speed 122; landing speed 60; climb 1100 ft. first min. at sea level; climb in 10 min. 7500 ft.; ceiling 16,000 ft.; gas cap. 80 gal.; oil cap. 8 gal.; cruising range at 15 gal. per hour was 600 miles; price at the factory field was $12,500.

The fuselage framework was built up of welded chrome-moly steel tubing in round and square section, heavily faired to a well-rounded shape and fabric covered. The pilot's cockpit was roomy and well protected by a large built-up windshield of shatter-proof glass; convenient steps and a hand-hold were placed for easy entry to the cockpit. An adjustable seat provided better visibility over the nose for take-offs and landing, the high-profile turtle-deck had built-in head-rest and the pilot had a separate baggage compartment for his personal equipment. The large cargo area ahead of the cockpit was fire-proof and weather-proof with easy access through three metal hatch covers; the rear compartment just ahead of the cockpit, was converted for carrying 2 passengers on the prototype model (PA-8S) with room up ahead for over 600 lbs. of baggage and cargo. Cockpit heat and ventilation were also provided. The high aspect ratio wing panels with well-rounded tips were built up of box-type spars in front and solid spruce spar beams in back; the box-type

Fig. 201. Unusual view shows prototype PA-8 during test.

spars were of spruce caps with plywood webs and the girder-type wing ribs were of spruce members reinforced with plywood gussets. The leading edges were covered with dural metal sheet and the completed framework was covered in fabric. Ailerons were on the lower panels only and the lower wing roots were faired-in at the fuselage junction. The robust out-rigger landing gear of 96 in. tread used oleo & spring shock absorbing struts; wheels were 32x6 and Bendix brakes were standard equipment. The swivel-type tail skid was mounted in rubber. The fabric covered tail-group was built up of welded chrome-moly steel tubing; the fin was ground adjustable and the horizontal stabilizer was adjustable in flight. A ground adjustable metal propeller, hand crank inertia-type engine starter, bonding for radio, Zerk lubrication on all movable fittings, a first-aid kit, flashlight, map case and 12 volt battery were standard equipment. Besides a normal complement of engine and flight instruments, a Daiber compass, Sperry artificial horizon, a compensated altimeter, bank and turn indicator, rate of climb

indicator, fuel level gauge, 8-day clock, direct and indirect instrument lighting and automatic fire extinguisher were also provided. Night-flying equipment was also available, with navigation lights, retractable landing lights and parachute flares. The next Pitcairn development was the PCA-2 "Autogiro" described in the chapter for ATC # 410.

Listed below are Pitcairn model PA-8 entries as gleaned from registration records:

X-10056; Model PA-8 (# 150) Wright R-975.
NC-10750;   ”   (# 161)   ”
NC-10751;   ”   (# 162)   ”
NC-10752;   ”   (# 163)   ”
NC-10753;   ”   (# 164)   ”
NC-10754;   ”   (# 165)   ”

Serial # 150 was prototype, tested both as PA-8M (mail) and PA-8S (sport); serial # 161 through # 165 operated by EAT; we have no record of PA-8 in civilian service after retirement by EAT, however, 2 were still available for restoration in 1966; PA-8 also available with 300 h.p. "Wasp Jr." engine but none on record.

A.T.C. #365
(9-13-30)
STEARMAN "CLOUDBOY," 6-A

Fig. 202. Stearman "Cloudboy" model 6-A with 165 h.p. Wright J6-5 engine.

Versatile and adaptable to a number of uses, "Stearman" airplanes in previous models had very often been used for pilot training, especially the popular C3B which was used by many schools for secondary phases. Though better known for their mail-planes, "Stearman" also produced varied models for business or sport and this popular line-up had been their mainstay of production up to this point. With the market for all craft of this type slowly dwindling away, the company was forced to look elsewhere to new horizons for its very existence. It was only logical then that Stearman should look hopefully to the armed services both here and abroad for a continuation of normal business. For an entry into this new field "Stearman" leaned heavily on past experience and developed a line of training planes which were to launch a program and new career that lasted into a span of some 15 years. Starting warily into this new field "Stearman" designed and developed the new 6-series biplane which was first planned as a commercial pilot-training craft but definitely slanted towards requirements of the Air Corps for a primary trainer. Submitted soon after to the Air Corps for test, the design was accepted and an order was received for four of the new trainers that became the YPT-9. The YPT-9 was powered with the 5 cyl. Wright J6 (R-540) engine of 165 h.p. and the commercial counter-part of this version was the model 6-A of which 3 or 4 were built. The new model 6-A was affectionately called the "Cloudboy" and it was also hoped that it might find a market amongst certain he-man sportsman-pilots, pilots that would be looking for a ship they could bang-around in and not worry about it breaking up or getting dirty. Built rugged and kept simple it was also hoped that the "Cloudboy" would find some favor amongst the civilian flying schools but the economic situation was such that a demand never did develop.

The tom-boy looking model 6-A was an open cockpit biplane seating 2 in tandem and was the first "Stearman" air plane to be specifically designed for pilot training throughout all its phases. Called the "Cloudboy" it is interesting to note that despite the slab-sided frame and angular lines that were devoted to simplicity and rugged character, the "Stearman" lineage that led to the development of this trainer seems to poke through here and there. The requirements for a true training airplane naturally influence its make-up and arrangement, so therefore a "trainer" hardly ever evolves as a thing of great beauty. Everything is primarily planned to provide a practical utility, an ease of service and maintenance, a care-free and dependable tenure on the flight-line and a rugged character that could withstand occasional abuse

Fig. 203. Rugged of frame and staunch of character, 6-A was ideal trainer and
excellent knock-about for sport flying.

or mis-use on the ground or in flight. Yet, none of this was to encroach upon the trainer's ability to handle easily, to help instill a proper "feel" to the fledgling pilot and still be ready, willing and able to do all sorts of nip-ups as the training progresses; almost without fault, the "Stearman 6" measured up to most of these requirements. Powered with the 5 cyl. Wright R-540 engine of 165 h.p. the model 6-A was not particularly blessed with speed but its performance was

Fig. 204. Army Air Corps version of 6-A was the YPT-9.

Fig. 205. Delivery flight of YPT-9 to San Antonio, Texas.

more than ideal for this type of work and there was yet some reserve left for an occasional tight spot. Flight characteristics were generally described as pleasant, general behavior was sprightly but by no means tricky, and response was quite eager if properly prompted. The type certificate number for the Stearman model 6-A was issued 9-13-30 and 3 examples of this model were manufactured by the Stearman Aircraft Co. at Wichita, Kansas.

Listed below are specifications and performance data for the Stearman "Cloudboy" model 6-A as powered with the 165 h.p. Wright J6 engine; length overall 24'8"; height overall (tail up) 9'6"; wing span upper 32'0"; wing span lower 28'0"; wing chord both 60"; wing area upper 148 sq. ft.; wing area lower 124 sq. ft.; total wing area 272 sq. ft.; airfoil N-22; wt. empty 1733 lbs.; useful load 667 lbs.; crew wt. 336 lbs.; baggage 86 lbs.; gross wt. 2400 lbs.; max. speed 110; cruising speed 90; stall speed 50; landing speed 45; climb 710 ft. first min. at sea level; climb in 10 min. 5600 ft.; ceiling 12,300 ft.; gas cap. 37 gal.; oil cap. 2.5 gal.; cruise range at 9 gal. per hour was 350 miles; price at factory field was $8,500, raised to $10,250 in 1931.

The fuselage framework was built up of welded chrome-moly steel tubing in a rigid truss form, lightly faired to shape and fabric covered. The cockpits were quite large and well accessible with the protection of 3-piece safety glass windshields; dual controls were provided and the metal bucket-type seats were adjustable for height. The robust wing framework was built up

of heavy-sectioned solid spruce spar beams with spruce and plywood truss-type wing ribs; the leading edges were covered with dural metal sheet and the completed framework was covered in fabric. The gravity-feed fuel tank was mounted in the center-section panel of the upper wing, which was supported from the fuselage by two N-type strut assemblies of streamlined steel tubing. The four Friese-type ailerons were connected together in pairs by streamlined push-pull struts. The wide tread split axle landing gear was of the out-rigger type with "Aerol" (air-oil) shock absorbing struts; wheels were 28x4 or 30x5 and wheel brakes were standard equipment. A swivel-mounted tail skid was attached to very end of fuselage for a longer wheel-base; later versions of the "Six" had tail wheel. The fabric covered tail-group was built up of welded steel tubing and the horizontal stabilizer was adjustable in flight. A metal propeller and wheel brakes were stand-are equipment. The next development in the "Stearman 6" trainer was the model 6-F described in the chapter for ATC #371 in this volume.

Listed below are "Stearman" model 6-A entries as gleaned from registration records:

X-786H; Model 6-A (#6001) Wright R-540
NC-787H;   ″   (#6002)   ″
NC-795H;   ″   (#6004)   ″

Serial #6001 also modified to model 6-F, 6-D, 6-L, at various times; serial #6002 modified to model 6-P on Group 2 approval 2-520; serial #6004 modified by Boeing School of Aeronautics into model 6-L with Lycoming 215 engine.

Fig. 206. Boeing "Monomail" model 221 with 575 h.p. "Hornet" B engine; craft was a break-through into new design philosophy.

Like a sign in the sky of things to come, an entirely new form in large transport aircraft, the revolutionary "Monomail" was directly responsible for many drastic changes in design, design ideas that led to a whole new concept for the air-liner. First flown in May of 1930 by E. T. Allen, the "Monomail" was introduced at first as an all-cargo version. While tests were still going on with the prototype airplane, a passenger transport sister-ship was already in the making and receiving its final grooming for an early maiden flight. More or less typical in its basic outer form, the "Model 221" version was comfortably fitted for carrying 8 passengers in the enclosed cabin up forward and the pilot was still aft in an open cockpit. Heavier when empty because of cabin appointments and other aids to passenger comforts, the Model 221 was penalized slightly by that amount in its useful load, however, there was still ample

payload allowance for 8 passengers and 575 lbs. of baggage-cargo. Used in test on Boeing's own transcontinental system from San Francisco to Chicago, the "221" underwent various modifications in an effort to study the numerous possibilities inherent in this design. Perhaps the most useful variation of the basic design was the Model 221-A, a craft that carried 6 passengers with a generous allowance for their baggage and still was able to carry up to 750 lbs. in cargo payload. In 1931, affiliate lines of the Boeing System and the national Air Transport merged together into a network called United Air Lines, spanning the breadth of this country from New York to San Francisco; continued tests of the 221 on this system laid the ground-work for better carriers. Because of the recent improvements in air-transports and considerable improvements in line operations, the one-way fare from the west coast to the east coast or vice-versa, was

Fig. 207. Retractable landing gear and cantilever wing of 221 were responsible for excellent performance.

soon reduced to $227.60. It takes a moment of meditation to actually realize what an important part the Boeing "Monomail" played in the advancement of future air-transportation.

The Boeing "Monomail" model 221 was a large low-winged cantilever monoplane of all-metal construction that was especially designed for the speedy transport of passengers. The cabin interior was arranged to carry 8 passengers with their baggage in the Model 221 and a modified version of this, called the Model 221-A, carried 6 passengers and baggage with ample allowance yet for cargo up forward. Feeling that the development of this particular design was more like opening the door onto many possibilities not yet dreamed of, the Boeing line did not stock up on these airplanes but only had the 2 "Monomail" for further studies. These studies paved the way for a

revolutionary twin-engined "bomber" and shortly after, the famous Model 247, the first of the "modern" air-liners. Powered with the 9 cyl. Pratt & Whitney "Hornet" B engine of 575 h.p., the Model 221 translated this power into exceptional performance for an airplane of this size, proving again that aerodynamic efficiency was the key to future gains in payload and speed. Flight characteristics of the 221 were good and the large craft was fairly easy to fly; singled out to herald the coming of a new day, the "Monomail" spent much of its time in test and demonstration along the transcontinental air-ways system. The type certificate number for the Model 221 was issued 9-16-30 and one example was initially built in this version; the cargo-carrying Model 200 (ATC # 330) was later converted to the 221 and both aircraft were then converted to the 221-A with special

Fig. 208. Cabin arrangements also allowed combination passenger-cargo loads; pilot's open cockpit heated for cold-weather flying.

Fig. 209. Model 221-A was tested with spatted cantilever landing gear.

variants of this version at different times. The "Monomail" was manufactured by the Boeing Airplane Co. at Seattle, Wash. Philip G. Johnson was president; C. L. Egtvedt was V.P. and general manager and C. N. Montieth was the chief of engineering.

Listed below are specifications and performance data for the Boeing "Monomail" model 221 as powered with the 575 h.p. "Hornet" B engine; length overall 41'2"; height overall 12'6"; wing span 59'2"; wing chord at root 12'4"; wing chord at tip 7'6"; total wing area 535 sq. ft.; airfoil "Boeing"; wt. empty 4990 lbs.; useful load 3010 lbs.; payload with 135 gal. fuel was 1935 lbs.; gross wt. 8000 lbs.; max. speed 158; cruising speed 137; landing speed 57; average take-off run 800 ft.; average landing run 700 ft.; climb 720 ft. first min. at sea level; climb to 10,000 ft. was 21.5 mins.; ceiling 14,700 ft.; gas cap. 135 gal.; oil cap. 12 gal.; cruising range at 32 gal. per hour was 540 miles; no price was announced.

The construction details and general arrangement of the Model 221 was typical to that of the Model 200 as described in the chapter for ATC # 330 of this volume, including the following. The cabin interior was neatly upholstered with seating arrangement for 8; four windows were provided down each side with a large door on the left side for entry. A broad wing-walk and hand-rails were provided for passengers to gain entry to the cabin and steps were provided to gain entry to the open cockpit. The cabin interior was modified from time to time with various appointments including ventilation, cabin lighting and cabin heat. In the 221-A version seating was arranged for 6 with allowance up forward for baggage and cargo. The pilot's open cockpit, a hold-over

from days gone by, was tested with various windshields, head-rest fairings and snap-on side-panels which were used for cold weather protection. The airframe was bonded and shielded for radio equipment with the radio gear stored in a compartment behind the pilot; full night-flying equipment included retractable landing lights and parachute flares. Separated from the main cabin, the cargo compartment was fire-proof for mail that was now carried in asbestos pouches; passenger baggage allowance averaged 30 lbs. per person. Normally the "Monomail" was equipped with a folding landing gear that retracted into wing wells on underside but a cantilever rigid gear, with a tread 121 inches, fully enclosed in large streamlined "spats," was also tested for comparison. The tail wheel was steerable and wheel brakes were provided for better handling. All movable control surfaces had slotted-hinge aerodynamic balance to lighten the control forces and allow easier manipulation by the pilot. In later tests the fuselage length was increased to improve stability and lengthen the pitching moment. Fairly bristling with innovations both aerodynamic and structural, the "Monomail" proved a good test-bed for a look into new concepts in transport aircraft design. The next Boeing development, discounting various military projects at this time, was the twin-engined Model 247 described in the chapter for ATC # 500.

Listed below are Model 221 entries as gleaned from registration records:
NC-725W; Model 221 (# 1153) Hornet B-575.
NC-10225;   "   (# 1154)   "
Serial # 1153 first as all-cargo Model 200; serial # 1154 first as 9-place Model 221, both aircraft later modified to 6-place model 221-A.

Fig. 210. Stinson "Tri-Motor" model SM-6000-A with three 215 h.p. Lycoming engines; economical operation permitted lowering of fares.

As described here earlier in the chapter for ATC #335, the model SM-6000 was the first version in the Stinson "Tri-Motor" series, an 11 place craft arranged primarily for coach-style service on the shorter routes; therefore, it lacked the maximum utility capable by this versatile transport. The model SM-6000-A, as will be described here, was introduced as an up-graded version of the earlier model with modifications now permitting somewhat more flexible interior arrangements with some aerodynamic improvements also to boost performance potentials. Some slight modifications to the interior now made possible a combination mail-passenger version for 8 or 9 passengers and cargo. A version completely stripped of seating was available as an all-cargo carrier and several "custom club" interiors were available on order. "Townend-ring" engine fairings were offered also as a part of this version for a gain in available speeds, first on the outboard engines only and later tests proved these "cowlings" beneficial for use on all 3 engines. Large tear-drop fairings over the 36x8 wheels were offered also to further boost the speeds and this combination of streamlining (wheel pants & engine cowlings) produced a gain of at least 8 m.p.h.; the maximum speed capable now was a good 146 m.p.h. These modified versions of the Stinson "Tri-Motor" were on occasion called the "Model T" and later versions yet to come were progressively improved to the point where they

were considered as one of the finest craft of this type. Quite unusual too is the fact that at least 25 of the SM-6000, in the various versions built, served actively for as long as 10 years after they were first brought into use.

The Stinson "Tri-Motor" model SM-6000-A (also Model T) was a high-winged cabin monoplane of the multi-motored type that had a standard seating arrangement for 10 in its spacious and comfortable interior. As an up-graded version of the earlier SM-6000, the modified SM-6000-A had interior arrangements that were now more flexible to enhance utility advantages for several types of service. Optional arrangements were now possible in the pilot's cabin to seat one or two pilots, with dual controls available if desired. The main cabin seating was arranged for 9 passengers but the space for one or two of the front seats could be converted to baggage and cargo bins for combination loads. With some major aerodynamic clean-up the SM-6000-A was blessed with ample speed and general all-round performance was of a caliber not usually found in a craft of this size and type. With no examples actually leaving the factory as brand-new model SM-6000-A, it must be assumed that this certificate of approval was the means to provide an up-grading for the earlier SM-6000 and was more than likely applied to all existing craft in this particular batch of airplanes. The best-known version in the Model T series was the improved SM-6000-B

Fig. 211. Interior of SM-6000-A showing baggage and mail compartments up front.

of which a good number were built for use on various lines and some were in use as executive transports in the field of big business. The type certificate number for the Stinson "Tri-Motor" model SM-6000-A was issued 9-19-30. The provisions of this approval probably affected no more than the first 14 airplanes of the SM-6000 series, because airplanes serial #5015 and upwards were covered by the certificate for the model SM-6000-B.

Listed below are specifications and performance data for the Stinson model SM-6000-A as powered with 3 Lycoming R-680 engines of 215 h.p. each; length overall 42'10"; height overall 12'0"; wing span 60'0"; wing chord 108"; total wing area 490 sq. ft.; airfoil Goettingen 398; wt. empty 5600 lbs.; useful load 2900 lbs.; payload with 200 gal. fuel was 1410 lbs.; payload with 120 gal. fuel was 1890 lb.; gross wt. 8500 lbs.; max. speed 146 (with 3 engine cowlings & wheel pants); cruise speed 122; landing speed 60; climb 1000 ft. first min. at sea level; climb in 10 min. 6500 ft.; ceiling 14,500 ft.; gas cap. max. 200 gal.; oil cap. 15 gal.; cruising range at 36 gal. per hour was 600 miles; price at factory field was listed as $25,000. for this version.

The construction details and general arrangement for the model SM-6000-A was typical to the model SM-6000 (as described in the chapter for ATC #335 in this volume), except for the following. As already mentioned, several interior lay-outs were now available for a greater variety of transport services. A coach-style transport with seating for 9 passengers and a pilot was the basic arrangement; by elimination of the two forward seats in the main cabin, this space could be converted into bins carrying 255 lbs. of baggage and cargo. To provide better access to pilot's cabin and cargo bins up forward, a door installation on the right side front was available. For service on the longer routes provisions were available for dual controls manned by a pilot and co-pilot; extra instruments in a lighted panel were optional. All windows were of shatter-proof glass and any window could be opened throughout length of the cabin; ventilators and main cabin lights were also provided. Custom interior decor was available and all cabin hardware was chrome-plated; over-stuffed seats, lounges and cabin heaters were optional. Engine fairings (a modified Townend-ring type) for all 3 engines were eventually standard equipment but metal wheel fairings were optional; standard wheel and tire size was 36x8 but 35x15-6 low-pressure tires were optional. Narrow fenders to protect whirling propellers from kicked-up debris were also optional. Long tail-pipes were fitted to the exhaust collector-rings and the complete assemblies were plated in chrome. Ground-adjustable metal propellers and electric direct-drive engine starters were standard equipment; landing lights, navigation lights, parachute flares, fire extinguishers and radio receiver equipment were optional. An unusual feature in the tail-group was the vertical fin which was adjustable in flight to compensate for uneven torque in the case of one engine's failure; the horizontal stabilizer was also adjustable in flight. The rudder and ailerons were aerodynamically balanced for lighter control forces. This approval was later amended to allow an 8600 lb. gross weight for the addition of extra equipment without penalizing the payload to any great extent. The next and most popular development in the Stinson tri-motored "Air Liner" (SM-6000 series) was the model SM-6000-B described in the chapter for ATC #420.

There were no available records to check on which of the early SM-6000 had been converted to the SM-6000-A but from various hints and photographs of the period, it is reasonable to assume that perhaps all of the serial numbers shown at appendage of chapter for ATC #335, had at one time or another at least, taken advantage of some or all the up-grading allowed by the provisions in ATC #367.

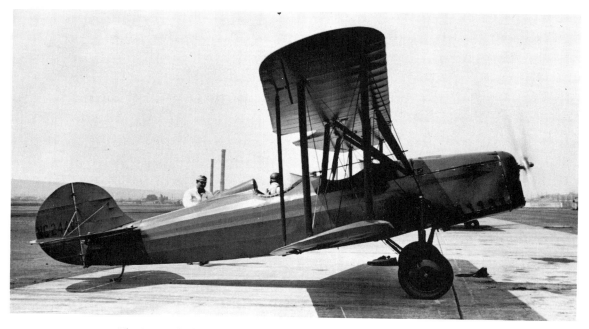

Fig. 212. Fairchild model KR-125 with 125 h.p. Fairchild 6-390 engine.

The Fairchild model KR-125 was a KR-21 series biplane in all respects except that it was modified as a flying test-bed for the newly developed Fairchild 6-390 in-line aircooled engine. Because of its robust frame and rugged character the KR-21 type was the logical choice for the in-flight testing and further development of this new powerplant. Early versions of this engine were loaned out to several aircraft manufacturers to get a cross-section of reaction and comments were very favorable. A Fairchild 6-390 engine of 120 h.p. was installed in the original Gee Bee "Sport" monoplane and flown by Lowell Bayles for some 6 months in air-races, stunting exhibitions and several fast cross-country hops. No trouble of any kind was experienced, the engine running smoothly at all r.p.m., often to over 3000 revs in power-dives. Bayles candidly claimed that the 6-390 was the sweetest-running engine he ever flew behind. Other pilots also voiced their approval in typical words. As adapted for the 6 cyl. inverted Fairchild engine, the model KR-125 would be rather hard to judge; considerably heavier than the KR-21-B which was comparable in power at least, the KR-125 was slightly shy in performance on most all counts, but certainly a suitable test-bed for the new engine.

The Fairchild model 6-390 engine as approved on powerplant ATC #57 (issued 7-14-30) was a 6 cyl. in-line inverted aircooled engine developing 120 h.p. at 2150 r.p.m. or 125 h.p. at 2250 r.p.m. A single under-head camshaft eliminated troublesome push-rod gear and an oil-bath cam cover provided continuous lubrication for the valve gear. Generously finned for adequate cooling, the cylinders were baffled for even operating temperature. Exceptionally smooth-running because of the inherent dynamic balance of the 6 cyl. in-line engine, the 6-390 had many other good features to its credit. Later known as the "Ranger" engine, with power output progressively increased to over 200 h.p., this engine was installed in many varied airplane models in the next 10 years.

The Fairchild model KR-125 was an open cockpit biplane, with seating for two in tandem, and was typical to the KR-21 series biplane in all respects except for its engine installation and whatever modifications were necessary for this combination. Because only one airplane of this type was built and because no record or lore was available, we can only surmise that its performance and general behavior would be more or less typical to the KR-21-B. The increase in weight over that of the KR-21-B is nowhere explained but changes in the center-section cabane, considerable change in the landing gear, and some 55 lbs. more in engine weight would account for most of the differ-

Fig. 213. KR-125 was basically a KR-21 modified for installation of in-line engine.

ence. With the extra weight acting as a slight detriment to its performance, there is however no doubt that the smooth-running nature of the 6-390 provided seemingly effortless power for enjoyable flight. The type certificate number for the Fairchild model KR-125 was issued 9-22-30 and only one example of this model was manufactured by the Kreider-Reisner Aircraft Co., Inc. at Hagerstown, Md. The 6-390 engine was manufactured by the Fairchild Engine Corp. at Farmingdale, Long Island, N. Y.

Listed below are specifications and performance data for the Fairchild model KR-125 as powered with the 125 h.p. Fairchild 6-390 engine; length overall 22'1"; height overall 8'6"; wing span upper 27'0"; wing span lower 24'6"; wing chord both 57" at root and 41" at tip; total wing area 193 sq. ft.; airfoil USA-45 modified; wt. empty 1295 lbs.; useful load 595 lbs.; payload with 30 gal. fuel 215 lbs.; gross wt. 1890 lbs.; max. speed 112; cruising speed 95; landing speed 52; climb 750 ft. first min. at sea level; climb to 5000 ft. was 9.7 min.; ceiling 14,000 ft.; gas cap. 30 gal.; oil cap. 4 gal.; cruising range at 7 gal. per hour was 380 miles; price at factory not announced.

The construction details and general arrangement of the model KR-125 was typical to that of the model KR-21-B as described in the chapter of ATC #363 of this volume, including the following. The most noticeable change in this model was of course the inverted in-line en-

Fig. 214. 6 cyl. Fairchild engine later became famous "Ranger" series.

gine installation, which was completely cowled in and baffled internally for proper cooling. Because no cylinders protruded from the cowling in front as they would in a radial type engine installation, visibility was much better around and across the nose. In general the wing cellule was typical of the KR-21 series biplane except for a different arrangement of the center-section struts which resembled a lop-sided N; it appears that exit or entry would be slightly more difficult. The landing gear was typical in most of its geometry but upper fastening points for the "oleo strut" shock absorber legs were raised considerably higher than that of the KR-21; it is questionable whether this arrangement was any improvement. The 24x4 wheels with high-pressure tires were equipped with brakes; a metal propeller was also standard equipment. The next model of this type, the KR-135, was an improved version of the model just discussed (KR-125) and is described in the chapter for ATC #415.

Listed below is the only known example of the model KR-125 as gleaned from registration records:

NC-244V; Model KR-125 (#1) Fairchild 6-390.
In 1929 an earlier version of the Fairchild engine, called the 6-375, was tested in a "Challenger" biplane (X-311H) model C-2 (KR-31); it is believed this engine was rated at approx. 100 h.p.

A.T.C. #369
(9-29-30)
CONSOLIDATED "FLEETSTER,"
17-2C

Fig. 215. Consolidated "Fleetster" model 17-2C mounted 575 h.p. Wright "Cyclone" engine.

The fast-flying "Fleetster" model 17 was a medium-capacity air-liner that was primarily designed for pioneer service on air-routes in South America. As more fully described in the chapter for ATC # 291 of U.S. Civil Aircraft, Vol. 3, the cigar-shaped monoplane was a revolutionary departure from the conventional types as normally built by Consolidated Aircraft. The earlier "Fleetster 17" was powered with the 575 h.p. Pratt & Whitney "Hornet" B engine and this new version under discussion here was powered with the big 9 cyl. Wright "Cyclone" (R-1820-E) engine of 575 h.p. As the model 17, Type 2-C, this new version was basically typical in most respects except for its powerplant installation and some minor modifications that were necessary to this combination. Content to stick with Pratt & Whitney engines for all of its larger designs, Consolidated was no doubt persuaded to develop this new Wright-powered sister-ship either on suggestion or upon special order. As a custom built craft the Model 17, Type 2-C was rather rare and its identity was perhaps best spotted by the large nose-type collector ring in front of the big "Cyclone" engine. Early operational history of this Cyclone-powered model reveals that NC-750V was first registered to Edwin Lowe, Jr. of San Francisco and was operated by the Pacific-International Airways in 1933 with base in Anchorage, Alaska.

The "Fleetster" model 17, Type 2-C was an "aerodynamically clean" high-winged cabin monoplane with seating for 8 in a circular all-metal fuselage; with its pure cantilever wing, its big engine shrouded in an NACA-type fairing

cowl and its wheels encased in long tear-drop fairings, the speedy "Fleetster" was almost a pure streamlined form. Void of external wing bracing and other drag-producing assemblies, the "Seventeen" could sustain a top speed of 180 m.p.h. with ease. The "Fleetster" handled quite nicely in the air and operated well in and out of the smaller fields but ground handling was somewhat of a bother. Powered with the 9 cyl. Wright "Cyclone" (R-1820-E) engine of 575 h.p., the model 17-2C had performance comparable to the "Hornet" powered Model 17 and its selection would only be dominated by a preference in engines. Slightly heavier when empty and a more generous allowance for useful load brought gross weight to 5600 lbs. but with little or no apparent detriment to overall performance. The first example in this series was approved as a 7-place carrier on Group 2 approval numbered 2-273 (issued 9-25-30) and it is assumed that this same craft was some 4 days later approved as an 8-place on ATC. The type certificate number for the Model 17, Type 2-C was issued 9-29-30 and no more than one or two examples of this model were manufactured by the Consolidated Aircraft Corp. at Buffalo, New York.

Listed below are specifications and performance data for the "Fleetster" model 17, Type 2-C as powered with the 575 h.p. Wright "Cyclone" engine; length overall 31'5"; height overall 9'2"; wing span 45'0"; wing chord at root 106"; wing chord at tip 68"; total wing area 313.5 sq. ft.; airfoil "Goettingen" 398 modified; wt. empty 3443 lbs.; useful load 2157 lbs.; payload with 145 gal. fuel was 1027 lbs.

Fig. 216. With seating for 8, Cyclone-powered 17-2C was rare version in "Fleetster" series.

(7 pass. at 161 lbs. each); gross wt. 5600 lbs.; max. speed 180; cruising speed 153; landing speed 64; climb 1100 ft. first min. at sea level; climb to 10,000 ft. was 14 min.; ceiling 18,000 ft.; gas cap. 145 gal.; oil cap. 12 gal.; cruising range at 32 gal. per hour was 675 miles; price at factory field approx. $28,000.

The construction details and general arrangement of the "Fleetster" model 17, Type 2-C was typical to that of the "Hornet" powered model 17 as described in the chapter for ATC # 291 of U.S. Civil Aircraft, Vol. 3, including the following. The circular metal monocoque fuselage of the 17-2C was typical but arranged differently; in this case, passengers sat in two full-width seats 3-across, all facing forward. Center seats folded out of the way to make an aisle and cabin entry door was further back to rear of cabin section. The pilot's cabin normally seated 2 but right-hand seat was removed for installation of radio equipment; no baggage was allowed with a full passenger

Fig. 217. 17-2C operated out of Alaska in 1933.

load. Cabin ventilation was provided and cabin heat was available by installation of tail-pipe and heater "muff." The long-legged landing gear of 96 in. tread used 32x6 wheels with Bendix brakes and these were encased in long teardrop metal fairings; 36x8 wheels without fairings were optional. A source of some annoyance on the "Seventeen" was the fixed tail-wheel which hampered ground maneuvering; most turns required a crew to swing the tail around. The pilot could "blast the tail up" and then kick rudder for the turn but this required particular finesse. Navigation lights were provided and retractable landing lights were built into under-side of the wing. A metal propeller, electric inertia-type engine starter, battery and generator were standard equip-

ment. Special interior arrangements and exterior color combinations were optional. The next Consolidated development was the "Fleetster" model 17-AF described in the chapter for ATC # 486.

Listed below is the only known Model 17, Type 2-C entry as gleaned from registration records:
NC-750V; Model 17-2C (# 6) Wright R-1820-E. This approval may also be for serial # 5 (registration number unknown); serial # 6 first on Group 2 approval numbered 2-273; just to account for balance of early "Fleetster" series, we mention that serial # 7 and# 8 were on Group 2 approval numbered 2-331; Model 17-AF as on ATC # 486 was a different series.

A.T.C. #370
(9-29-30)
IRELAND "PRIVATEER", P-2

Fig. 218. Ireland "Privateer" model P-2 with 110 h.p. Warner "Scarab" engine; shown here "on step."

Realizing that a large luxurious amphibian airplane such as the "Neptune" with its correspondingly high price-tag, was somewhat out of line now during these "depressed times" of 1930-31, Ireland Aircraft developed the small "Privateer" which could be owned and operated for just a little more than the average deluxe speed-boat. Of a different arrangement because of its vastly different purpose, the new "Privateer" was a trim monoplane of small proportion with the ability and agility to scamper in and out of tight places such as were likely to be found on the smaller lakes and harbors. Because of its comparatively low first-cost and reasonably low upkeep and maintenance costs, the "Privateer" was leveled at a market that was just beginning to be seriously explored. Being an amphibious airplane that could pick and choose its landing places, whether on some grassy strip or a small body of water, the trim "Privateer" could lay claim to a utility that would be equally suitable for the flying-services, for business or for sport; however its nominal seating, rugged construction and playful nature were attributes that made it a natural companion when flying just for the fun of it. Though the market was still quite limited for the sales of a strictly-for-the-fun-of-it airplane such as the "Privateer", there were still a good amount of people that could afford it even at this particular time.

Introduced early in 1930 and heralded as the lowest-priced amphibian airplane in the world ($5550) the compact "Privateer" was first developed and flown with the 4 cyl. Wright "Gipsy" engine of 90 h.p., flown in a prototype (Model P-1) that was a good deal lighter than subsequent production models. Light of frame, agile as a trim skiff and just right for two people, the early "Privateer" was a barrel of fun but its overall utility was somewhat shy of what might be expected from the clientele that would be interested in owning and flying a ship of this type. The top speed was about 80 m.p.h., the cruising speed was about 65, it landed at 40, its initial climb-out was near 700 ft. per min. and the cruising range could be stretched to some 290 miles. In order to provide a little more practical utility, the useful load was increased, wheel brakes, dual controls, a metal propeller, engine starter and other such called-for necessities were offered as optional equipment; as a consequence, the gross weight soared to an increase of nearly 400 lbs. and more horsepower was now in order. In a redesign altering several aspects of the original configuration, the Warner "Scarab" engine became the standard installation for the improved model P-2.

The Ireland "Privateer" model P-2 was a compact wire-braced monoplane of the flying boat type with sporty accommodations for two

Fig. 219. Pilot hauls anchor prior to flight in "Privateer" P-2.

in an open cockpit placed just ahead of the shoulder-mounted wing. Perched atop a tripod mount, in a rather precarious position over the cockpit, was a 7 cyl. Warner "Scarab" engine of 110 h.p.; mounted backwards as a "pusher" the engine's whirling propeller was only scant feet from the occupants who learned to be mindful of its presence. Seated in a forward position that offered panoramic vision, occupants were treated to the best view and were in a position best suited for mooring and docking; maneuverability in the water was excellent even in close quarters and required only the minimum in seamanship. Though not blessed with a thundering reserve of power, the P-2 was of sprightly nature, it was comfortably stable and operated from land or water with ease. Flight characteristics and general behavior were quite pleasant, its frame was able to absorb considerable punishment, and in view of the fact that the "Privateer" had such a tremendous potential as the ideal airplane for sport, it behooves one to mentally wager that it certainly would have been seen in far greater number had economic conditions been more favorable. The type certificate number for the Ireland "Privateer" model P-2 was issued 9-29-30 and some 10 or more examples of this model were built by Ireland Aircraft, Inc. at Garden City, Long Island, N.Y. In the refinancing and reorganization of Ireland Aircraft during the latter part of 1930, the firm name had been changed to Amphibions, Inc. Bertram Work was president; D. J. Brimm, Jr. was V.P. and chief engineer, later succeeded by T. P. Leaman and P. H. Spencer was sales manager. Some of the "Privateers" were owned and flown by noted sportsmen and one was operated by the Aeronautics Branch of Dept. of Commerce.

Listed below are specifications and performance data for the Ireland (Amphibions) "Privateer" model P-2 as powered with the 110 h.p. Warner engine; length overall 28'0"; height overall 8'4"; wing span 38'0"; wing chord 72"; total wing area 198 sq.ft.; airfoil Curtiss C-72; wt. empty 1403 [1450] lbs.; useful load 562 [650] lbs.; payload with 30 gal. fuel was 195 [283] lbs.; gross wt. 1965 [2100] lbs.; figures in brackets are for amended allowances; max. speed 95; cruising speed 78; landing speed 42 [50]; climb 650 [550] ft. first min. at sea level; climb in 10 min. 5800 [4500] ft.; climb to 10,000 ft. was 18 [28] min.; ceiling 10,500 [10,000] ft.; figures in brackets are for 2100 lb. gross wt.; gas cap. 30 gal.; oil cap. 2.75 gal.; cruising range at 7 gal. per hour was 300 miles; price at factory was $5800.

The hull framework was built up of spruce and ash longerons, spruce and ash bulkheads heavily gusseted with plywood into a semi-monocoque arrangement that was covered with "Alclad" metal sheets screwed to the frame. The short hull had a single step with a flat bottom. An open cockpit was located just ahead of the wing's leading edge with ample room for two; life-preserver seat cushions were fitted and dual controls were available. The wing framework was built up of solid spruce spar beams with wing ribs stamped out of Allegheny metal; the leading edges were covered with Alclad sheet and the completed framework was covered in fabric. The shoulder-mounted wing had 5 deg. of dihedral and 3.25 deg. of sweepback in an aerodynamic arrangement that promoted excellent stability. Metal covered floats were fitted to a point near the wing tips to keep wing from digging into the water. The Warner engine was mounted atop a steel tube frame high above the

Fig. 220. Although basically a "boat", "Privateer" could operate off land also.

cockpit; fuel and oil tanks were mounted in the streamlined nacelle. Mounted backwards as a "pusher", the engine was shrouded in a Townend speed-ring cowling. Wing bracing consisted of streamlined steel wire in heavy section; attach points were from the wing to a point on the engine nacelle and from the wing to a point on lower part of hull. The main axle tube of the semi-cantilever landing gear passed through the hull, with wheels attached to short stub legs on its outer end; wheels were raised or lowered by a lever in the cockpit that rotated the axle which then locked in either the up or down position. Low-pressure "airwheels" absorbed all the shocks and were suitable for soft ground or sandy beaches. The fixed surfaces of

Fig. 221. "Privateer" was flying fun for two.

the tail group were built up of spruce spars, with a combination of Allegheny metal formers and Alclad ribs, covered in fabric; the movable surfaces were built up of welded steel tubing and also covered in fabric. The tail-group was then mounted on a steel tube outrigger boom attached to several points on the hull with a decided upsweep to position it high above the water. The horizontal stabilizer was adjustable on the ground with additional trim possible from the cockpit through two movable airfoiled struts that also acted as braces. A swiveling tail-wheel was attached to the stern-post of the hull and was fitted with a small fin to act as a water-rudder. Normal equipment included a Curtiss-Reed metal propeller, Heywood air-operated engine starter, 10 lb. anchor with 50 ft. of rope, fire extinguisher, first-aid kit and cockpit and engine covers. Wheel brakes were $100 extra, dual controls were $50 extra, running lights were $15 extra and any custom two-color paint combination was $25 extra. All model P-2s were also eligible with 125 h.p. improved Warner engine for a slight increase in overall performance. Serial # 61 was eligible with 1965 gross wt. with allowance for 21 lbs. of baggage that included tools, anchor, rope and bilge pump. Serial # 62 through # 65 eligible with 2003

lbs. gross wt. with allowance for 59 lbs. of baggage. Serial # 66 and up eligible with 2100 lb. gross wt., a metal propeller and allowance for 70 lbs. of baggage that included tools, anchor and rope, 2 parachutes at 20 lb. each and other loose equipment. The next development in the "Privateer" series was the model P-3 as certificated on a Group 2 approval numbered 2-449.

Listed below are Ireland (Amphibions) "Privateer" model P-2 entries as gleaned from registration records:

NC-92K; Model P-2 (# 61) Warner 110.
NC-93K;   "   (# 62)   "
NC-94K;   "   (# 63)   "
NC-95K;   "   (# 64)   "
NC-96K;   "   (# 65)   "
NC-793Y;   "   (# 66)   "
NC-794Y;   "   (# 67)   "
NC-795Y;   "   (# 68)   "
NS-15;   "   (# 69)   "
NC-797Y;   "   (# 70)   "

X-376N was prototype (Model P-1) for "Privateer" series; registration number for serial # 64 unverified; serial # 66 first owned by Lamont DuPont, Jr.; serial # 69 operated by Department of Commerce.

A.T.C. #371
(9-30-30)
STEARMAN "CLOUDBOY," 6-F

Fig. 222. Stearman model 6-F with 165 h.p. Continental A-70 engine, another version in trainer series.

Planned to offer interesting variety in a series of airplanes specifically designed for pilot training, the model 6-F was another development in the "Stearman Six." With a stout frame designed to easily absorb up through 300 h.p. it is interesting to note the number of modifications that finally appeared in this one basic design. It has always been a source of interest, and sometimes even amazement, the changes that can take place in a basic airplane design simply by the change of engines, and perhaps some slight modifications necessary to a particular installation. More likely than not the airplane always emerged as a slightly different personality and in some cases the changes were nearly amazing. The model 6-F under discussion here differs only slightly from the model 6-A discussed previously (ATC # 365) but models in this series released later show increasing differences, in personality at least. The 6-F, as shown, gives off a faint difference but otherwise it was typical except for certain weights which were mainly influenced by engine installation differences. Had we to pick a likely model in this series for use by civilian flying schools the choice would probably be slanted to the model 6-F. The final development of

this series in the commercial "Cloudboy" version went into 5 more engine installations, with power ranging up to 300 h.p., so anticipation of discussing these later models is viewed with interest. Of the four model 6-A that were delivered to the Air Corps as YPT-9, one was converted to the YPT-9A which was the same as the model 6-F, two were converted into YPT-9B with 200 h.p. Lycoming engines (similar to 6-L) and one was converted to the YPT-9C with 170 h.p. Kinner engine (similar to 6-H). Further modifications changed one YPT-9 into a YBT-3 with 300 h.p. Wright engine and another YPT-9 was changed into a YBT-5 with 300 h.p. "Wasp Jr." engine. Those at the Air Corps presumably had an interesting time creating and re-creating these different models and it certainly speaks well for the versatility of the "Stearman 6" design.

The "Stearman" model 6-F was an open cockpit biplane seating two in tandem and was the second in a series specifically designed for pilot training. In a drawing board game of give and take, some performance potential was naturally sacrificed for more desirable features; therefore we find that the model 6-F was not particularly fast and somewhat over-weight, but it did have ample strength and a good solid feel. Other outstanding features were ease of service, ease of maintenance, dependability on the flight-line and a nature that cooperated well with all those concerned. Powered with the 7 cyl. Continental A-70 engine of 165 h.p. the model 6-F was no snorting charger but its performance was ample and well suited to its particular work. Though only 10 examples in the "6-series" were built and some were equally as rare in number as the 6-F, the model 6-F seemed to be the lesser known model of them all. The type certificate number for the "Stearman" model 6-F trainer was issued 9-30-30 and records indicate that no more than 1 or 2 examples of this model were built by the Stearman Aircraft Co. at Wichita, Kan. Lloyd Stearman never did like to "fly a desk," he was much happier in the shop or at the drawing board so it is understandable that in Feb. of 1931 he vacated the presidency to become consulting engineer, a technical advisor and a member of the company board. The new president was then Walter P. Innes, Jr. who certainly knew the ins and outs at Stearman Aircraft and also acted as treasurer; Mac Short continued his fine performance as V.P. and chief engineer; J. E. Schaefer was V.P. in charge of sales.

Listed below are specifications and performance data for the "Stearman" model 6-F as powered with the 165 h.p. Continental A-70 engine; length overall 24'8"; height overall (tail up) 9'6"; wing span upper 32'0"; wing span lower 28'0"; wing chord both 60"; wing area upper 148 sq. ft.' wing area lower 124 sq. ft.; total wing area 272 sq. ft.; airfoil N-22; wt. empty 1727 lbs.; useful load 673 lbs.; crew wt. 342 lbs.; baggage 86 lbs.; gross wt. 2400 lbs.; max. speed 110; cruising speed 90; stall speed 50; landing speed 45; climb 710 ft. first min. at sea level; climb in 10 min. was 5600 ft.; ceiling 12,300 ft.; gas cap. 37 gal.; oil cap. 2.5 gal.; cruise range at 9 gal. per hour was 350 miles; price at factory field was $7945, raised to $9460 later in 1931.

The construction details and general arrangement of the model 6-F was typical to that of the model 6-A as described in the chapter for ATC # 365 in this volume, including the following which is also typical for both models. Two parachutes at 20 lbs. each were part of the 86 lb. baggage allowance; a small baggage compartment was located high in the fuselage just ahead of the front cockpit with access door in top cowling on right hand side. A direct-reading fuel gauge was visible from either cockpit. Metal bucket-type seats had wells for parachute pack and were adjustable for height. The excessive stagger of the lower wing in relation to the upper wing, plus a substantial cut-out in the center-section panel, provided ease of entry to either cockpit and also provided excellent visibility to occupants of either seat. Ailerons on early models were metal-framed and fabric covered but later modifications had metal-framed ailerons covered with dural metal sheet that was corrugated for added stiffness. The unusual aileron actuation was transmitted through push-pull struts that were fastened to a point ahead of the aileron hinge line; somewhat unusual but quite positive. Hand-holds were provided in the lower wing to assist in ground maneuvering. Interplane struts and all interplane bracing was of heavy gauge to withstand stresses of increases in power and acrobatic flying. The rudder was aerodynamically balanced to lighten pedal pressures and both rudder and elevators were actuated by stranded steel cable. The wide tread out-rigger landing gear used "Aerol" (air-oil) shock absorbing struts; wheels were 28x4 or 30x5 and brakes were standard equipment. A metal propeller, hand crank inertia-type engine starter, fire extinguisher and first-aid kit were standard equipment. The next development in the "Stearman 6" trainer was the thundering model 6-D of 300 h.p. described in the chapter for ATC # 402.

Listed below are "Stearman" model 6-F entries as gleaned from registration records: NC-786H; Model 6-F (# 6001) Continental A-70. NC-788H;   "   (# 6003)   "
Serial # 6001 was first as model 6-A, later modified as 6-F; serial # 6003 later modified to models 6-P and 6-L.

LINCOLN "ALL-PURPOSE," AP-B5

Fig. 223. Lincoln "All-Purpose" model AP-B5 with 125 h.p. Kinner B5 engine.

Feeling quite certain that the death-knell for the 3-place open cockpit general purpose biplane, such as they themselves had built for several years, would not be too long in coming, Lincoln Aircraft backed up this belief by designing and developing the AP (All-Purpose) monoplane series. All factors weighed carefully, this proved to be a practical series of aircraft which were definitely constructed and arranged for "all-purpose" work, as usually carried out with the classic biplane lay-out heretofore — with the bonus offer, however, of several advantages in the cabin monoplane that would make it the more logical choice for the average flying-service operator. Profitable payloads, comfortable, roomy interiors and a distinct improvement in performance were among the many features offered by this new line of airplanes; but, in spite of all the new Lincoln AP-series had in its favor they could not at this time attract any great switch to the closed-in monoplane for this type of general usage. Proud of the new AP monoplane, as well they should have been, the hopes of country-wide popularity and good quantity production run for this new model withered away sadly in the creeping paralysis caused by the "general depression." Flying service operators the country over were certainly investigating the obvious merits of such new models as the Lincoln AP monoplane, as a boon to better business; however, the open

cockpit biplane, still around in great numbers, could be bought for about one-half as much. All things considered, this was still a deciding factor in its continued popularity. Like most all aircraft manufacturers, Lincoln was overly-enthusiastic about their new monoplane series and tended to over-emphasize its performance capabilities and upon deducting about 5 percent for factory pride it is apparent that Lincoln still had plenty to brag about.

The Lincoln model AP-B5 was a rather plain-looking high winged cabin monoplane with seating for 3 in a roomy arrangement that placed the pilot up forward and two passengers were seated side-by-side in the rear. Versatile and not choosy as to chore, the AP-B5 was also fitted with dual controls for advanced pilot-training. The more powerful of the 2 models in this series, the model AP-B5 was powered with the 5 cyl. Kinner B5 (R-440) engine of 125 h.p. and a fairly low power-loading translated itself into an exceptionally good all-round performance. Loaded to its gross weight, the AP-B5 could break ground easily in about 450 ft., a spirited climb-out allowed operations out of the smallest fields and landing runs were generally less than 300 ft. Inherent high-altitude characteristics were excellent to the point that a factory pilot nudged the AP-B5 to an unofficial record of 28,000 ft. but normal service ceiling was more like 18,000 ft. Well arranged in long and easy aerodynamic proportions the AP was credited with gentle nature and good behavior, an easy-to-get-along-with airplane that asked no favors. Due to its scarcity it failed to make much of an impression in flying circles or leave behind much lore, but it is reasonable fact that had it a better chance to prove its actual worth the AP monoplane would surely have been seen in far greater number. The type certificate number for the Lincoln "All-Purpose" monoplane model AP-B5 was issued 10-1-30 and records indicate that no more than 2 examples of this model were built by the Lincoln Aircraft Co. at Lincoln, Neb.

Listed below are specifications and performance data for the Lincoln monoplane model AP-B5 as powered with the 125 h.p. Kinner B5 engine; length overall 26'0"; height overall 8'3"; wing span 37'0"; wing chord 72"; total wing area 206 sq. ft.; airfoil USA-35B; wt. empty 1352 lbs.; useful load 828 lbs.; payload with 36 gal. fuel was 412 lbs. (2 passengers at 170 lb. each & 72 lb. baggage); gross wt. 2180 lbs.; max. speed 120; cruising speed 103; landing speed 50; climb 800 ft. first min. at sea level; climb in 10 min. was 7400 ft.; service ceiling 18,000 ft.; gas cap. 36 gal.; oil cap. 4 gal.; cruising range at 8 gal. per hour was 450 miles; price at factory first quoted at $5,995, lowered to $4,495 and then to $4,395 in May of 1931.

The fuselage framework was built up of welded chrome-moly steel tubing, lightly faired to shape and fabric covered. The cabin interior, of ample proportion, was upholstered neatly and seats were arranged for plenty of leg-room; the front and rear seats were adjustable for height. A large cabin door on the right-hand side with two convenient steps provided easy entry;

a large baggage bin with allowance for 72 lbs. was in back of the rear seat and all windows were of "Duolite" shatter-proof glass. Dual controls were optional for pilot training and the rear seat was easily removed to make space for bulky cargo. The robust wing framework, in two halves, was built up of solid spruce spars that were routed to an I-beam section with basswood truss-type wing ribs; the leading edges were covered with dural metal sheet and the completed framework was covered in fabric. Wing braces were parallel struts of streamlined steel tubing built into a rigid truss with the landing gear. The landing gear of fairly wide tread used oil-draulic shock absorbing struts; low pressure semi-airwheels were fitted with Air Products brakes. A novel lever engaged the wheel brakes for engine run-ups and parking. The tail skid was of the steel spring-leaf type, mounted in rubber; a tail wheel was optional. The fabric covered tail-group was built up of welded chrome-moly and 1025 steel tubing; the fin was ground adjustable and the horizontal stabilizer was adjustable in flight. A Fahlin wooden propeller was standard but a metal propeller was optional; a Heywood air operated engine starter was also optional. The lower-powered model AP-K5 was a companion model in this series as described in the chapter for ATC #373 in this volume.

Listed below are the only known entries of the Lincoln model AP-B5 as gleaned from registration records:
NC-424V; Model AP-B5 (#401) Kinner B5.
NC-954N;   "   (#402)   "
Serial #401 first tested and approved as a model AP-K5, soon after modified to model AP-B5.

Fig. 224. Lincoln "All-Purpose" model AP-B5 with 100 h.p. Kinner K5 engine.

Bearing the unmistakable touch of Ensil Chambers, versatile design-engineer for Lincoln Aircraft, the model AP-K5 as shown, was the first in a series of 3-place monoplanes designed specifically to replace the time-honored all-purpose biplane. Familiar with the peculiar requirements necessary for handling the chores of the average flying-service operator, Lincoln studied the cabin monoplane concept in this light. They felt it would be entirely suitable for these varied duties, even in a more practical and profitable manner. Not motivated to create a "beauty" of appealing line or proportion, but only striving to formulate a design that was meant to replace the open biplane for general service, Lincoln craftsmen dealt only in predetermined requirements that were to be adapted to the characteristics of the monoplane, with an eye for doing the job much better. After all design factors were translated to lines on paper and eventually into a finished airplane, the form was more or less straight-forward and quite plain; but a sturdy character and amiable nature held priority above all else so no compromise was even considered. The preliminary tests of this craft had the flavor of a dream come true and the AP-K5 posed right then and there as an answer to the prayers of the flying-service operator, or even the private-owner pilot who hauled family and friends for the fun of it. Fitting all the basic qualifications for a practical and profitable general-purpose airplane to a tee, and in this case with extra features not usually found in an airplane of this type, one would guess that

the AP-series should have taken the country by storm in a great back-log of orders. Sad to say, this was not the case. The flying-folk who really needed this type of airplane, the one-man flying service operator and the private-owner pilot who generally flew just for the fun of it, just had not the way or the means to invest $4000. in a new airplane. With the market fairly glutted with last-year models at slashed prices and used bargains at a give-away, the new-production airplane had very slim chances. Fate and the circumstances of the times had a habit of dealing harshly with the dreams of many good men, and thus this particular period of aviation history was being shaped.

The Lincoln model AP-K5 was a straight-sided and very honest-looking high wing monoplane with seating arranged for 3 in the confines of roomy cabin comfort. Especially designed to handle the varied chores of all-purpose service, the AP monoplane was in many ways the ideal for charter trips off the beaten path, for pilot training, light cargo hauling, or as just a sportplane for the owner-pilot. Inherently reliable because of its robust frame and inherently stable because of good aerodynamic arrangement, the AP-K5 was almost fool-proof to operate. The more economical of the 2 models in this series, the AP-K5 was powered with the 5 cyl. Kinner K5 engine of 100 h.p. and rendered an all-round performance well above the norm for this type of airplane. Loaded to its gross weight, the AP-K5 broke ground easily in about 650 ft., climbed out at a rather generous rate and a landing run usually amounted to around 250 ft. Seated in the roomy comfort of this gentle airplane, the owner-operator could easily relax even while on a busy schedule and certainly didn't need to act or look the part of the "intrepid aviator". Of the several monoplanes designed to replace the open biplane for general-service, the Lincoln AP (All-Purpose) was perhaps the best approach thus far. The type certificate number for the Lincoln model AP-K5 was issued 10-1-30 and according to record, only one example of this model was built by the Lincoln Aircraft Co., Inc. at Lincoln, Neb. Victor Roos was president and general manager and Ensil Chambers was chief of design and engineering. Feeling the pinch from a steady loss of sales through 1930, plans had been made to form a merger between the Lincoln Company and the American Eagle Aircraft Corp. of Kansas City, a move calculated to re-finance and strengthen the structure of both companies. In the re-shuffle of company executives, Victor Roos became president of the new merger and E. E. Porterfield, Jr. the president of American Eagle, became sales manager. The Lincoln facilities, more modest of the two, were moved to the American Eagle plant where a continued production was planned for the combined line of models manufactured under 15 different approved type certificates (ATC); most of the emphasis was placed on the building of the little "Eaglet", the Lincoln PT training biplane and the Lincoln "All-Purpose" monoplane. Somewhere along in 1931, Lincoln developed an ultra-light "pusher" monoplane called the "Playboy" with seating for 2 and powered with a modified version of the Wright-Morehouse "twin" engine called the Lincoln "Rocket". Outside of the development of this one prototype, very little else is known of this project or why it was dropped.

Listed below are specifications and performance data for the Lincoln model AP-K5 as powered with the 100 h.p. Kinner K5 engine; length overall 26'0"; height overall 8'3"; wing span 37'0"; wing chord 72"; total wing area 206.4 sq.ft.; airfoil USA-35B; wt. empty 1320 lbs.; useful load 828 lbs.; payload with 36 gal. fuel was 412 lbs. (2 pass. at 170 lb. each & 72 lb. baggage); gross wt. 2148 lbs.; max. speed 110; cruising speed 93; landing speed 48; climb 580 ft. first min. at sea level; climb in 10 min. was 4500 ft.; ceiling 12,000 ft.; gas cap. 36 gal.; oil cap. 4 gal; cruising range at 7 gal. per hour was 450 miles; price at factory first quoted at $4395, reduced to $3995 in May of 1931.

The construction details and general arrangement of the Lincoln AP-K5 was typical to that of the AP-B5 as described in the chapter for ATC # 372 of this volume, including the following. Cabin seats were of the bucket type with wells for parachute pack and adjustable for height. Center panels of cabin windows were split to slide open for rapid ventilation; an exhaust-muff cabin heater was optional. The horizontal stabilizer was adjusted by a lever and screw mechanism to obtain a "finer trim" for varied loads. The all-weather engine nose-cowling was provided with adjustable louvers that could be operated by pilot to regulate engine temperature. Gravity-feed fuel tanks were mounted in the root end of each wing half and fuel gauges were provided for each. The fuselage and wing were provided with numerous inspection plates for inspection and maintenance of control mechanism and other vital gear. A Fahlin wooden propeller and wiring for navigation lights were standard equipment. A metal propeller and Heywood engine starter were optional.

Listed below is the only known example of the Lincoln model AP-K5 as gleaned from registration records:
X-42N; Model AP-K5 (# 901) Kinner K5.
X-424V was originally tested as model AP-K5 but soon after modified to model AP-B5 with 125 h.p. Kinner B5 (R-440) engine.

Fig. 225. Fleet model 7 with 125 h.p. Kinner B5 engine; added power and revised tail-group were main improvements.

By this particular time in its development, the lean and brawny "Fleet" biplane had well established itself as one of the most practical pilot-trainers of this period; practical in the sense that it was specifically arranged for flight training, and was therefore easy to get in and out of, even with strapped-on parachute. The visibility was excellent from any angle and the rugged frame was able to withstand more than the average abuse, whether that be on the ground or cavorting in the air. Reliable and economical, the "Fleet" was ever popular with operators of flying schools because it provided them a craft that easily convinced the would-be pilot he would be flying a ship with exceptionally stout heart, and its constant readiness on the flight-line afforded the operator a reasonable profit. Several hundred of the "Fleet" had already been built by this time and most all were in service with flying schools that were scattered through all parts of the country; literally thousands of pilots had learned to fly in the "Fleet." An occasional "Fleet" was used strictly for sport-flying and those pilots who indulged in acro-batic type flying reveled in its maneuverability, and enjoyed that comforting feeling of knowing that it would really take some doing to break this ship apart. The type of flying that was possible with the "Fleet" was only limited by

the pilot's capabilities and his fortitude. As an up-graded development in this popular series, Fleet Aircraft now introduced the Model 7 which was still fairly typical to the earlier mod-els 1 and 2 but contained several improvements in a new trainer version; a utility model, a sea-plane version and a version more slanted to the tastes and requirements of the sportsman-pilot who flew just for fun.

As shown here in various views, the "Fleet" model 7 was an open cockpit biplane seating two in tandem and with a form of lean and brawny lines that made hardly any effort to hide the muscle that lie underneath. Bearing an in-crease in power, the "Seven" was arranged for more utility and consequently was available in a variety of models. A landplane was avail-able as a trainer, or with belly-tank holding extra fuel to extend the cruising range, or with a coupe-type canopy for those that didn't like the breezy open cockpits. The "Seven" was al-so available as a two-place seaplane on twin-float gear or as a one-place seaplane with cruis-ing range greatly extended for special services like harbor or forest patrol, or "spotting" schools of fish out in open waters. In any version the "Fleet 7" served well but it was naturally better noted as a pilot-trainer. The "Seven" was not particularly docile but even though somewhat

Fig. 226. Fleet 7 on Edo floats, flying patrol for Washington Dept. of Fisheries.

sassy and eager it did cooperate nicely and soon gave the fledgling pilot that grand feeling he was getting along just fine and would soon become a hot-rod pilot. On occasion a "Fleet" would get wound-up in a "spin" that flattened out as it went and sometimes recovery was either a hairy incident or sometimes no recovery was possible at all; this of course caused considerable concern. A study revealed that proportion and areas of the tail-group caused inadequate control for prompt recovery; more fin and rudder area in a slightly different proportion allowed a more normal recovery. Some of the "Fleet" looked outlandish with their new tail-feathers but that's what it took to do the job so no one seemed to mind. The "Fleet" was never noted for its speed but it cruised along happily at about 90 m.p.h. and was always completely ready to do as bidden. Flight characteristics were quite pleasant, maneuverability was exceptional, and its general behavior kept it popular the country over. Its rugged structure and its continued popularity naturally promoted longevity so it is no wonder that nearly 275 of the "Fleet" biplanes were still flying actively some 10 years later. In recent years it has become very popular for restoration by those that like to fly for the fun of it.

Due to the fact that "Fleet" designation, as generally known, had jumped from the Model 2 to the Model 7, it might be well to explain a little of what went on in-between. The grand-daddy of the "Fleet" biplane was of course the Consolidated "Model 14" also known as the "Husky Junior" (see ATC # 84 in U. S. Civil Aircraft, Vol. 1) and then came the "Fleet 1" which was quite similar in early examples (see ATC # 122 in U.S. Civil Aircraft, Vol. 2). A

companion model powered with the Kinner K5 engine was introduced as the "Fleet 2" and it was by far the most plentiful of the series (see ATC # 131 in U. S. Civil Aircraft, Vol. 2). The "Fleet 3" was powered with the 5 cyl. Wright J6 engine of 150-165 h.p. and the "Fleet 4" was powered with the 6 cyl. Curtiss "Challenger" engine of 170 h.p., both of these were experimental versions about which very little is known. The "Fleet 5" was powered with the 6 cyl. Brownback (Light) "Tiger" of 90 h.p., a model which was slated for production but cancelled due to a sag in the market. The "Fleet 6" was apparently a Model 2 that was rigged up with a pick-up apparatus in its top wing to test in-flight hook up to a dirigible. The "Fleet 7", as approved on this certificate (ATC # 374), was powered with the 5 cyl. Kinner B5 engine of 125 h.p. and was available as the models 7, 7-C and 7 Deluxe, with an amendment to this certificate some time later that also covered approval of the Model 10. The Model 7 was the basis for the XPT-6 that was developed as an Air Corps primary trainer. The type certificate number for the "Fleet" model 7 was issued 10-4-30 and some 30 or more examples of this model were manufactured into 1932 by the Fleet Aircraft, Inc. at Buffalo, New York. Lawrence D. Bell was president; Ray P. Whitman was V.P. and Joseph Marr Gwinn, Jr. was V.P. in charge of engineering. Reuben H. Fleet, president of the Consolidated Aircraft Corp. was reported to be on the west coast at this time looking for a suitable site to build a branch factory; Fleet must have been impressed with the possibilities of aircraft manufacture on the coast because the whole operation was finally moved to San Diego, Calif. just a few years later.

Fig. 227. Shown here os 7-C, Fleet was also available with coupe-type canopy.

Listed below in separate paragraphs are specifications and performance data for the "Fleet" models 7, 7-C, 7 Deluxe and model 10 as powered with the 125 h.p. Kinner B5 engine. For "Fleet 7" as two-place landplane; length overall 21'6"; height overall 8'0"; wing span upper & lower 28'0"; wing chord both 45"; wing area upper 100 sq. ft.; wing area lower 95 sq. ft.; total wing area 195 sq. ft.; airfoil "Clark Y-15"; wt. empty 1146 lbs.; useful load 594 lbs.; payload with 23.5 gal. fuel was 260 lbs.; crew wt. (2) 340 lbs.; baggage 50 lbs. including 2 parachutes at 20 lbs. each; gross wt. 1740 lbs.; max. speed 115; cruising speed 95; landing speed 47; climb 800 ft. first min. at sea level; ceiling 14,000 ft.; gas cap. 23.5 gal.; oil cap. 3 gal.; cruising range at 7 gal. per hour was 300 miles; price at factory was $4485.

For Model 7 as a two-place seaplane with Edo I-1835 floats; wt. empty 1279 lbs.; useful load 594 lbs.; payload with 23.5 gal. fuel was 260 lbs.; crew wt. 340 lbs.; baggage 50 lbs.; gross wt. 1873 lbs.; max. speed 105; cruising speed 88; landing speed 52; climb 710 ft.; ceiling 12,850 ft.; gas cap. 23.5 gal.; oil cap. 3 gal.; range at 7 gal. per hour was 265 miles; price approx. $5500. For the Model 7 as a one-place seaplane on Edo I-1835 floats; wt. empty 1319 lbs.; useful load 632 lbs.; payload with 48.5 gal. fuel was 148 lbs.; which could include 128 lbs. baggage and 20 lb. parachute; crew wt. (1) 170 lbs.; gross wt. 1951 lbs.; max. speed 105; cruising speed 88; landing speed 55; climb 680 ft.; ceiling 12,500 ft.; gas cap. 48.5 gal.; oil cap. 3 gal.; range was 600 miles; price at factory approx. $5500.

For the "Model 7 Deluxe" as a two-place landplane with 48.5 gal. fuel; empty wt. 1187 lbs.; useful load 744 lbs.; payload with 48.5 gal. fuel was 260 lbs.; crew wt. 340 lbs.; baggage 50 lbs.; 2 parachutes 40 lbs.; gross wt. 1931 lbs.; max. speed 112; cruising speed 92; landing speed 55; climb 690 ft. first min. at sea level; ceiling 12,500 ft.; gas cap. 48.5 gal.; oil cap. 3 gal.; range was 600 miles; price at factory $5185. For the Model 7-C as a two-place landplane with coupe-top canopy & 48.5 gal. fuel; wt. empty 1216 lbs.; useful load 715 lbs.; payload with 48.5 gal. fuel 230 lbs.; crew wt. (2) 340 lbs.; baggage 60 lbs. which included 2 parachutes at 20 lb. each; gross wt. 1931 lbs.; max. speed 115; cruising speed 98; landing speed 55; climb 690 ft.; ceiling 12,500 ft.; gas cap. 48.5 gal.; oil cap. 3 gal.; range 620 miles; price at factory $5300, lowered to $5185.

For the Model 10 as a two-place landplane with 48.5 gal. fuel; wt. empty 1221 lbs.; useful load 710 lbs.; payload with 48.5 gal. fuel 225 lbs.; crew wt. (2) 340 lbs.; baggage 55 lbs.; gross wt. 1931 lbs.; max. speed 115; cruising speed 95; landing speed 55; climb 690 ft.; ceiling 12,500 ft.; gas cap. 48.5 gal.; oil cap. 3 gal.; range 620 miles. For the Model 10 as a two-place seaplane on Edo I-1835 floats with 24 gal. fuel; wt. empty 1354 lbs.; useful load 597 lbs.; payload with 24 gal. fuel 259 lbs.; crew wt. (2) 340 lbs.; baggage 55 lbs.; 2 parachutes 40 lbs.; gross wt. 1951 lbs.; max. speed 110; cruising speed 90; landing speed 55; climb 680 ft.; ceiling 12,500 ft.; gas cap. 24 gal.; oil cap. 3 gal.; cruising range 300 miles. Dimensions were typical for all models.

The construction details and general arrangement for the "Fleet" model 7 series and the Model 10 was typical to that of the models 1 and 2 as described in the chapters for ATC # 122 and # 131 of U.S. Civil Aircraft, Vol. 2, including the following. The "Fleet 7" was nearly identical to previous models except for some change in cockpit shape, with larger plastic

Fig. 228. Formerly a "Fleet" 2, this craft was modified into model 7 and mounted on Edo floats.

windshields to offer better protection and a revised tail group for better control at lower speeds. The 125 h.p. Kinner B5 engine, as mounted in the "Seven" and the "Ten", usually was fitted with a nose-type collector ring so there was a slight change in the engine cowling; baggage compartments were located one ahead of the front cockpit and one behind the rear cockpit for a total allowance of 50 lbs. The landing gear was typical but wheels and tires were changed to 26x5. Fittings for the float-gear were part of the fuselage structure and the model 7 was also available with an externally mounted

belly-tank installation for 25 gal. extra capacity to extend the range. The 7 Deluxe was also fitted with a belly-tank for a total fuel capacity of 48.5 gal.; the belly-tank installation weighed 40 lbs. The Model C was fitted with a coupe-type canopy that weighed 30 lb. and required additional fin area; a belly-tank was optional. The Model 10 was introduced in 1932-33 and several changes were incorporated into the otherwise basic configuration. The cockpit cowling profile was raised for better protection and the turtle-back was also raised to fair out the fuselage in a somewhat larger cross-section; the tail-group was also re-

Fig. 229. Fleet 10 had high cockpit cowl and revised tail-group; engine was 125 h.p. Kinner B5.

Fig. 230. Fleet model 10 on Edo floats.

vised to a more pleasant proportion. The landing gear on the "Ten" was typical except for elimination of the cross-axle arrangement; axles from the lower vee were linked in the center to a pyramidal structure, 6.50x10 low pressure semi-airwheels were standard equipment and a 25 gal. belly-tank was also available. As a seaplane, the "Ten" required additional rudder area. Standard equipment for all models included a wooden propeller, spring-leaf tail skid and one set of instruments in the front cockpit. A metal propeller, wheel brakes, 6.50x10 semi-airwheels, a coupe-top canopy, dual controls, dual set of instruments, head-rest for rear cockpit, tail wheel, belly-tank, seaplane float gear, navigation lights, metal panel cover for front cockpit, Heywood engine starter, speed-ring engine cowling and wheel streamlines were all optional as extra equipment. The next development in the "Fleet" biplane was the models 8 and 9 as described in the chapter for ATC # 428.

Listed below are "Fleet" model 7 and 10 entries as gleaned from registration records:

NC-446K; Model 7 (# 169) Kinner B5.

NC-682M;	"	(# 230)	"
NC-649M;	"	(# 233)	"
NC-684M;	"	(# 234)	"
NC-716V;	"	(# 311)	"
X-756V;	"	(# 337)	"
X-757V;	"	(# 338)	"
NC-762V;	"	(# 343)	"
NC-763V;	"	(# 344)	"
NS-700Y;	"	(# 345)	"
NC-765V;	"	(# 346)	"
NC-769V;	"	(# 350)	"
NC-780V;	"	(# 361)	"
NC-782V;	"	(# 363)	"
NC-784V;	"	(# 365)	"
NC-786V;	"	(# 367)	"
NC-787V;	"	(# 368)	"
NC-788V;	"	(# 369)	"
NC-789V;	"	(# 370)	"
NC-790V;	"	(# 372)	"
NC-794V;	"	(# 375)	"
NC-776V;	"	(# 380)	"
NC-62V;	"	(# 401)	"
NC-86V;	"	(# 404)	"
NC-87V;	"	(# 405)	"
NC-88V;	"	(# 406)	"
NC-89V;	"	(# 408)	"
NC-90V;	"	(# 409)	"
NC-702V;	Model 10 (# 262)		"
NC-751V;	"	(# 411)	"

Serial # 169-230-234 first as Model 2; ser. # 233 first as Model 1; ser. # 338 was 7 Deluxe, later as Model 10-G; ser. # 344 first as Model 2; ser. # 346 first as Model 1; ser. # 350 as 7 Deluxe; ser. # 363 as seaplane; ser. # 367 as 7 Deluxe; ser. # 345 with N.Y.C. Police Dept.; ser. # 380 as 7 Deluxe; ser. # 401 as seaplane with Edo Corp.; ser. # 262 first as Model 2, later converted to Model 10.

This certificate for serial # 169-230-233-234-311-337-338-343-344-345-346 and up. Aircraft eligible for Model 10 were ser. # 262, 411 and up.

A.T.C. #375
(10-8-30)
SIKORSKY "SPORT AMPHIBION"
S-39-B

Fig. 231. Improved Sikorsky model S-39-B with 300 h.p. "Wasp Jr." engine.

Introduced some 3 or 4 months previous as the 4-place model S-39-A, the Sikorsky "baby amphibian" was now being offered in an improved version called the S-39-B. Evolving from the experience gained in the testing of two different prototype airplanes (refer back to ATC #340), the S-39-A emerged as quite a success in general; however with the first flush of excitement in readying the new model for production, a few annoying short-comings did crop up here and there. As a result, the full capability of the design was misjudged, misjudged to the extent that its total potential was not put to good use. Not long in finding these things out, the third ship off the line was ear-marked for modification into the new model S-39-B, an improved version that was to embody all the necessary chances to obtain better results. A substantial increase in the gross loaded weight now allowed the seating for 5 instead of 4, with also a handsome increase in the baggage allowance. By aerodynamic refinement and some rearrangement, the same amount of power (300 h.p.) was translated into more efficient flight, and despite the increases in gross loading there was some performance improvement. Basically a little-sister version of the popular S-38 design, the S-39-B was called upon to perform all manner of chores, some of which were an unrelentless test of an airplane's

true character. As a truly versatile airplane, well able to operate with equal proficiency from land or the water, the S-39 was to prove its reliability conclusively in more than 60,000 miles of exploratory flight in the dark continent of wild Africa with Martin Johnson's expedition. In some of the most difficult terrain the world has to offer, the fully loaded amphibious craft often operated at 6,500 ft. or more above sea-level from in-land waters, and sometimes operating at about 7500 ft. above sea-level from high mountain landing strips. Even in 1935, although now a veteran of years of service but not yet outmoded by normal progress, an S-39 in stock trim, climbed to an altitude of 18,641 ft. for a class C3 record. An international speed record for the same class was also set over a 1000 kilometer course at 99.95 m.p.h. Primarily designed to handle the varied chores of all-purpose service, the S-39 in all its variants worked nearly unheralded as an air-taxi for men of business, shuttled back and forth in air-ferry service, complied to the whims of sportsman-pilots, and served in a short tour of duty at the West Point academy as the YIC-28. Some were even marshalled to work extra hard at emergency rescue during the troubled years of World War 2. Built in rather small number and under the prophecy of being "a very good airplane at a very bad time," the S-39 sport amphibian cer-

Fig. 232. S-39-B now seated 5, other improvements increased the performance.

tainly piled up a meritorious service record in its every-day life.

The Sikorsky "Sport Amphibian" model S-39-B was basically a "parasol" monoplane of the flying-boat type with its tractor-mounted engine high in the leading edge of the elevated wing. The slightly modified all-metal hull was a seaworthy structure to which was added a wheeled landing gear, a gear that was raised clear for operating on water or lowered into position for operating on land. Equally at home on land or water, the "amphibian" was never too far from a landing place and it selected either by extending or retracting its wheels; the wheels were pumped up or down hydraulically by a hand-pump in about 10 seconds. The short boat-type hull had excellent water character-

istics with a shape that allowed no water spray into the propeller or over the windshield. The landing gear wheels could be lowered singly or together while in the water to use for braking, maneuvering or beaching. In the water, the large steerable tail wheel (18"x3") acted as a water rudder. Water maneuvering in a high wind was pretty much of a problem, but dropping a wheel on either side for a focal point to turn on, helped considerably. Generally, the visibility out of the S-39 was very good but restricted somewhat to forward because of the rather long bow. Under certain operating conditions it was possible and even advisable to remove the landing gear and tail wheel to make up a true "flying boat" capable of carrying some 200 lbs. extra in useful load. The "Amphibion's"

Fig. 233. S-39-B sport "Amphibion" used for business and pleasure; many served for ten years or more.

behavior on land was admirable despite its somewhat unwieldy appearance; effective wheel brakes and a large steerable tail wheel made taxiing and other ground maneuvering a fairly easy chore. Comfortable and exceptionally roomy, the cabin now had seating for 5 with a wide full-length roof hatch for easy loading and unloading. Powered also with the 9 cyl. P&W "Wasp Jr." engine of 300 h.p., performance of the S-39-B was in some cases improved and more than adequate for a craft of this type; whether on land or water, the S-39 landed gently, take-offs were short and clean and the climb-out left good margin for operating out of smaller fields. Airborne, the S-39-B responded well, was stable in all directions and a very pleasant airplane to fly. The type certificate number for the 5-place model S-39-B was issued 10-8-30 and some 9 examples of this model were built, although most all of the earlier S-39-A were later modified to the new specifications. By this time in the latter part of 1930, the keel for the big 17-ton 4-motored model S-40 was being laid; ordered by Pan Am, it was the biggest craft Sikorsky had ever built until now and was the original of the famous Pan American "Clipper Ships."

Listed below are specifications and performance data for the Sikorsky model S-39-B as powered with the 300 h.p. "Wasp Jr." A (R-985) engine; length overall 32'2"; height on wheels 11'8"; wing span 52'0"; wing chord 85"; total wing area 350 sq. ft.; airfoil Sikorsky GS-1; wt. empty 2678 lbs.; useful load 1322 lbs.; payload with 65 gal. fuel 717 lbs.; (4 passengers at 165 lbs. each and 57 lbs. baggage); gross wt. 4000 lbs.; max. speed 119; cruising speed 100; landing speed 54; climb 750 ft. first min. at sea level; climb in 10 min. was 6000 ft.; ceiling 18,000 ft.; gas cap. normal 65 gal.; oil cap. 6 gal.; cruising range at 15 gal. per hour was 400 miles; price at factory was $20,000 as of May 1931.

The construction details and general arrangement of the model S-39-B was typical to that of the S-39-A as described in the chapter for ATC # 340 of this volume, including the following. The spacious cabin interior was now arranged for the seating of five; the two forward seats were the individual type and the rear seat was a wide bench type for seating 3 across. A baggage allowance of 157 lbs. included anchor and rope at 22 lbs., a tool kit at 15 lbs. and 5 life-preserver jackets at 9 lbs.; a "speed-ring" engine cowling accounted for another 27 lbs. of the baggage allowance, leaving some 84 lbs. for the actual baggage load. Normal fuel capacity was 65 gal. in 2 tanks of 32.5 gal. each; 2 extra tanks of 12.5 gal. each were also eligible to increase fuel capacity to 90 gal. All fuel tanks were mounted in the center-section panel of the wing. Serial # 902 was eligible with 105 gal. max. fuel cap. The outer wing panels were detachable from the center-section portion for storage; with floats still attached, the total width was then only 18 ft. & 2 in. The single rudder on the S-39-A proved inadequate for proper maneuvering in water or in the air so on the S-39-B it was enlarged in area with a sub-rudder extension, increasing the overall length by some 3 inches. Because of its overhead placement, the engine was rather difficult to work on, so factory mechanics devised a small scaffold that fitted across the hull to reach the engine in a comfortable attitude, and to prevent tools from falling through the windshield below. Despite its spidery arrangement the S-39 was quite rugged in frame and character and most examples served usefully for 15 to 20 years. Two other versions of the S-39 were the model S-39-C as built under Group 2 approval numbered 2-391 and the model S-39-CS as built under Group 2 approval numbered 2-436. The next Sikorsky development was the twin-motored 16 passenger model S-41-A described in the chapter for ATC # 418.

Listed below are Sikorsky model S-39-B entries as gleaned from registration records:

NC-888W; Model S-39-B (# 902) Wasp Jr. A
NC-896W; " (# 906) "
NC-50V; " (# 912) "
NC-51V; " (# 913) "
NC-53V; " (# 915) "
NC-54V; " (# 916) "
NC-55V; " (# 917) "
AC32-411; " (# 919) "
NC-58V; " (# 920) "

Serial # 900X-901-903-904-905-907-908-909-910 first as model S-39-A, all were modified as S-39-B; ser. # 911-918 modified to S-39-C; ser. # 914 modified to model S-39-CS; ser. # 919 as Army Air Corps YIC-28 at Wright Field & West Point Academy; ser. # 920 later modified to S-39-C Special.

A.T.C. #376
(10-17-30)
FAIRCHILD, MODEL KR-34-D

Fig. 234. Fairchild model KR-34-D with 165 h.p. Comet 7-E engine.

The Kreider-Reisner KR-34 series biplane had long been a favorite for all-purpose work and had by this time been introduced in several different versions. This model in discussion here was the last of the type, that is, discounting a few "experimentals" that were later used mostly as engine test-beds and for various other tests. More or less an improvement over the earlier model C-4-A, which was actually the first of the KR-34 series (refer to ATC # 88 of U.S. Civil Aircraft, Vol. 1), the Fairchild (Kreider-Reisner) model KR-34-D was still basically typical; however it harbored several modifications in the total airframe, both inside and out. Powered also with a 7 cyl. "Comet" engine but now with the much-improved model 7-E of 165 h.p., the KR-34-D had advantage of the extra muscle in a 35 h.p. increase and utilized it mainly to gain a better all-round performance. Upgraded in general to keep pace with ever-changing demands, the new D-version also offered several pilot aids, and some conveniences that were lacking, in the earlier C-4-A (KR-34-B). Considerably heavier when empty (184 lbs.) than the early version, the KR-34-D took the penalty in its useful load in order to offer that worthwhile increase in performance. Strangely enough, this compromise was not necessarily a bad move because the average owner or operator had actually come to appreciate better performance, more than the ability to carry a larger paying load. In all honesty, it seems the only justification for the KR-34-D as a better airplane than

its earlier sister-ship, was hardly more than the increase in h.p. for a better performance, and a few added doo-dads to make its operation a bit easier. Introduced at a time when the market for this type of airplane was at its lowest ebb, this model found very few customers and was built only in small number. As a consequence, Fairchild found themselves with several new "Comet" engines to dispose of; these were advertised for quick sale in 1931 at rock-bottom prices.

The Fairchild model KR-34-D was an open cockpit biplane with seating for 3 and was primarily offered as an all-purpose airplane that could handle the many chores of the average flying-service operator or as an air-taxi for the businessman who did his own flying. Typical to other models in the KR-34 series, the D-version had several slight improvements in its make-up. Powered with the 7 cyl. "Comet" 7-E engine of 165 h.p., its performance was comparable to that of the Wright "Five" powered KR-34-C (ATC # 162), and it was offered at a much better price. In view of the progressive advances made in this type of craft in the few preceeding years, the KR-34-D was on the brink of being old-fashioned with its high-lift aerodynamic arrangement, surely a detriment to its ability to develop more speed; however when it came to operating out of miserably small and rough fields with anything like a decent payload aboard, the KR-34-D and its like, were very hard to beat. Reasonably gentle and

Fig. 235. KR-34-D was last production model in KR-34 series.

well-behaved, the KR-34 biplanes were pleasant to fly and were actually preferred over similar types in other makes by many pilots. The type certificate number for the Fairchild model KR-34-D was issued 10-17-30 and this supersedes a Group 2 approval numbered 2-250 that had been issued for this version several months earlier. We have no accurate count on the number built but it appears that at least 10 examples were manufactured by the Fairchild Airplane Mfg. Corp. Throughout the coverage we've given the KR-34 series up to this point, the reader must certainly have noticed the confusion in the designations of the various models. Inconsistent factory releases and inaccurate listings

in various old records have been the instigators of this confusion. The following list as based more or less on averages, appears to be the most accepted list of designations for the KR-34 series; ATC # 88 as KR-34-B; ATC # 162 as KR-34-C; ATC # 208 as the rare KR-34-A and ATC # 376 as the KR-34-D.

Listed below are specifications and performance data for the model KR-34-D as powered with the 165 h.p. "Comet" 7-E engine; length overall 23'0"; height overall 8'9"; wing span upper 30'1"; wing span lower 28'9"; wing chord both 63"; wing area upper 154 sq. ft.; wing area lower 131 sq.ft.; total wing area 285 sq.ft.; airfoil Aeromarine 2A modified; wt. empty 1515

Fig. 236. KR-34-D harbored several improvements, note oleo landing gear and upper wing cut-out.

lbs.; useful load 855 lbs.; payload with 45 gal. fuel 389 lbs. (2 pass. at 170 lb. each & 49 lb. baggage); gross wt. 2370 lbs.; max. speed 122; cruising speed 102; landing speed 45; stall speed 48; climb 785 ft. first min. at sea level; climb in 10 min. 5830 ft.; climb to 10,000 ft. 21 min.; ceiling 15,200 ft.; gas cap. 45 gal.; oil cap 3.5 gal.; cruising range at 9.5 gal. per hour 430 miles; price at factory $5675.

The construction details and general arrangement of the model KR-34-D was typical to that of the KR-34-B (ATC # 88), the KR-34-C (ATC # 162) and KR-34-A (ATC # 208) including the following. The gravity-feed fuel tank of 45 gal. capacity was mounted high in the fuselage just ahead of the front cockpit; an easy-reading fuel gauge projected up through the forward cowling. A baggage compartment behind the rear cockpit had a metal panel locking-type door and allowance for 49 lbs. The new-series "Comet" engine was provided with an unusual venturi-type exhaust collector ring that helped to smooth out the air-flow in front; a long tail-pipe deflected gases well below the cockpit level. The landing gear of 72 in. tread was still typical but now used oleo-spring shock absorbing struts; large roly-poly Goodyear "airwheels" were equipped with brakes. The center-section panel of the upper wing on some examples of this model had a large trailing edge cut-out for better visibility upward. The tail-group was typical

except for some added area to the vertical fin. A rubber-mounted tail skid swiveled through a large range for better ground handling. A Curtiss-Reed metal propeller, navigation lights, altimeter, oil pressure and temperature gauges, air-speed indicator, compass, booster magneto, tachometer, dual controls, engine and cockpit covers, log books, first-aid kit, tool kit, fire extinguisher, tie-down ropes and engine instruction manual were all standard equipment. An Eclipse engine starter was optional. The next Fairchild (Kreider-Reisner) development was the "Fairchild 22" parasol monoplane described in the chapter for ATC # 408.

Listed below are Fairchild model KR-34-D entries as gleaned from registration records:
NC-634E; Model KR-34-D (# 263) Comet 7-E.
NC-245V;   "   (# 803)   "
NC-249V;   "   (# 804)   "
NC-949V;   "   (# 805)   "
NC-950V;   "   (# 806)   "
This approval for ser. # 803, ser. # 804 to # 812 (previously on Group 2 approval numbered 2-250) also eligible; information on ser. # 801-802 unknown but there was a ser. # S-801 as model "140" to flight-test "Ranger" V-12 engine; registration no. for ser. # 807-808-809-810-811-812 unknown; ser. # 263 (formerly a KR-34-B) also eligible as KR-34-D when changed to conform.

Fig. 237. The beautiful Laird "Speedwing" model LC-RW300 with 300 h.p. "Wasp Junior" engine.

The beautiful Laird "Speedwing," although a rare example in actual numbers built, was nevertheless a most colorful airplane that was quite well known in flying circles far and wide, leaving a rather indelible impression in the annals of sport flying. Stemming directly from a model called the "Speedster," which was built especially for contest in 1928, the "Speedwing" incorporated refinement of design through several different models of distinctive form and character; to perhaps reach its peak of development in the "Wasp Junior" powered model LC-RW300 as discussed here. Carefully custom built for those who desired and could afford to buy the finest, there was absolutely no room for mediocrity in any part of the Laird airplane. With special attention to every requirement as either laid down by "Matty" Laird or perhaps suggested by a customer, speed, response, dependability and mechanical beauty received their maximum expression in the "Speedwing" model LC-RW300. E. M. "Matty" Laird, pioneer flyer and builder who struggled hard to make his way, was fiercely proud of his handsome "Speedwing" series. He nurtured each example through production and assembly almost personally so that it would be the best aircraft possible to make. Sad to say, not many of the "Speedwing" biplanes were built but those few were a shining beacon for "Laird" performance and quality. Although not offered in regular production beyond 1934, the deluxe "Speedwing," in any of several models, could have been ordered and assembled from new parts on hand as late as 1940.

The Laird "Speedwing" model LC-RW300 was normally a 3-place open cockpit biplane of exceptionally high performance that was specifically leveled at the sportsman pilot; a proud breed of man with discriminate tastes who would appreciate and enjoy an airplane that had the heart and courage of a thoroughbred, with a performance and muscle to match. Assembled to mechanical excellence in a design that was developed through several years, one got the subtle feeling by just walking up to it, that the "Speedwing" was proud of what it was, and eager to show it. The "Speedwing" biplane though primarily a sporting airplane, was designed to offer utility also and normally carried three; when off on a jaunt with pilot only, the front cockpit could be closed off for better airflow across the fuselage for a considerable gain in cruising speeds. With the ability to carry a fair-sized payload and the fuel capacity for a high cruising range, the LC-RW300 posed as ideal for longer cross-country jaunts, and was normally fitted with the very best in instruments and other pilot aids. Flight characteristics, although demanding much more than the uncertain directing of an amateur pilot, were obedient and quite pleasant under the guidance of a loving firm hand; sensitive to feeling and the tempo, the "Speedwing" shared the joy of every flight. Being an airplane of almost special category, the LC-RW300 was custom built to order; this imposed limits on its number so only a few were built. Because of the scant market for an airplane of this type, Matty Laird's development of this series was slowed consid-

Fig. 238. Front cockpit of LC-RW300 shown closed off to gain speed.

erably by 1931 and the LC-RW300 was the last Laird airplane to receive an approved type certificate (ATC). This does not mean that Laird quit building airplanes because many fine airplanes left the plant for several years afterwards, but most were racing or special-purpose craft that were built in limited number on Group 2 approvals. The type certificate number for the Laird "Speedwing" model LC-RW300 was issued 10-24-30 and no accurate account was available, but it seems likely that only one or two examples of this model were built by the E. M. Laird Airplane Co. at Chicago, Ill. Emil Matthew Laird was president and general manager; Lee Hammond was V. P. and sales manager, with Raoul Hoffman as chief en-

gineer; R. L. Heinrich had replaced Hoffman as chief engineer by 1935, and in a few more years Laird was to vacate his plant to make room for the new DGA monoplanes as built by Benny Howard.

Listed below are specifications and performance data for the "Speedwing" model LC-RW300 as powered with the Pratt & Whitney "Wasp Jr." (R-985) engine of 300 h.p.; length overall 22'7"; height overall 9'0"; wing span upper 28'0"; wing span lower 24'0"; wing chord upper 54"; wing chord lower 52"; wing area upper 121 sq. ft.; wing area lower 81 sq. ft.; total wing area 202 sq. ft.; airfoil "Laird" (modified M-6); wt. empty 1922 lbs.; useful load 1088 lbs.; payload with 78 gal. fuel 388 lbs.; (2 pass. at 170

Fig. 239. Sporting nature of LC-RW300 drew it to all the air-meets, where it was consistent winner.

Fig. 240. "Speedwing" fostered design of "Solution", winner of
1930 Thompson Trophy Race.

lb. each & 48 lb. baggage); gross wt. 3010 lbs.; max. speed 180 (front cockpit covered & 27" General "streamlined" tires); cruising speed 150; landing speed 60; climb 1500 ft. first min. at sea level; ceiling 20,000 ft.; gas cap. 78 gal.; oil cap. 8 gal.; cruising range at 16 gal. per hour 650 miles; price at factory first at $15,500, lowered to $14,250 in 1931. For a resume of the Laird models produced to this point, we suggest reference to ATC #86 of Civil Aircraft, Vol. 1 and ATC #152 and #176 of Vol. 2.

The construction details and general arrangement of the "Speedwing" model LC-RW300 was typical to later versions of the model LC-R300 as described in the chapter for ATC # 176 of U.S. Civil Aircraft, Vol. 2, including the following. The fuselage framework was built up in the same manner and deeply faired to shape to streamline the large diameter of the NACA type engine cowling; the fuselage was covered with removable metal panels to rear edge of pilot's cockpit and portion aft of that was covered in fabric. The front cockpit had footsteps and a small door to aid entry and windshield folded down to allow covering of the front pit with a metal panel for gains in speed. The pilot's cockpit was deep and well protected by a large windshield of Triplex shatter-proof glass; the interior was richly upholstered in leather providing several handy pockets for maps, gloves, etc. The pilot's seat had a deep well for parachute pack and was adjustable for height; base of controls were covered with snug leather boots to avoid draft and dust up through the floor-boards. The pilot's panel was loaded with all sorts of instruments and other pilot aids; heating and ventilating were also provided. The wing framework was typical, with closely spaced wing ribs to preserve the airfoil form across the span; 4 ailerons, one in each wing panel, were of the Friese balanced-hinge type. The distinctive "I" shaped wing struts were a laminated plywood sandwich shaved and shaped to a streamlined form; all interplane struts, wires and fittings were carefully streamlined to reduce drag. A large fuel tank was mounted high in the fuselage ahead of the front cockpit with extra fuel in a tank mounted in the center-section panel of the upper wing. The split-axle landing gear of 70 in. tread now used Gruss oleo-spring shock absorbing struts and low pressure semi-airwheels with AP brakes; 27 in. General "streamlined" tires were also available. A large baggage compartment with allowance for 50 lbs. was behind the rear seat bulkhead and access was through a locking-type metal panel door. A metal propeller, electric engine starter, wheel brakes, navigation lights, cockpit clock, thermocouple with indicator for cylinder head temperature, oil cooling radiator, battery, generator, fire extinguisher and fuel gauges were standard equipment.

Listed below are the only known entries of the "Speedwing" model LC-RW300:
NC-10591; LC-RW300 (# 180) Wasp Jr. R-985.
NC-14803;   "   (#   )   "
Serial # 180 also eligible with Wasp Jr. T3A engine of 400 h.p.; serial number for NC-14803 unknown; approval expired 7-1-32.

A.T.C. #378
(10-27-30)
DETROIT-LOCKHEED, "SIRIUS"
DL-2

Fig. 241. Detroit-Lockheed "Sirius" model DL-2 with 420 h.p. "Wasp" engine, a rare type that led a diversified life.

Among the bright and shining stars in the Lockheed "constellation" was a rare gem called the "Sirius" model DL-2. A different "Sirius" though, not quite like all the others, because of its gleaming all-metal fuselage that was mated to the standard wooden wing. Having the know-how and the experience for metal airplane construction within its organization, the Detroit Aircraft Corp. had already designed and built an all-metal monocoque fuselage for the speedy "Vega." Now, they were launching further experiments in the same type of fuselage, incorporated into the configuration of a "Sirius." Flown successfully as a 2-place sport plane and also as a one-place mail plane, (with either open cockpits or cockpits covered by a sliding transparent canopy), the DL-2 was also used in experiments for developing a practical retracting landing gear. With its low-mounted wing the "Sirius" was a logical choice for experiments on the folding gear and these developments finally led to the creation of the speedy "Altair" and "Orion." As fitted with the newly developed flush-type retractable undercarriage, the "Sirius" DL-2 became the model DL-2A. Upon successful demonstration, it was sold to the Army Air Corps as a command transport to be used by the Asst. Sec. of War, F. Trubee Davison. Operating for less than 8 months as

the DL-2 with a "panted" fixed landing gear, the DL-2A conversion (YIC-23) led a very exciting and diversified life in the Air Corps through 1938.

The Detroit-Lockheed model DL-2 was a low-winged monoplane of typical "Sirius" configuration, except for its all-metal fuselage and other modifications necessary to this combination. Seating two in tandem as a sport-plane with range enough for extended flights, the DL-2 also had ample capacity for large amounts of baggage and personal equipment. When arranged as a single-place mail-cargo carrier the forward compartments had capacity for up to 1100 lbs. As a highly specialized craft, the lone DL-2 was not immediately attracted to interested customers so it remained for a while as a demonstrator at Detroit Aircraft, and served in furthering the developments of a retracting landing gear. Powered with the 9 cyl. Pratt & Whitney "Wasp" SC engine of 450 h.p., that was shrouded in a big NACA type low drag cowling, performance was more or less comparable to that of the all-wood "Sirius" (see ATC #300) in most respects but slightly less than that of the vivacious "Vega." With no especial merits of its own to cling to, it however played its part in the transition to the speedy "Altair" and the outstanding "Orion."

Fig. 242. Metal fuselage seated two with cargo space forward.

The type certificate number for the metal-fuse-laged DL-2 was issued 10-27-30 and only one example of this model was built in the shops of the Detroit Aircraft Corp. at Detroit, Mich., Robert J. Woods was project engineer for the Detroit corporation on several interesting developments and Vance Breese, noted airplane designer and test-pilot, did much of the test flying during this period.

Listed below are specifications and performance data for the "Sirius" model DL-2 as powered with the 450 h.p. "Wasp" engine; length overall 27'10"; height overall 9'2"; wing span 42'10"; wing chord at root 102"; wing chord at tip 63"; total wing area 294 sq. ft.; airfoil "Modified Clark Y"; wt. empty 2958 lbs.; useful load 2212 [2242] lbs.; payload with 140 gal. fuel 1122 [1152] lbs.; gross wt. 5170 [5200] lbs.; figures in brackets are for 1-place cargo-carrier; max. speed 175; cruising speed 145; landing speed 65; climb 1100 ft. first min. at sea level; ceiling 18,000 ft.; gas cap. normal 140 gal.; oil cap. 10-15 gal.; cruising range at 22 gal. per hour 870 miles; price at factory first quoted at $22,000.

The construction details and general arrange-

Fig. 243. Fixed landing gear of DL-2 replaced with retractable type on DL-2A, forerunner of famous "Altair".

ment of the model DL-2 was typical to that of the Lockheed "Sirius" model 8 as described in the chapter for ATC #300 of U. S. Civil Aircraft Vol. 3, including the following. The fuselage was an all-metal structure typical to that of the "Vega" DL-1 (see ATC #308 in this volume) except that the wing was mounted on the bottom side and tandem cockpits were arranged in the aft portion. The cantilever wing, faired neatly into the fuselage was an all-wood structure typical to that of the "Sirius" model 8; two fuel tanks were mounted in the wing flanking the fuselage. A baggage compartment of 50 cu. ft. capacity was placed forward of the front cockpit; the mail-cargo version had extra compartments in this same area. Originally with two open cockpits, the DL-2 later had a sliding transparent canopy for better weather protection. The divided landing gear of 144 in. tread used oleo shock absorbing struts; wheels were 32x6 or 9.50x12 and brakes were standard equipment. The landing gear vees were faired to a

solid section and the wheels were encased in large tear-drop fairings; a tail skid or a tail wheel was optional. The all-wood tail group was of cantilever construction and of somewhat larger area than that of the earlier "Sirius" 8. Quoted empty weight of the DL-2 includes NACA engine cowling, wheel streamlines, navigation lights, battery, landing lights, electric engine starter and generator. A metal propeller, dual controls, fire extinguisher and first-aid kit were also standard equipment. The next Lockheed development was the "Vega" model 5-C described in the chapter for ATC #384 of this volume.

Listed below is the only entry for the Detroit-Lockheed "Sirius" model DL-2:

NC-8494; Model DL-2 (#165) Wasp 450 Serial #165 converted 10-17-31 to model DL-2A (metal-fuselaged "Altair") on Group 2 approval numbered 2-386; sold to Army Air Corps and operated as YIC-23 command transport.

Fig. 244. The Swallow "Sport" mounted several different engines, the model HW
had 165 h.p. Wright J6-5 engine.

On the face of what we can see, the new "Sport" design (also called "Special") was more or less a conventional open cockpit biplane with seating for three; in a working arrangement called the all-purpose type, because of its adaptability to varied uses. Comparatively speaking, the new design shows careful planning and it stands well apart from anything that Swallow Aircraft had ever produced before this. With "faster wings" now smaller of unequal chord and of unequal span, the configuration bears marked evidence throughout of the attempt to rearrange and regroup the various items making up a useful load within the confines of a much smaller airplane. As a result, this allowed for a compact airframe that was coupled up much shorter, yet without loss of leg room or any other of the normal conveniences; more efficient flight, with considerable gains in performance per horsepower, were not then results of paring down the airplane strictly in the interests of less drag and more speed. With intent to allow the prospective buyer a choice of several fine powerplants in a comparable power range, the "Sport" was also introduced in a companion model to the HA (see ATC #341) that mounted the 5 cyl. Wright J6 (R-540) engine of 165 h.p. as the model HW. Fashioned with a flair to pri-

marily appeal to the sporting type of flyer, a flyer who appreciated bonuses in performance but was limited to operate on a more realistic budget, the "Sport" model HW had all in its favor except that the group of potential customers for an airplane of this type had not the means to buy airplanes at this particular time.

The Swallow "Sport" model HW, although we have no photo to show it specifically, was an open cockpit biplane with seating for 3, and was of course similar to the previously discussed model HA except for its engine installation, and some minor modifications necessary to this particular combination. Powered with the 5 cyl. Wright J6 (R-540) engine of 165 h.p., the performance of the HW was proportionately better than that of the 150 h.p. model HA. As compared to other airplanes of this type with this amount of power, the Swallow "Sport" was just about on top of the list. Fashionable trim, light in weight and of rather small size, the HW had a substantial power reserve that translated into a rapid climb-out and a better than average speed range. It was blessed with all the requisites for good response and maneuverability with nothing more than a light touch. Because of its scarcity, handed-down lore for this airplane was not to be found (and still lurks somewhere in the past) so the flight characteristics and general behavior are somewhat in question; however, it is safe to assume just by reckoning, that its aerodynamic arrangement must have been conducive to lively reaction, with a positive response to good command. Being some 36 lbs. lighter when empty than the Axelson-powered model HA, the HW had the advantage of carrying this much more in its useful load and with gross weight kept to the same figure, the 15 extra horsepower was there to add to the "snap and flash." It would have been quite interesting to interview a pilot from this era, a pilot that had the opportunity to fly and become acquainted with any of the Swallow "Sport" models, because one just gets the feeling that this must have been a spirited airplane that offered practical sport-flying at its best. The type certificate number for the Swallow "Sport" (Special) model HW was issued 10-28-30 and only one example of this model was built by the Swallow Airplane Co. at Wichita, Kan. The airplane used to make up the model HW for certification tests was the model HA just 3 months or so previous; slight modifications and a change of engines transformed this airplane into the newer model and the manufacturers serial number was also changed to better identify the new series.

Listed below are specifications and performance data for the Swallow "Sport" model HW as powered with the 165 h.p. Wright J6 engine; length overall 22'3"; height overall 8'2"; wing span upper 31'0"; wing span lower 23'0"; wing chord upper 60"; wing chord lower 48"; wing area upper 150 sq. ft.; wing area lower 90 sq. ft.; total wing area 240 sq. ft.; airfoil (NACA) M-12; wt. empty 1380 lbs.; useful load 820 lbs.; payload with 40 gal. fuel 372 lbs. (2 pass. at 165 lbs. each & 42 lbs. baggage); gross wt. 2200 lbs.; max. speed 135; cruising speed 115; landing speed 42; normal landing run 370 ft.; climb 1200 ft. first min. at sea level; climb to 10,000 ft. was 9 min.; ceiling 16,000 ft.; gas cap.

40 gal.; oil cap. 5 gal.; cruising range at 8 gal. per hour 500 miles; price at factory was $5,350, lowered to $4,895 in May of 1931.

The construction details and general arrangement for the Swallow model HW was typical to that of the model HA, (as described in the chapter for ATC #341 of this volume), including the following, which naturally applies to all 3 models in this series. A baggage allowance of 42 lbs. was stowed in a bin under the front seat with another small baggage locker in the dash-panel of the front cockpit. The cockpits were rather deep, well upholstered in auto-type leather and well protected by large streamlined windshields. The upper wing panels were joined together on the center-line and a 20 gal. gravity-feed fuel tank was mounted in the root end of each half. Friese-type ailerons were in the upper wing and were actuated through rods and bellcranks by two push-pull tubes coming up from the rear cockpit. The split-axle landing gear was a stiff-legged structure of welded chrome-moly steel tubing and no shock absorbers were normally used. Large squashy "airwheels" carrying only 7 lbs. air pressure were calculated to absorb all the shocks, however, oleo-spring shock absorbers were available if desired and wheel brakes were optional. The tail-skid was of the normal steel spring-leaf type with a removeable hardened shoe. The normal color scheme was believed to be black and white but other two-tone combinations were available on order. The next development in the Swallow "Sport" was the Continental-powered model HC described in the chapter for ATC #399 of this volume.

Listed below is the only known example of the Swallow model HW as gleaned from registration records:

NC-109V; Model HW (#2002) Wright R-540 Serial #2002 was previously a model HA as mfgrs. ser. #102.

A.T.C. #380
(11-18-30)
AMERICAN EAGLE "EAGLET",
230

*Fig. 245. The American Eagle "Eaglet" model 230 with 30 h.p. Szekely engine;
shown here in an early example.*

Perhaps no development was watched any closer by certain aircraft manufacturers than the development and introduction of the Aeronca C-2. It had almost immediate acceptance in private-flying circles that had taken to sitting around on their hands, wishing they could afford to fly. Success of the Aeronca (see ATC # 351 in this volume) and the curtailing effects of the so-called business depression, had together spurred the movement to lighter low-cost airplanes. These airplanes would be a lot easier to buy and they could be operated for so much less. Altogether this was to permit many of the flying folk to keep on flying who otherwise would be hampered by high expenses and either not get to fly at all or only on rare occasion. It didn't take Ed Porterfield very long to see the handwriting on the wall so, spurring everyone into action at American Eagle Aircraft, he had the new "Eaglet" built and flying in somewhere near record time. With the first one already under construction by June of 1930, the "Eaglet" was introduced soon after as a rather conventional looking airplane in contrast to the "Aeronca"; nevertheless it was a true flivver-plane that one could enjoy flying for just a few dollars an hour. Larry Ruch and Ralph Hall were company test-pilots, so it was proba-

bly they who put the "Eaglet" through all the testing in two different versions and by Nov. of 1930 it received its approval for manufacture of the two-place version with 30 h.p. It was the only airplane licensed to carry two people on 30 h.p. Refinanced in Sept. of 1930 for some $500,000 to launch the building and marketing program of the "Eaglet" model 230, it wasn't too long before "Eaglets" began slowly rolling off the assembly floor. Due to the fact that things were rather unsteady at American Eagle during some of this time and an unsettled household did have its hampering effects, it is commendable that a dozen or so were finally built by the spring of 1931. Exhibited at the Detroit Air Show for 1931, the "Eaglet" fostered a lot of interest and orders were pouring in by the hundreds; for a change, the future was beginning to look much brighter. A certain amount of readjustment had been necessary, refinancing also brought on its obligations, and before long there was talk of a merger between American Eagle and the Lincoln Aircraft Co. of Nebraska to form the American Eagle-Lincoln Aircraft Corp. The Lincoln Aircraft firm was moved in to the American Eagle factory, company officials were all scrambled together and reassigned new positions, some were let go and some had taken

Fig. 246. Ralph Hall, test pilot, flying "Eaglet" 230.

their own leave; in general, there was some more readjustment and a loss in some of the former harmony. Internal conditions at the factory caused set-backs and too much idle time to allow taking the full advantage of the "Eaglet's" sales potential. Eventually, the little monoplane began to sell by mid-1931 and were soon in all parts of the country but its number built was far from the bright-eyed calculations that were first predicted.

The development of the "Eaglet" was first centered around the 2 cyl. Cleone engine of 25 h.p., in a craft that was limited to a pilot only, but it had carried two on several occasions with only a slight loss in performance. There were great hopes for this feather-weight airplane at first because of its economy and low first-cost ($995); however the 2-cycle engine was not entirely satisfactory and a single-place airplane was judged to have its limitations too, so this Model 130 served mostly as a test-bed for the brand-new airframe. Switching to the 3 cyl. Szekely of 30 h.p., the "Eaglet" was now able to carry two with a fairly good performance. It was more in keeping with the type of airplane that would fit the new trend in private-flying, and also with the fixed-base operator who was anxious to cut his costs so that he could lower prices. Struggling through a period of unrest and readjustment, the "Eaglet" model 230 finally began rolling off the line to take its place on the American scene. Powered also with the little 9 cyl. Salmson AD-9 (French) engine of 40 h.p., several were built as the model 231 and the performance gains with the extra 10 h.p. prompted a new version for 1931-32, known as the model B-31. Basically similar, it was powered with a Szekely SR-3-0 engine of 45 h.p. A version powered with the 4 cyl. Continental A-40 engine was also built as the model A-31 but its development was discontinued. In general, the "Eaglet" remained quite popular through the years, holding up well under all kinds of service; one was even shipped to Madras, India and of the total that were built, at least half were still flying in active service by 1939.

The American Eagle "Eaglet" model 230 was a conventional looking parasol type monoplane with seating for two in tandem and because of its configuration it didn't have the look of a "flivver", even though it was in just about every sense. With its empty weight of less than 500 lbs., it could be easily lifted by its tail and pulled around by one man; it was easy to get in and out of and though everything about it was toy-like by comparison, pilots were eager to fly it and proclaimed their enjoyment with a big grin. Powered with the 3 cyl. Szekely SR-3-L of 30 h.p. or the SR-3 of 35 h.p. the "Eaglet" was certainly not a "tiger" by any means, but its performance was quite adequate for the purpose and it afforded an hour's flying pleasure for two on just a few dollars. Because of its light weight, its low power and its ample proportion, it did take a certain technique to fly it properly but it didn't ask too much of a pilot and most of them learned to fly it well in just a few hours. Every light plane of course had its limitations and so did the "Eaglet", but if it was flown within these limitations the satisfaction per dollars invested was proportionately quite high. Flight characteristics and general behavior of the "Eaglet" were light and well-mannered and its rugged character both in frame and in nature allowed a certain amount of mistreatment, so on the average, it was easy to keep her on the flight-line. By late 1931, several other builders were introducing light airplanes and it is commendable at least that quite a few were using the practical lines of the "Eaglet" to go by. The first 33 examples of the "Eaglet" model 230 were awarded a Group 2 approval numbered 2-303; the 34th example and up were awarded

Fig. 247. Early "Eaglet" tested with 2 cyl. Cleone engine of 25 h.p.; pilot beams his pleasure but engine was not reliable.

the type certificate number (# 380) that was issued 11-18-30. The first few of the model 130 with the 2 cyl. Cleone engine were eventually modified into the model 230 and in total, some 80 or more examples of the "Eaglet" were manufactured by the American Eagle Aircraft Corp. on Fairfax Field in Kansas City, Kan. Of course the later models were built by the new company known as the American Eagle-Lincoln Aircraft Corp. Victor Roos was president of the new combine and E. E. Porterfield was sales manager. Not entirely satisfied with the workings of the new arrangement, Ed Porterfield left the firm to pursue his own interests; John Carroll Cone, Jack E. Foster and others had also left to pursue varied interests.

Listed below are specifications and performance data for the "Eaglet" model 230 as powered with the Szekely SR-3-L engine of 30 h.p.; length overall 21'6"; height overall 7'9"; wing span 34'4"; wing chord 60"; total wing area 164 sq.ft.; airfoil (NACA) M-series; wt. empty 467 lbs.; useful load 400 lbs.; payload with 8.5 gal. fuel 172 lbs.; gross wt. 867 lbs.; max. speed 80; cruising speed 70; landing speed 25-35; climb 650 ft. first min. at sea level; ceiling 14,000 ft.; gas cap. 8.5 gal.; oil cap. 1.5 gal.; cruising range

at 2 gal. per hour 280 miles; price at factory first announced as $1395, raised to $1475 in Oct. of 1930. Also eligible with Szekely SR-3 of 35 h.p. at 900 lb. gross wt. for slight performance gain.

The fuselage framework was built up of welded chrome-moly and 1025 steel tubing and covered in fabric. Two metal-framed seats were mounted in tandem and the fuselage was 23 in. wide in the cockpit area; a large door was on the left side for entry and exit. Dual joy-stick controls were provided and all movable controls were operated by stranded steel cable. The wing framework of one continuous section was built up of solid spruce spar beams with spruce and plywood truss-type wing ribs; the leading edges were covered with dural metal sheet and the completed framework was covered in fabric. The wing was perched atop an N-type cabane of steel tubing in a "parasol" fashion and braced to the fuselage longerons by parallel steel tube struts; with the wing mounted high in relation to the occupants, the visibility was good in all directions. The 8.5 gal. gravity-feed fuel tank was mounted high in the fuselage just behind the metal firewall; there were no provisions for baggage. The split-axle landing gear of 54 in. tread used 16x7-4 low pressure airwheels and the shock absorbers were Rusco rubber donut-rings in compression; the steel tube tail skid was snubbed with several wraps of rubber shock cord. The fabric covered tail group was built up of welded steel tubing; the fin was rigidly mounted and the horizontal stabilizer was adjusted for load trim on the ground only. A Flottorp wooden propeller, Pyrene fire extinguisher bottle, first-aid kit and log books were standard equipment. The next development in the "Eaglet" was the 45 h.p. model B-31 described in the chapter for ATC # 450.

Listed below is partial listing of "Eaglet" model 230 entries as gleaned from registration records:

NC-295N- Model 230 (# 1007) SR-3L-30.

NC-454V;	"	(# 1013)	"
NC-475V;	"	(# 1034)	"
NC-480V;	"	(# 1035)	"
NC-484V;	"	(# 1037)	"
NC-482V;	"	(# 1039)	"
NC-485V;	"	(# 1040)	"
NC-486V;	"	(# 1041)	"
NC-487V;	"	(# 1044)	"
NC-483V;	"	(# 1045)	"
NC-489V;	"	(# 1046)	"
NC-490V;	"	(# 1047)	"
NC-491V;	"	(# 1048)	"
NC-492V;	"	(# 1049)	"
NC-494V;	"	(# 1050)	"
NC-495V;	"	(# 1051)	"
NC-496V;	"	(# 1052)	"
NC-497V;	"	(# 1053)	"
NC-498V;	"	(# 1054)	"
NC-499V;	"	(# 1055)	"
NC-504Y;	"	(# 1057)	"
NC-513Y;	"	(# 1058)	"
NC-515Y;	"	(# 1060)	"

Registration listing continues at least to ser. # 1078; this approval for ser. # 1007-1013-1034 and up; ser. # 1036 & # 1043 were model 231 with Salmson 40 engine on Group 2 approval numbered 2-387; reg. no. for ser. # 1038-1042-1056-1059 unknown; ser. # 1001 through # 1033 were on Group 2 approval numbered 2-303.

Fig. 248. Sharing the light-plane market with two other popular makes, the "Eaglet" was more normal in its make-up.

Fig. 249. Northrop "Alpha" 2 with 420 h.p. "Wasp" C engine; model 2 seated 6 passengers and a pilot.

Introduced as the first in a long line of "Northrop" high-performance airplanes, the sleek "Alpha" light commercial transport was viewed with great interest by those in the aviation industry as the first practical, all-metal, low winged cabin monoplane in this country. Fairly bristling with innovations in aircraft design and construction, despite its simple business-like appearance, the "Alpha" quietly launched part of a new day and set a bright new pattern for transport airplanes of the immediate future. Seating a pilot and 6 passengers as a high-speed transport, or with pilot only as an all-cargo carrier, capable of carrying some 1060 lbs. of paying load, the "Alpha" had a definite future in scheduled air-line work. The first "Alpha" transport, gleaming in its shiny metal skin, had been readied for tests and was successfully flown by May of 1930; tested quite thoroughly thereafter in the next few months for its government approval, several orders were already on hand. Commencing a production run at the close of the year, they completed and delivered one "Alpha 2" to the NAT line (National Air Transport), 5 of the "Alpha" were delivered to TWA (Transcontinental & Western Air) by mid-1931 and 3 were being completed for the Army Air Corps as a 5-place (YC-19) personnel transport; the one example delivered to NAT was a 7-place "Alpha 2" and the 5 aircraft delivered to TWA were the 3-place combination mail-passenger type designated the "Alpha 3". One example of

the "Alpha" had also been delivered for the personal use of Col. Clarence M. Young, the Asst. Sec. of Commerce for Aeronautics, who was also an accomplished pilot. Used during 1931 on portions of the TWA line to haul mixed loads of mail and passengers, most of the "Alpha" were then modified into the improved models 4 and 4-A with increases in gross loading and slightly higher operating speeds.

John K. Northrop, a true pioneer and a professed self-made engineer who designed the first of the outstanding "Vega" series, left Lockheed in 1928 to make detailed studies in all-metal aircraft construction and to toy with some ideas he had on a "flying wing" design. Forming the Avion Corp., Northrop experimented with multi-cellular all-metal construction and the practical application of smooth metal "stressed skin" covering. Built and successfully tested too, was a small airplane to study various aspects of flying-wing design. In the fall of 1929, Wm. E. Boeing, noted airplane builder, became avidly interested in the development of the Northrop "Flying Wing". He arranged to take over the Avion Corp. as a nucleus for the proposed Northrop Aircraft Corp. which was to be another division of the United Aircraft & Transport Corp. With a proposed transport monoplane that was to employ many of Northrop's new ideas in all-metal construction already in the works, a small factory and a laboratory were erected on United Airport in Burbank,

Fig. 250. "Alpha" 3 seated 3 passengers and pilot, balance of payload was in mail and cargo.

Calif. With plans now forming fast, this factory building was soon enlarged to provide room for quantity production of the new "Alpha" transport. By Jan. of 1931, the prototype airplane had already been built, tested and approved, with 20,000 sq.ft. of factory space already tooled up for the production of an order from the TWA line. From the handfull that started a year previous, the working force was already up to 80 men familiar with the various metal crafts. Completed and flown too by now was a small sport-model, a swift low-winged all-metal monoplane for two, called the "Beta"; construction principles were similar to the larger "Alpha" and high performance was its cardinal feature. In an effort to enforce economy, a move brought about by a sagging aircraft market, the Northrop facility was consolidated with that of Stearman Aircraft (another division of the UA&T Corp.) by Sept. of 1931; consequently, all assets, machinery (some of which were specially designed hydraulic presses) and all materials were moved to the affiliated plant in Wichita, Kan. Somewhat unhappy with the move and miffed with the ensuing circumstances, John K. Northrop left shortly after for Calif. to continue his studies of new methods in all-metal construction, and to adapt several new-found design principles to an entirely new line of "Northrop" transport airplanes.

The Northrop "Alpha 2" was a low-winged cabin monoplane of cantilever all-metal construction with interior arranged for 6 passengers, or cleared of all seating to haul large loads of mail and cargo; an open cockpit for the pilot was placed aft of the cabin section. A combination mail-passenger version called the "Alpha 3" was offered with seating for 3 passengers and a cargo load of some 465 lbs. Embodying such principles as a sturdy cantilever wing, a cigar-shaped monocoque fuselage and a deep-chord NACA type low-drag engine fairing, the "Alpha"

was also arranged in good aerodynamic proportion for high efficiency and the maximum in performance. Powered with the 9 cyl. Pratt & Whitney "Wasp" C (R-1340-C) engine of 420 h.p., the "Alpha" used every available horsepower wisely and delivered ample utility with high performance. Loaded take-off runs were in the average of some 550 ft. and landing runs averaged some 600 ft. without using brakes, proving the "Alpha" well capable to operate out of the average small airfield. Climb-out from take-off was very good and the service ceiling was high enough to top the highest mountain. Flight characteristics were described as excellent and its general behavior was cooperative and compatible whether in flight or on the ground. Easy to service and easy to maintain, the "Alpha" was a good example of an airplane that could work day-in and day-out at top capacity in high performance with a fair return in profit. The type certificate number for the "Alpha 2" was issued 11-22-30 with added provisions to include the combination mail and passenger carrying "Alpha 3"; some 10 or more examples were built in these two versions by the Northrop Aircraft Corp. a division of the United Aircraft & Transport Corp. The factory was first located on United Airport in Burbank, Calif., the beautiful air-depot of the Boeing System. Wm. E. Boeing was chairman of the board; W. K. Jay was president and general mgr.; John K. Northrop was V.P. and chief of engineering; and Don R. Berlin was the chief engineer.

Listed below are specifications and performance data for the Northrop "Alpha" as powered with the 420 h.p. "Wasp" R-1340-C engine; length overall 28'5"; height overall (tail up) 9'0"; wing span 41'10"; wing chord at root 100"; wing chord at tip 66"; total wing area 295 sq.ft.; airfoil Clark Y; wt. empty 2679 [2590] lbs.; useful load 1821 [1910] lbs.; payload with 100 gal.

Fig. 251. "Alpha" 2 used by Col. Clarence M. Young, Asst. Sec. of Commerce for Aeronautics.

fuel 975 [1060] lbs. (7 pass. & no baggage or 3 pass. & 465 lbs. mail-cargo); figures in brackets for all-cargo version; gross wt. 4500 lbs.; max. speed 170; cruising speed 145; landing speed 60; climb 1350 ft. first min. at sea level; climb in 10 min. was 9500 ft.; ceiling 19,300 ft., gas cap. max. 100 gal.; oil cap. 10 gal.; cruising range at 22 gal. per hour 600 miles; price at factory field $21,500; empty wts. shown do not include engine starter, generator, and battery, a total of 46 lbs.; wts. of all other extra equipment also deducted from payload.

The fuselage framework of monocoque construction, was built up of duralumin ring-shaped bulkheads of channel section and longitudinal stiffeners that were covered with a smooth "Alclad" metal stressed skin; of circular section, the outer metal skin as supported by rings and stiffeners, carried nearly all of the fuselage stresses. Seating for six was arranged on 2 wide seats with 3 across and facing each other; with one seat removed, the seating was arranged for 3 and the cargo bin allowed up to 465 lbs. of mail, baggage and cargo. With all cabin seats removed, the cabin interior provided 120 cu.ft. capacity with allowance for over a half-ton of cargo. Padded and well upholstered, sound-proofed and insulated, the interior was provided

Fig. 252. Several "Alpha" 3 served Transcontinental and Western Air routes.

Fig. 253. "Alpha" 2 on Edo floats.

with ventilators and a cabin heater; a large door on the right in the aft section, provided convenient entry or exit. The pilot's open cockpit, placed aft of the cabin section, was deep and well protected by a large windshield; a sliding canopy for this cockpit was offered as an option on the "Alpha" but apparently none were so equipped. The cantilever wing framework in 3 sections, built up entirely of "Alclad" metal sheet, was a multi-cellular stressed-skin construction; the center portion of a constant chord was built as an integral part of the fuselage with the tapered outer panels bolted to ends of the stub wing. Two fuel tanks were mounted in the stub-wing, flanking each side of the fuselage. The split-axle landing gear of 102 in. tread was fitted to underside of the stub-wing and provided with oleo-rubber shock absorbing struts; wheels were 32x6 and Bendix brakes were standard equipment. Low pressure semi-airwheels were optional. Early experiments in a retractable landing gear proved somewhat heavy and unreliable so this mechanism was shelved for further study; later versions of the "Alpha" were fitted with metal streamlined boots over separate cantilever legs and proved to be nearly as efficient as a retractable gear. The cantilever tail-group was an all-metal multicellular construction covered with smooth "Alclad" metal sheet. A metal propeller, NACA engine cowling, navigation lights, wheel brakes, swiveling

tail-wheel, a hand crank inertia-type engine starter and fire extinguisher were standard equipment. Low pressure semi-airwheels, landing lights, parachute flares, generator, battery and electric engine starter were optional; actual wts. of these units to be deducted from the payload. The next development in the "Alpha" series was the improved "Alpha 4" described in chapter for ATC # 451.

Listed below are "Alpha" model 2 and 3 entries as gleaned from registration records:

X-2W;	Alpha 2 (# 1) Wasp 420.	
X-127W;	" 2 (# 2)	"
NC-11Y;	" 2 (# 3)	"
NC-999Y;	" 3 (# 4)	"
NC-933Y;	" 3 (# 5)	"
NC-942Y;	" 3 (# 6)	"
NC-947Y;	" 3 (# 7)	"
NC-961Y;	" 3 (# 8)	"
NC-966Y;	" 2 (# 9)	"
NC-993Y;	" 3 (# 16)	"

Serial number for NS-1 unknown; ser. # 2 later as Alpha 4; ser. # 3 later as Alpha 4-A, also on floats; ser. # 4-5-6-7-8 later as Alpha 4; ser. # 9 later as Alpha 3; ser. # 10-11-12 as Alpha 4 on ATC # 451; ser. # 13-14-15 probably delivered to Air Corps; ser. # 17 as Alpha 4; all ser. # 2 through # 17 eligible as Alpha 2-3-4-4A when converted to conform to certain specifications.

A.T.C. #382
(11-24-30)
BRUNNER-WINKLE "BIRD", BW

Fig. 254. "Bird" model BW with 110-125 h.p. Warner "Scarab" engine, a happy combination of fine engine and excellent airplane.

Although one of the more rare versions in the popular "Bird" biplane line-up, the model BW was one of the most interesting. Merged in combination with the smooth-running Warner "Scarab" engine, an engine that packed a good solid punch with grace, the model BW not to be outdone, brought out its best "Bird" behavior also and together they made into a most delightful machine. It is almost certain that had the price-tag on the BW been more in keeping with what the flying-folk could afford during these times, it surely would have been seen in far greater number around the country. Reorganized in Sept. of 1930 as the Bird Aircraft Corp. with a better capital structure, the new company spent a good amount in promotion and tried to make buying a "Bird" almost as easy as buying a car. Through their own finance plan requiring only 40% down, balance was in 12 monthly payments with a low interest rate; several operators paid for their airplanes from profit earned while making their payments. Proud of their product and rightfully so, Bird Aircraft kept the flying populace well aware of the many good points to be found in the "Bird" biplane; one cute reminder went as follows: "Be popular with the ladies, it is no trouble at all getting them in or out of a "Bird" and they'll enjoy it". In fact, there were quite a few good lady-pilots who enjoyed getting in and out of a "Bird". With its glide ratio of about 10 to 1, the "Bird" biplane could often be floated to

distant safety during an emergency requiring a hasty landing; its good control at the lower airspeeds and its reluctance to fall off into an honest-to-goodness "spin" was also subject for conversation at many hangar sessions. Good-natured, obedient and sensitive to the guiding mood, the "Bird" was not an aggressive airplane but it most always showed up at the various air shows, air races, air-derbies, and most always did well too in the competition. Although not as rabid in enthusiasm as some, "Bird" owners were a relaxed and happy lot and there was always a great bond of friendship between man and machine. Of the 200 or so examples that were built in all since 1924, some 120 of the "Bird" were still flying actively in 1939. It is not surprising that several of the "Bird" biplanes have been carefully restored to fly again, even at this late date, and many barns and dusty hangars are being checked for a rebuildable specimen.

The handsome "Bird" model BW was an open cockpit biplane of the classic general purpose type with seating arranged for 3. As an all-purpose airplane adaptable to many and varied uses, this model was not particularly slanted towards the commercial operator but more towards the private-owner pilot who would be using his airplane mostly for sport or personal transportation. Typical in configuration to all other "Bird" models, the BW was also arranged in good aerodynamic proportion for inherent

Fig. 255. Wm. Beard flying "Bird" BW, Beard later presented ship as a present to his wife, Melba.

safety and pleasant flight characteristics. Because of the sesqui-plane arrangement and because of the large amount of stagger of the wings in relation to cockpit placement, visibility was excellent in flight, on take-offs and landings and in ground handling. Well equipped and neatly fitted, the graceful BW was one of the finest examples in an airplane of this type. Powered with the 7 cyl. Warner "Scarab" engine of 110 h.p., this model used each horsepower wisely and to its fullest, to deliver an excellent performance in the broadest range and with good economy. Not often exposed to national competition, the model BW however could muster up some fancy flying when the occasion arose and distinguish itself with top-notch performance. Melba Beard, an accomplished aviatrix and a loyal "Bird" enthusiast to this very day, owned a BW in 1933 and later won several acrobatic contests with it. The type certificate number for the "Bird" model BW was issued 11-24-30; 4 examples of this model were built and 3 were converted into the BW from earlier models, by the Bird Aircraft Corp. at Glendale, Long Island, N.Y. Just for the sake of record it might be appropriate to say that nearly 400 of the Warner "Scarab" had been built and sold going into 1930; the engine's increasing popularity pointed to an exceptionally good future.

Listed below are specifications and performance data for the "Bird" model BW as powered with the 110 h.p. Warner engine; length overall 23'0"; height overall 8'6"; wing span upper 34'0"; wing span lower 25'0"; wing chord upper 69"; wing chord lower 48"; wing area upper 184 sq.ft.; wing area lower 82 sq.ft.; total wing area 266 sq.ft.; airfoil USA-40B (Goettingen) modi-

fied; wt. empty 1235 lbs.; useful load 785 lbs.; payload with 37 gal. fuel 370 lbs. (2 pass. at 170 lb. each & 30 lb. baggage); gross wt. 2020 lbs.; max. speed (with speed-ring engine cowling) 117; cruising speed 98; landing speed 36; climb 750 ft. first min. at sea level; ceiling 15,000 ft.; gas cap. 37 gal.; oil cap. 3 gal.; cruising range at 6.5 gal. per hour 500 miles; price at factory first as $4250, later raised to $4395.

The construction details and general arrangement of the "Bird" model BW was typical of other models as previously described, including the following. The two open cockpits were deep and well protected, with a large door on the right side for entry to the front seat; a baggage compartment with allowance for 30 lbs. was accessible through a locking-type metal panel in the top cowling ahead of the front windshield. The fuel tank was mounted forward, high in the fuselage, just behind the engine section with a direct-reading fuel gauge projecting through the cowling. The Warner "radial" engine was usually shrouded with a Townend-type "speed ring" cowling. As on all "Bird" models, the ailerons in the upper wing were mounted on a canted hinge line and effective area varied across their span for more positive control, expecially at the lower air-speeds. The cross-axle landing gear of 66 in. tread was fitted with 7.50x10 semi-airwheels and wheel brakes; tail skid with a hardened removable shoe was of the steel spring-leaf type. The fin was ground adjustable and the horizontal stabilizer was adjustable in flight. A wooden propeller, navigation lights, Heywood air-operated engine starter, dual controls, wheel brakes, cockpit covers and a first-aid kit were standard equipment. A metal propeller, and "speed ring" engine cowling were optional. Several color combinations were available; the 'plane shown in heading of this chapter was reported to be that sickly orange fuselage with silver wings and tail. The next development in the "Bird" biplane series was the model C described in the chapter for ATC # 387 of this volume.

Listed below are "Bird" model BW entries as gleaned from registration records:

X-892W; Model BW (# 3001) Warner 110.
NC-723N;   "   (# 3002)   "
NC-724N;   "   (# 3003)   "
NC-725N;   "   (# 3004)   "
NC-170H;   "   (# 1024)   "
NC-836W;   "   (# 1084)   "
NC-734Y;   "   (# 2043-23)   "

Serial # 3002 operated by Warner Motors as demonstrator; ser. # 3003 registered to Melba Beard in 1933; ser. # 1024 and # 1084 first as model A with Curtiss OX-5 engine; ser. # 2043-23 first as model BK with Kinner K5 engine.

Fig. 256. The "Spartan" model C4-300 with 300 h.p. Wright J6-9 engine; buxom beauty was blessed with high performance.

With performance somewhat penalized because of above-average gross weight, brought on by stout construction, aerodynamic arrangement of long moment arms, deluxe appointments and extras in equipment, the model C4-225 by "Spartan" (see ATC # 310) was a bit short in get up and go. The only logical solution for this predicament was of course more power, so this same design was slightly revised to mount an engine of 300 h.p. Had the general layout been left alone and modified only to take the 300 h.p. Wright J6 engine, a substantial increase in performance would surely have been the gain, with only the additional weight of the larger engine and its installation. But, to this was added the weight of a low drag Townend-ring engine cowling and large tear-drop wheel streamlines so just the empty weight increase came to 242 lbs.; on top of this was the increase of 208 lbs. in useful load. The engine fairing and the wheel streamlines added measurably to the available speed and the greater useful load no doubt added to the utility; therefore now we have a gross load that has just about absorbed the extra in power, and some of the performance capabili-

ties have dropped nearly to the former levels. However, the payload jumped to some 660 lbs., which would now allow the seating of four large people with full gas load and yet an allowance of 100 lbs. for baggage; in this light a definite advantage was gained and the model C4-300 earned a useful spot in the new "Spartan" line-up. With a price-tag that was rather high for these times, even though there was good value for the money, the C4-300 was left begging for customers and only one example appears in any of the records.

The "Spartan" model C4-300 was a high-winged cabin monoplane of rather large proportion with more than ample seating for 4, in accommodations of good taste, comfort and high quality. One can easily agree that the "Spartan" monoplanes were beautiful airplanes with a distinctly feminine character that were leveled at the owner who could well afford the best. Spartan Aircraft spared very little expense in the make-up of this series to provide the appropriate finery and deluxe equipment that would put them in a class somewhat above the average. As a further accommodation to the type of clientele that would be buying this sort of airplane, the "C-Four" was quite easy to fly, pleasantly stable and did not require a high degree of piloting technique for satisfactory operation. Powered with the 9 cyl. Wright J6 (R-975) engine of 300 h.p. the power advantage over the C4-225 was just about cancelled by increases in weight, so the performance would be nearly typical in most cases except for some gains in available speed. Cleaned-up in the aerodynamic sense and of good aerodynamic proportion too, the C4-300 was not left behind by too many other 300 h.p. craft and whatever shortcomings it did have in certain performance maximums were well overcome by its well-behaved character and pleasant nature. The type certificate number for the "Spartan" monoplane model C4-300 was issued 11-26-30 and there seems to be evidence of only one example built by the Spartan Aircraft Co. at Tulsa, Okla. Some reports indicate that the C4-300 was available also as a single-place cargo-hauling transport but it is doubtful if any were built in this arrangement; this would actually be a modification that could be performed on any existing C4-300.

Listed below are specifications and performance data for the model C4-300 as powered with the 300 h.p. Wright R-975 engine; length overall 32-6"; height overall 9'0"; wing span 50'0"; wing chord 80"; total wing area 299 sq.ft.; airfoil Clark Y; wt. empty (includes Townend ring) 2567 lbs.; useful load 1398 lbs.; payload with 85 gal. fuel 660 lbs.; gross wt. 3965 lbs.; max. speed (with wheel pants) 143; cruise speed 121; landing speed 56; climb 890 ft. first min. at sea level; ceiling 14,500 ft.; gas cap. 85 gal.; oil cap. 7.8 gal.; cruising range at 16 gal. per hour 605 miles;

price at factory $11,500 fully equipped (as 4-place or mailplane), later raised to $11,700.

The construction details and general arrangement of the model C4-300 was similar to that of the model C4-225 (as described in the chapter for ATC # 310 of this volume) including the following. With the use of long moment arms for a gentler response, balanced controls for lighter forces and a graceful wing of high aspect ratio (relation of chord to span) for better efficiency, the "Spartan" monoplane was literally loaded with aerodynamic aids to make it an airplane that would be a pleasure to fly. A large baggage compartment of 9.7 cu.ft. capacity, with allowance for 100 lbs., was located behind the rear seat; cabin dimensions were 40x84x51 in. The so-called Freighter version had ample clear floor area for hauling bulky cargo; crew wt. was 170 lbs. and actual payload was 660 lbs. Tread of the outrigger landing gear was 120 inches; wheels were 32x6 or 9.50x10 low pressure semi-airwheels encased in streamlined fairings. Bendix brakes were standard equipment. A metal propeller, Eclipse electric engine starter, navigation lights and tail wheel were also standard equipment. The Towend-ring engine cowl was more or less a part of the airplane but it was actually offered as optional. The next development in the "Spartan" monoplane series was the model C5-301 described in the chapter for ATC # 389 in this volume.

Listed below is the only known model C4-300 entry as gleaned from registration records: NC-981N; Model C4-300 (# E-1) Wright R-975.

NC-981N was first registered to Wright "Ike" Vermilya c/o Metal Aircraft Co. of Cincinnati, Ohio.

A.T.C. #384
(12-19-30)
LOCKHEED "VEGA", 5-C

Fig. 257A. Lockheed "Vega" model 5-C, one of the last examples built.

The Lockheed model 5-C was the last and perhaps the finest example of the illustrious "Wooden-Vega" line. That it should be the finest and the best in this series of airplanes is only logical development because since the very first example of 1927, every "Vega" that was built spent an exciting life of one kind or another, probing every facet of record-breaking flight. Because of its very nature and its basic endowment, the "Vega" was called upon to lead a life usually steered to gaining maximums in performance or in achievement; this naturally tended to improve the breed. Stressing a craft to its maximums repeatedly, whatever they may be, naturally discloses what it is capable of but also tends to reveal any of its little shortcomings. These shortcomings certainly became obvious in time to those guiding the destiny and development of these craft at the Lockheed factory. Building the fastest and the finest and vowing to keep it that way if humanly possible, "Vega" development engineers kept the basic design continually up-graded to meet its fullest demands, and then too eliminated the little shortcomings as they would appear. The final result now in the model 5-C, was a better airplane with flawless lines of near-perfect beauty that pleased and amazed everyone with its load-toting ability. Its brilliant performance while bearing this sizable load and its swift-

ness up into and through the air was certainly not impaired a bit. Some of the most famous "Vega" in history were the model 5-C or else they had been earlier models that were modified and up-graded to conform to the 5-C specification. Of the 30 or so examples that were listed as the model 5-C, some 24 had been modified from earlier models and only 6 had originally come off the assembly line in this latest specification. Nearly all of these 30 were famous airplanes that recorded exciting entries into the annals of aviation history and with such a stellar line-up it would surely be hard to award top-billing to any one of these. Although now having to share the lime-light with other Lockheed stars such as the "Sirius", the "Altair" and the "Orion", the "Vega" was a romantic favorite and remains with its head above all.

The Lockheed "Vega" model 5-C was an all-wood high winged cabin monoplane with seating arranged for 7, allowing 165 lbs. each for 6 passengers and 90 lbs. for baggage. The normal payload of 1080 lbs. was also listed as "variable" which translated into less payload when carrying extra fuel and oil for longer range or for a greater allowance in baggage. Although the only visible difference in the model 5-C, as compared to earlier "Vega", was the larger fin and rudder for better directional stability and better control, there was beyond that a certain subtle

Fig. 257B. This "Vega" was first model 5-A, modified to 5-C for Braniff Airways.

Fig. 258. This was first "Vega" built at factory as 5-C; flown on pleasure trip in Europe, Africa and Near East.

Fig. 259. Shell Oil Co. "Vega" 5-C was official "press plane" at 1933 National Air Races.

Fig. 260. Vega 5-C on Edo floats.

quality that was slightly different. Fuselage framing was slightly heavier at points of severe bending and torsional stresses, and the canti- lever wing structure was also beefed-up to with- stand the added stresses of higher gross loads. Heavier when empty because of this added muscle, the 5-C had allowable gross weight to compensate for this higher empty weight, also an increase in the payload and weight allow- ance yet for added equipment. Powered with the 9 cyl. Pratt & Whitney "Wasp" SC1 or S3D1 engines of 450 h.p., performance was compara- ble to the earlier model 5-B; converted at vari- ous times, clear up into the late thirties, some versions of the 5-C mounted a controllable pitch propeller which allowed blade-pitch selection for a sizeable increase in the all-round perform- ance. Normally fitted with a deep-chord NACA type engine cowling and large tear-drop wheel pants as a landplane, the model 5-C was also eligible as a seaplane on Edo K or model 4650 twin-float gear. Serving on the fastest air-lines both here in the U.S.A. and abroad, serving in

private ventures and adventures of all sorts too, the 5-C also saw some World War II service as the UC-101 utility transport. To W. P. Fuller of "Fuller Paints" goes the honor of purchasing the very last "Vega" that was built; built in June of 1934, this bright and shiny 5-C was per- haps the finest of its type and served for more than 10 years at varied duties. The type certifi- cate number for the "Vega" model 5-C was is- sued 12-19-30 and some 30 examples of this model were manufactured or converted by the Lockheed Aircraft Corp. at Burbank, Calif.

Listed below are specifications and perform- ance data for the Lockheed "Vega" model 5-C as powered with the 450 h.p. "Wasp" engine; length overall 27'6"; height overall 8'6"; wing span 41'0"; wing chord at root 102"; wing chord at tip 63"; total wing area 279 sq.ft.; airfoil at root Clark Y-18; airfoil at tip Clark Y-9.5; wt. empty 2850 lbs.; useful load 1900 lbs.; payload with 96 gal. fuel 1080 lbs; gross wt. 4750 lbs.; max. speed 180 [190]; cruising speed 155 [168]; landing speed 62; climb 1200 [1300] ft. first min. at sea level; ceiling 18,000 [19,000] ft.; figures in brackets with controllable pitch propeller; gas cap. normal 96 gal.; gas cap. max. 184 gal.; oil cap. 10-15 gal.; cruising range at 22 gal. per hour [96 gal] was 620 miles; price at factory field first as $18,985, raised to $19,000 in March of 1931. Following figures are for seaplane wt. empty 3150 lbs.; useful load 1730 lbs.; payload with 96 gal. fuel 910 lbs.; gross wt. 4880 lbs.; max. speed 170 [180]; cruising speed 145 [165]; landing speed 65; climb 1000 [1100] ft.; ceiling 16,000 [17,000] ft.; figures in brackets with controllable pitch propeller; cruise range

Fig. 261. The last "Vega" manufactured in Burbank, built for W. P. Fuller early in 1934.

Fig. 262. "Vega" 5-C flown to Hawaii by Amelia Earhart.

575-650 miles; all other figures same as land-plane.

The fuselage framework was of wooden cigar-shaped monocoque construction built up of two plywood half-shells that were assembled over spruce annular rings. The walls of the cabin interior were lined with heavily-padded pigskin which gave the interior a sense of elegant richness; there were 4 single adjustable seats to the front and one double seat to the rear, with cabin entry door on the right side. The cabin was fitted with dome lights, 5 windows down each side provided ample vision and the cabin was also fitted with ventilation and heat; a baggage compartment of 12 cu.ft. capacity was to the rear of the cabin section with allowance for up to 180 lbs. The seaplane version was allowed 84 lbs. of baggage. The cantilever wing framework was built up of spruce and plywood box-type spar beams with spruce and plywood web-type wing ribs; the completed framework was covered with 1/8 and 3/32 in. plywood sheet. Two fuel tanks were mounted in the wing, flanking each side of the fuselage; two extra fuel tanks of 44 gal. capacity each could be mounted outboard of the regular tanks for extended range. The center section of the wing, which was actually the cabin roof, could be fitted with a baggage bin for allowance up to 300 lbs. or another extra fuel tank of 48 gal. capacity. The long-legged landing gear used oleo shock absorbing struts and was quickly detachable to mount twin-float seaplane gear; wheels were 32x6 and Bendix brakes were standard. The cantilever tail-group, now slightly larger in area, was of all-wood construction similar to that of the wing; the horizontal stabilizer was adjustable in flight. A metal propeller, electric inertia-type engine starter, NACA type engine cowling, navigation, cockpit and cabin lights, a battery, cabin and cockpit heaters, swivel tail wheel, Pyrene fire extinguisher and first-aid kit were standard equipment. Landing lights, 9.50x12 semi-airwheels with brakes, oil-cooling radiator, extra fuel tanks, bonding and shielding for radio equipment, custom colors and a controllable pitch propeller were optional at extra cost. The next Lockheed development was the low-winged "Orion" model 9 described in the chapter for ATC # 421.

Listed below are "Vega" 5-C entries as gleaned from registration records:

NC-6526;	Vega 5-C (# 9)	Wasp 450.	
NC-7953;	"	(# 23)	"
NC-195E;	"	(# 26)	"
NC-433E;	"	(# 49)	"
NC-434E;	"	(# 50)	"
NC-2875;	"	(# 60)	"
NC-625E;	"	(# 63)	"
NC-898E;	"	(# 72)	"
NC-891E;	"	(# 73)	"
NC-306H;	"	(# 76)	"
NC-975H;	"	(# 96)	"
NC-47M;	"	(# 99)	"
NC-48M;	"	(# 100)	"
NC-49M;	"	(# 101)	"
NC-32M;	"	(# 102)	"
NC-538M;	"	(# 107)	"
NC-539M;	"	(# 108)	"
NC-540M;	"	(# 109)	"
NC-104W;	"	(# 121)	"
NC-105W;	"	(# 122)	"
NC-107W;	"	(# 124)	"
NC-161W;	"	(# 127)	"
NC-934Y;	"	(# 138)	"
NC-997N;	"	(# 139)	"
NC-972Y;	"	(# 160)	"
NC-959Y;	"	(# 170)	"
NC-965Y;	"	(# 171)	"
NC-980Y;	"	(# 191)	"
NC-13705;	"	(# 203)	"
NC-14236;	"	(# 210)	"

Serial # 138-160-170-191-203-210 were originally built as model 5-C; all other serial numbers shown, converted to 5-C from earlier models; ser. # 99 and # 101 operated as seaplanes.

A.T.C. #385
(12-13-30)
NICHOLAS-BEAZLEY, NB-4L

Fig. 263. Nicholas-Beazley model NB-4L with 90 h.p. Lambert engine.

In nearly two years of service now in various parts of the country at jobs usually falling within the sphere of all-purpose service, the "Barling" NB-3 (Nicholas-Beazley) monoplane was finally scheduled for a face-lifting and some modification. No extensive changes had been planned and the modifications were only meant to improve the craft in general. This is not to say that the NB-3 actually needed any serious modification but overall demands on an airplane were fast becoming more stringent with hardly any tolerance for certain limitations; and tastes in styling were, of course, changing too. Thus, the basic changes created in the new NB-4 series centered about the increase in horsepower and a judicious use of cross-sectional area to provide a bit more room and also to do away with the somewhat gaunt look of the earlier NB-3 airframe. In general, the new NB-4 was an improvement but in its effort to conform to new demands it almost grew out of its original concept as designed by Walter Barling. Because of the deviation from original form it became heavier, more buxom and bigger and a slightly more expensive airplane, one similar now to other airplane models all vying for the same piece of the market. As designed by Tom Kirkup, the new NB-4 series, still adhering closely to the earlier design basics in general, were better arranged now for the occupant's comfort. They were probably a little better looking to most people and the expected performance increase, though small, was of course gained by the 30% increase in horsepower. The earlier NB-3, in its various versions, was quite popular and built in some 100 examples; it is quite likely

that the improved NB-4 would certainly have done almost as well had it not been immediately confronted with a sagging in the aircraft market. In flight-test by May of 1930, announced formally in Oct. of 1930 and first nationally shown at the Detroit Air Show for 1931 held in April, the NB-4 attracted many onlookers who paused with interest; however it seems that most all walked away with their hands in their pockets.

The model NB-4L was a low-winged open cockpit monoplane with seating for 3, and in general was still quite typical of earlier series except for the increase in power and several obvious changes to the basic airframe form. Using cross-sectional area in the fuselage to relieve that gaunt look, the NB-4L was faired-out more deeply, but in a slab-sided form that somehow had the appearance of being much too functional for a craft that was to appeal also to the private owner-flyer. A higher fuselage profile offered better protection to all occupants and the interior offered a bit more stretch room. The long-legged oleo landing gear was not quite as finicky to miscalculation or abuse but an increase in wheel tread would have been surely helpful. Powered with the 5 cyl. Lambert R-266 engine of 90 h.p., the NB-4L had inherited a better performance throughout the entire range and its economy per seat-mile was still one of its better features. Both fuselage and wing were of a sturdy all-metal structure that offered durability, ease of maintenance and field repairs to damaged sections; designed for service normally performed by the classic 3-place open cockpit biplane, the NB-4L however posed as a more

Fig. 264. Improved NB-4L was larger and heavier than earlier NB-3.

efficient, practical machine for a more profitable operation. The model NB-4L was first approved on a Group 2 approval numbered 2-260 issued 8-15-30 for serial # 103 and up. The type certificate number for the model NB-4L was issued 12-13-30 and only one example was built by the Nicholas-Beazley Airplane Co. at Marshall, Mo. At a later date, a few of the model NB-4W were converted to the NB-4L type by a change in powerplants. Russell Nicholas was president; Chas. M. Buckner was V.P.; Howard Beazley was secretary and Tom A. Kirkup was chief engineer. Tom Kirkup had replaced Walter H. Barling who had originally designed the NB-3 series and then left N-B employ to design and develop several new designs on his own, one of which was a 6-place cabin monoplane with a remarkable performance on only 165 h.p. Jack Whitaker by now also had replaced Harold Speer as sales manager.

Listed below are specifications and performance data for the model NB-4L as powered with the 90 h.p. Lambert engine; length overall 23' 6"; height overall 6'11"; wing span 32'8"; wing chord 62"; total wing area 159 sq.ft.; airfoil "Barling" 90-A; wt. empty 828 lbs.; useful load 683 lbs.; payload with 24 gal. fuel 350 lbs.; gross wt. 1511 lbs.; max. speed 105; cruising speed 87; landing speed 40; climb 800 ft. first min. at sea level; climb in 10 min. was 6300 ft.; ceiling 16,000 ft.; gas cap. 24 gal.; oil cap. 2.5 gal.; cruising range at 5 gal. per hour was 4.4 hours or 390 miles; price at the factory was $3900.

The fuselage framework was built up of welded chrome-moly steel tubing, faired to a deep cross-section in slab-sided form, then covered in fabric. The cockpits were deep and well protected by large windshields with added width and length in cockpit area providing more

Fig. 265. Wide long-leg landing gear took some of the worry out of landings and take-offs.

leg and shoulder room; seat cushions and seat-backs were air-filled for comfort. A baggage compartment of 2.5 cu.ft. capacity was in the top cowl section behind the front seat with max. allowance for 10 lbs. The thick cantilever wing in 3 sections, of an all-metal construction, was built up of U-section box-type spar beams with stamped-out duralumin wing ribs; the leading edges were covered with dural metal sheet and the completed framework was covered in fabric. Central portion of the wing was straight and flat with end panels of 5 ft. 6 in. length set at a dihedral angle of 6.5 degrees. The gravity-feed fuel tank of 24 gal. cap. was mounted high in the fuselage just ahead of the front cockpit. The long-legged landing gear now of more robust proportion used oleo shock absorbing struts; wheels were 24x4 and semi-airwheels with brakes were later optional. The fabric covered tail-group was built up of welded chrome-moly steel tubing in a larger profile for more effective

area; the vertical fin was ground adjustable and the horizontal stabilizer was adjustable in flight. A wooden propeller and 24x4 hi-pressure tires were standard equipment. Low pressure semi-airwheels with brakes, an Eclipse inertia-type engine starter and navigation lights were optional. The next development in the NB-4 series was the Warner-powered model NB-4W, described in the chapter for ATC # 386 of this volume.

Listed below are model NB-4L entries as gleaned from registration records:
NC-427V; NB-4L (# 103) Lambert 90.
NC-430V;   ”   (# 104)   ”
NC-431V;   ”   (# 105)   ”
Serial # 104 first as model NB-4G with 80 h.p. "Genet" engine as built on Group 2 approval numbered 2-280, later modified to NB-4L with Lambert 90 engine; ser. # 105 first as model NB-4W, later modified to NB-4L.

A.T.C. #386
(12-13-30)
NICHOLAS-BEAZLEY, NB-4W

Fig. 266. Nicholas-Beazley model NB-4W with 90 h.p. Warner "Scarab Jr." engine.

Working on the premise already learned, that a certain amount of airplane buyers would prefer to make a choice from several different engine combinations, Nicholas-Beazley also offered the model NB-4W as a companion model. This model was quite typical of the NB-4L just described, except for its powerplant installation and a change in the fairing-out of the fuselage form. In contrast to the slab-sided form as used on the NB-4L, the NB-4W was faired out to a more rounded cross-section and consequently was the more handsome of the two. Use of the fairing structure and a difference in engine weights added some 32 lbs. to the gross weight but the NB-4W was not noticeably penalized by this extra burden. The model NB-4W was powered with the 5 cyl. Warner "Scarab Junior" engine of 90 h.p. and putting the difference in fuselage form aside for the moment, a selection of the NB-4W over that of the NB-4L would largely be motivated by the preference in engines. That the NB-4W did have more appeal is demonstrated by the fact that a greater number were built in this version. An unverified report stated that an NB-4 was ordered by a mining company in Mexico to use as a courier to bring in mail, the pay-roll and medical supplies to scattered mining camps isolated in the high mountains. In Mexico, this type of work over almost impassable terrain called for a maximum in good all-round performance; that an NB-4 should even be considered for this job is testimony enough of its good reputation. Having two models now in a brand-new line that showed some promise, Nicholas-Beazley

waited patiently for a surge of buying that never came; in hopes of discovering what it was that people would buy, N-B experimented with several different types in the light and ultra-light class, searching for that magic formula. In the meantime, N-B kept busy filling an Army Air Corps contract order for aircraft skis built up of ash wood, duralumin and moly-chrome steels.

The Nicholas-Beazley model NB-4W was an open cockpit low-winged monoplane, with seating arranged for 3, and was generally typical of the NB-4L except for a change in engines and a welcome rounding out of the fuselage form. Especially leveled at the flying-service operator or the private owner-flyer, the NB-4W was well able to perform all chores that would come up in general-purpose service, with the added bonus of comparable or better performance, in an economy that was bound to earn a higher profit. Powered with the 5 cyl. Warner engine of 90 h.p., the NB-4W offered excellent performance through the whole range and was particularly adaptable for use out of the smaller sod-covered airfields. Its nature and its general behavior was best described as compatible and pleasant and its durability and comparative simplicity offered many hours of trouble-free service. That it was not seen about the country in greater number is only a quirk of fate and a circumstance of the times. The model NB-4W was first submitted for approval in Feb. of 1930 and awarded a Group 2 certificate numbered 2-183 which was superseded by approval numbered 2-264, issued 8-27-30. On 10-7-30 the 80 h.p. "Genet" powered version was awarded a

Fig. 267. NB-4W shown here on flight-line of flying school.

Group 2 approval numbered 2-280 as the NB-4G. The type certificate number for the model NB-4W was issued 12-13-30 and some 6 or more examples of this model were built by the Nicholas-Beazley Airplane Co. at Marshall, Mo. Disappointed of course, but not totally discouraged by the failure of the NB-4 to sell, Nicholas-Beazley developed other models in an effort to yet get a piece of the market. The NB-4 type was also powered with an 80 h.p. "Genet" (British) engine as the model NB-4G but its further development was dropped. A mysterious NB-PG type was powered with a 2 cyl. "Aeronca" engine but other than the fact it must have been a flivver-type airplane, other details are unknown. Another flivver-plane for 2 was powered with the new 4 cyl. Continental A-40 engine of 37 h.p. and was called the NB-7; more detail than this is also unknown. The last

of this experimental bunch was a 2-place parasol type monoplane called the NB-8; powered in early tests with a 3 cyl. Szekeley engine of 45 h.p., it was also modified to take the 5 cyl. "Genet" engine of 80 h.p. and Nicholas-Beazley did have a fair amount of success with this particular version.

Listed below are specifications and performance data for the model NB-4W as powered with the 90 h.p. Warner engine; length overall 23'4"; height overall 6'11"; wing span 32'8"; wing chord (constant) 62"; total wing area 159 sq. ft.; airfoil "Barling" 90-A; wt. empty 860 lbs.; useful load 683 lbs.; payload with 24 gal. fuel 350 lbs.; gross wt. 1543 lbs.; max. speed 105; cruising speed 87; landing speed 40; climb 800 ft. first min. at sea level; climb in 10 min. 6300 ft.; ceiling 16,000 ft.; gas cap. 24 gal.; oil cap. 2.5 gal.; cruising range at 5 gal.

Fig. 268. NB-4W modified with installation of 110 h.p. Warner "Scarab" engine.

per hour was 4.4 hours or 380 miles; price at factory $3,900.

The construction details and general arrangement of the model NB-4W were typical to that of the NB-4L as described in the chapter for ATC #385 of this volume, including the following. The fuselage framework was faired out deeper to a more rounded cross-section by wooden formers and fairing strips, then fabric covered. Metal cockpit cowling panels were also rounded out and the high profile of the turtleback was now dropped down and fitted with a streamlined head-rest. Well padded cockpits were deep and well protected by unusually large windshields; cockpit interiors were neatly upholstered and the seat cushions and seatbacks were filled with air for comfort. The longlegged landing gear of some 5 ft. tread was fitted with low pressure semi-airwheels and brakes; the tail skid was a rubber-snubbed steel tube but some later versions used the steel spring-leaf type. The fabric covered tail group had slightly more area in the vertical fin and the horizontal stabilizer was adjustable in flight. A wooden Fahlin propeller was standard equipment. A metal propeller, navigation lights, and Eclipse hand crank inertia-type engine starter were optional. The next Nicholas-Beazley development was the model NB-8 a "parasol" monoplane described in the chapter for ATC #452.

Listed below are model NB-4W entries as gleaned from registration records:

NC-431V;   Model NB-4W (#105) Warner 90.
NC-433V;   ”   (#106)   ”
NC-951N;   ”   (#107)   ”
NC-967N;   ’   (#108)   ”
NC-437V;   ”   (#109)   ”
NC-501Y;   ”   (#110)   ”

Serial #105 later converted to model NB-4L; ser. #110 later fitted with 110 h.p. Warner "Scarab" engine.

A.T.C. #387
(12-20-30)
BIRD, MODEL C

Fig. 269. "Bird" model C with 165 h.p. Wright J6-5 engine was rare version in popular series.

With not much to go on but some factory released data and a few scattered hints about its actual existence, there is very little else of any interest to discuss about the "Bird" model C. As an all-purpose 3-place biplane, similar to other "Bird" models in most respects, we notice that it mounted an engine of 165 h.p. and wonder just what purpose it was intended to serve. Judging from its utility arrangement, with the performance that was the result of a more powerful combination, it immediately comes to mind that it was probably leveled more at the sportsman-pilot. Brought out in a period of time when the aircraft market was beginning to dwindle away at an alarming rate, Bird Aircraft, like so many others, was extending their line of airplanes into many different models. In this way they hoped to get at least a little piece of the business. Not having a model for the flying sportsman who could afford to pay a little more for a fine airplane with high performance, it is logical then to assume that the Model C was groomed as an airplane to fill this need. As a potential top-ranking star in the "Bird" line-up for 1931, a more concerted effort in its promotion probably would have eventually paid off, but the company seemed to be very quiet about this model. Stacked against other airplanes of this type, by any comparison, the "Bird" C stands nearly at the top of the list.

Had the model C been built in any number, it would have been interesting to note the reaction from pilots who would have hardly believed that so much performance was possible on just 165 h.p. As it was, the model C had no formal debut, nor a chance to make friends, so was left begging for customers.

Bird Aircraft was proud of their product and the spontaneous testimony of better than 100 satisfied owners already was enough to bear them up in any boast, casually reminding one to "Ask the pilot who flies one". Bird was so confident of their comparative excellence in performance, they offered a direct challenge to those in the industry with no holds barred. At the 1931 National Air Races they offered to match a "Bird" biplane against any airplane, be it equipped with rotors, vanes, wing slots, flaps or variable camber. There were no official tryouts to settle this question, as far as we know, but rumored incidents always had the "Bird" coming out on top. With a fair breeze to bolster the lift, it is entirely conceivable that the "Bird" could put an autogiro to shame with its ability to get in and out of a small 'field. Many seasoned pilots who had the experience to make a comparison, endorsed the "Bird" biplane as the safest airplane built, regardless of type. Wiley Post, famous round-the-world flyer who was not particularly concerned with safety but well able to recognize it, singled out the "Bird" as the safest airplane built for the average private owner-pilot. No airplane is perfect, any airplane has traits that need to be understood; even the "Bird" had some faults of operational nature that had to be overcome by study and practice. Because of its aerodynamic arrangement the top.heavy "Bird" often came to grief while on the ground but it was hard to get into any trouble while in the air.

The "Bird" model C was an open cockpit biplane with seating arranged for 3 and was typical of other models in this series except for its engine installation and some minor modifications necessary to this particular combination. Powered with the 5 cyl. Wright J6 (R-540) engine of 165 h.p., the model C poses as the most powerful version offered in this series of all-purpose airplanes. Lacking any opinion or recorded information concerning the characteristics and behavior of the Model C, we resort to a little guesswork, but this is not out of line because adding power to the basic airframe of the "Bird" biplane points only to predictable advantages. Quicker, shorter take-offs, a faster, steeper climb-out, and snappier response in the more difficult maneuvers would be obvious gains, with the slight increase in gross weight being of very little consequence to overall performance. Because of its aerodynamic arrangement the "Bird" biplane was reluctant in translating extra horsepower into very much more speed so the gains were small but still within the range of the best average. Pondering on the qualifications of the Model C as a whole, it leaves one with the impression that this certainly would have been an excellent airplane for the type of owner that is usually dubbed sportsman-pilot. The type certificate number for the "Bird" model C was issued 12-20-30 (sometimes listed as 12-23-30) and only one example of this model was built by the Bird Aircraft Corp. at Glendale, Long Island, N.Y. Maj. Thomas G. Lanphier was president; Wm. E. Winkle was V.P.; James Phelan was secretary; Michael Gregor was chief engineer, assisted by J. F. Lindstrom on detail design and drafting. It is not known if Bird Aircraft had any company pilots but Lee Gehlbach and Wm. Lancaster were often credited with promotion in the field. Lanphier, Winkle and Gregor were all accomplished pilots and no doubt also handled many of the company's piloting chores.

Listed below are specifications and performance data for the "Bird" model C as powered with the 165 h.p. Wright J6 engine; length overall 22'7.5"; height overall 8'6"; wing span upper 34'0"; wing span lower 25'0"; wing chord upper 69"; wing chord lower 48"; wing area upper 184 sq.ft.; wing area lower 82 sq.ft.; total wing area 266 sq.ft.; airfoil USA-40B modified; wt. empty 1425 lbs.; useful load 825 lbs.; payload with 45 gal. fuel 355 lbs.; gross wt. 2250 lbs.; max. speed 125; cruising speed 105; landing speed 38; climb 950 ft. first min. at sea level; ceiling 18,000 ft.; gas cap. 45 gal.; oil cap. 4 gal.; cruising range at 9 gal. per hour 500 miles; price at factory $5870.

The construction details and general arrangement of the Model C was typical to that of other "Bird" models, particularly as described in the chapter for ATC #101 of U.S. Civil Aircraft, Vol. 2 and ATC #239 of Vol. 3, including the following. The cross-axle landing gear of 66 in. tread was fitted with oleo-spring shock absorbing struts faired with metal streamlined cuffs; wheels were first 30x5 with Bendix brakes, later changed to 7.50x10 semi-airwheels with AP brakes. Some listings give upper wing span as 33 ft. and upper wing area as 175 sq. ft. for a total area of 257 sq. ft. for the Model C but this hardly seems likely. Navigation lights were first of the streamlined-egg type mounted externally and later changed to flush-type mounted internally. It appears that retractable landing lights were also later fitted to underside of upper wing for night flying. There have been scattered hints that Bird was grooming the Model C as an economy-type mail-cargo carrier but this could not be confirmed. A metal panel cover for the front cockpit was probably optional and bettered the top speed by some 5 m.p.h. The baggage allowance was 15 lbs. and this included a 10 lb. tool kit that was stored in the engine compartment. A metal propeller, navigation lights and low pressure semi-airwheels were standard equipment. The next development in the "Bird" biplane was the 4-seated model CK described in the chapter for ATC #388 of this volume.

Listed below is the only known example of the "Bird" model C as gleaned from various records:

X-876W; Model C (#5001) Wright R-540.

Several changes appeared on this model after being registered as NC-876W; some time later this same ship was also listed as a Model A, whether this was an error or if the R-540 engine was replaced with a Curtiss OX-5 we cannot determine.

A.T.C. #388
(12-26-30)
BIRD, MODEL CK

Fig. 270. "Bird" model CK with 125 h.p. Kinner B5 engine; front cockpit seated 3.

With an airplane that had the inherent capability to haul a sizeable load on nominal power, the "Bird" design-team realized the potential of adding an extra passenger to the payload of their versatile all-purpose biplane. Using the popular model BK (ATC # 239) as a basis for the new conversion, the front cockpit opening was enlarged just a bit to seat 3, and the installation of the 125 h.p. Kinner B5 engine was just enough power to offer a good performance in still the strictest economy. As the model CK, this new 4-seater was offered on wheels or floats and it soon became one of the more popular versions in the "Bird" biplane line-up. Many operators, forced to operate during these times on a limited budget for some profit return, were attracted by the extra seating offered, a payload that could mean extras in revenue; they therefore chose the CK for service ranging from charter-taxi and joy-rides to student instruction. Although the hard-working CK appealed mostly to the economy-minded owner who was trying to make a living, it also had some appeal to those that had plenty to spend; Bernarr McFadden and Powell Crosley, Jr. each harbored a CK in their hangars. It has also been said and it sounds reasonable too, that the 4-seated CK was the inspiration for several small air-lines in the east that offered shuttle service from small outlying towns to points on the main air-lanes. With every "Bird" model being of the same mold, so to speak, the model CK was of course typical and

perhaps it inherited all of the better qualities that has made this airplane a perennial favorite. Extremely versatile as a landplane, the seaplane version was particularly outstanding amongst craft of its type and could carry a useful load nearly equal to its empty weight. Operating in nearly all parts of the U.S.A., one CK even found its way down to Brazil in So. America. The rugged character, the inherent safety and the popularity of this model naturally promoted its longevity, so it is not surprising that at least 28 out of the 42 or so that were built, were still flying in active service by 1939.

The "Bird" model CK was an open cockpit biplane with seating arranged for 4; three passengers were normally seated in the front cockpit and it was also possible to fold up the front seat to install dual controls for flight instruction. It was possible too to remove all seating and load the front pit with up to 540 lbs. of cargo; it was quite unusual for an airplane of such nominal power to haul such a hefty load and not be restricted in its base of operations. Any normal strip would do. As a seaplane on Edo twin-float gear, the CK was loaded for test to a 2800 lb. gross and broke clean from the water in 15 seconds; not greatly penalized by the bulky floats, performance of the seaplane version was very good. Powered with the 5 cyl. Kinner B5 (R-440) engine of 125 h.p., the model CK nurtured every single horsepower to best advantage and by virtue

Fig. 272. Prototype CK shows excellent arrangement for visibility and stable flight.

of its sensible design, was able to operate nimbly and efficiently with best economy. It would not be unusual to toss laurels at the wheels of any "Bird" because each model in some way would be deserving, but it is felt that perhaps the model CK had earned the better right. Let us also say here, that Michael Gregor, creator of the "Bird" line in its infancy, watched over each successive development carefully and every example was a flying tribute to his apparent genius. The type certificate number for the "Bird" model CK was issued 12-26-30

and at least 42 examples of this model were manufactured by the Bird Aircraft Corp. at Glendale, L.I., N.Y. We might add that two of the "Bird" CK were flown to 6th and 7th position in the National Air Tour for 1931 by Wm. Lancaster and Lee Gehlbach, averaging better than 100 m.p.h. for the long grind.

Listed below are specifications and performance data for the model CK as powered with the 125 h.p. Kinner B5 engine; length overall 23'0"; height overall 8'6"; wing span upper 34'0"; wing span lower 25'0"; wing chord upper 69"; wing

Fig. 273. "Bird" CK shown during 1931 National Air Tour; two entries finished 6th and 7th.

Fig. 274. 4-seated "Bird" CK was popular as air-taxi.

Fig. 275. "Bird" CK on Edo floats; craft was excellent seaplane.

chord lower 48"; wing area upper 184 sq. ft.; wing area lower 82 sq. ft.; total wing area 266 sq. ft.; airfoil USA-40B modified; wt. empty 1350 lbs.; useful load 985 lbs.; payload with 42 gal. fuel 540 lbs.; gross wt. 2335 lbs.; max. speed 118; cruising speed 100; landing speed 38; climb 700 ft. first min. at sea level; ceiling 16,000 ft.; gas cap. 42 gal.; oil cap. 3 gal.; cruising range at 7 gal. per hour 530 miles; price at factory $4395. Following figures are for seaplane as mounted on Edo M twin-float gear; wt. empty 1602 lbs.; useful load 1005 lbs.; payload with 42 gal. fuel 560 lbs.; gross wt. 2607 lbs.; max. speed 112; cruising speed 95; landing speed 42; climb 620 ft.; ceiling 14,500 ft.; price at factory $6255. CAA approved wts. later quoted for landplane as 1318 lbs. empty, 985 lbs. useful load and 2303 lbs. for gross wt.; this would have a slight improving effect on the performance.

The construction details and general arrangement of the "Bird" model CK was typical to that of other models, including the following. The front cockpit opening was enlarged to accomodate an extra passenger and entry was by way of a large door on the right side; the extra seat could be folded up when not in use. A fair-sized baggage compartment was in the turtle-back section just behind the rear cockpit; the landplane was allowed 30 lbs. in this bin and the seaplane was allowed 30 lbs. also, plus an anchor and rope (20 lbs.) which was stored in the front cockpit. The fuel tank was mounted high in the fuselage ahead of the front cockpit and extended partially into the nose-section. The cross-axle landing gear of 66 in. tread used oleo-spring shock absorbing struts; wheels were 7.50x10 of the low pressure type and brakes were available. Metal formed cylinder head fairings were optional for an improved air-flow around the cylinders and later examples were often shrouded with a special Townend-type "speed ring" cowling. A metal propeller, navigation lights, dry-cell battery, wheel brakes and Heywood engine starter were later offered as standard equipment. A wooden propeller was

optional. The next development in the "Bird" biplane series was the Jacobs-powered model CJ described in the chapter for ATC #419.

Listed below are "Bird" model CK entries as gleaned from registration records:

X-787Y;	Model CK (#4001) Kinner B5	
NC-790Y;	" (#4002	"
NC-791Y;	' (#4003)	"
NC-914V;	" (#4004)	"
NC-915V;	" (#4006)	"
NC-726Y;	" (#4007)	"
NC-725Y;	" (#4008)	"
NC-845W;	" (#4009)	"
NC-818W;	" (#4010)	"
NC-850W;	" (#4012)	"
NC-916V;	" (#4014)	"
NC-917V;	" (#4015)	"
NC-918V;	" (#4016)	"
NC-919V;	" (#4017)	"
NC-933V;	" (#4018)	"
NC-758Y;	" (#4019)	"
NC-759Y;	" (#4020)	"
NC-931V;	" (#4021)	"
NC-782Y;	" (#4022)	"
NC-94V;	" (#4023)	"
NC-95V;	" (#4024)	"
NS-34;	" (#4025)	"
NS-35;	" (#4026)	"
NC-96V;	" (#4027)	"
NC-97V;	" (#4028)	"
NC-98V;	" (#4029)	"
NC-972M;	" (#4031)	"
NC-976M;	" (#4032)	"
NC-980M;	" (#4035)	"
NC-981M;	" (#4036)	"
NC-726N;	" (#4037)	"
PP-TAU;	" (#4042)	"

Serial #4003 also on floats; ser. #4004 registered to Bernarr McFadden; reg. no. for ser. #4005-4011-4013-4030-4033-4034-4038-4039-4040-4041 unknown; ser. #4029 registered to Powell Crosley, Jr.; ser. #4039 as model RK with 160 h.p. Kinner R engine on Group 2 approval numbered 2-502; ser. #4042 operated in Brazil; approval expired 3-3-1933.

A.T.C. #389
(12-26-30)
SPARTAN, MODEL C5-301

Fig. 276. Spartan model C5-301 with 300 h.p. "Wasp Jr." engine; craft seated 5.

As a companion model to the C4-series monoplane, Spartan Aircraft developed the C5-series which were basically typical in most respects, except for a rearrangement in the interior capacity to allow the seating of a pilot and 4 passengers. Broad, tall, and long, there was actually ample room in the basic layout of the C4-series interior so it was no great chore to utilize part of the huge baggage compartment in the rear for a fifth seat, and still have room enough to one side for 125 lbs. of baggage. With the ability to carry the extra payload, and cabin room already available to locate the extra seat, "Spartan" engineers were not greatly taxed to come up with a new monoplane series that was powered with the choice of 2 different engines. The model C5-300 was to be powered with the 9 cyl. Wright "Whirlwind Nine" (R-975) engine of 300 h.p. and the model C5-301 was powered with the 9 cyl. "Wasp Junior" engine, also of 300 h.p. Performance of either model would be more or less identical, so selection of either would be but a matter of engine preference; however the C5-301 could be considered the more popular of the two in regard to the number built. Of proportion and dimension fairly typical of the 4-place model C4, the model C5 somehow at least, appeared to be a somewhat bigger airplane; most of its performance maximums were slightly penalized in comparison, by the extra in gross weight. No ready market was available for this new craft but those few that were built were in use by mid-western oil companies for business promotion. A rather unusual vocation for the big "Spartan" monoplane (model un-

known) was the contract received from the U.S. Weather Bureau by Johnny Starr, a Tulsa pilot. The story relates that early each morning the "Spartan" was flown on a daily weather check to an altitude of 14,000 ft. above the Omaha, Neb. station; careful readings were taken for wind velocity, humidity, air temperatures at different levels and other related weather data. A "Spartan" biplane (model C3-225) was also known to be engaged in weather reporting at about this same time in another area.

The "Spartan" model C5-301 (C5-300) was a high-winged cabin monoplane of large and well-rounded proportion that had seating for five in accommodations of comfort and high quality. Thoroughly equipped with just about all of the so-called extras, the C5-series carried a rather high price-tag for these particular times, but in its favor was the fact that owners would not be plagued with extras to buy after taking delivery. Carefully planned to instigate efficient aerodynamic forces, the C5 was pleasantly stable, rather easy to fly and did not require the maximum in piloting technique for a satisfactory operation; one need not be the intrepid airman to fully appreciate the joys of relaxed flight in a craft of this character. Offered as the C5-300 with the 300 h.p. Wright J6 engine and as the C5-301 with the 300 h.p. Pratt & Whitney "Wasp Jr." engine, the model C5-301 seemed to be the only one appearing in any registration records; it is believed that one example of the C5-300 was built but this cannot be confirmed by record. Despite the heavy loadings the C5 had ample performance for a craft of this type and

Fig. 277. C5-301 operated by Skelly Oil Co., exploiting "The Air Adventures of Jimmie Allen", a popular radio show in the thirties.

the owner could be confident that his ship was purposely stout for the added safety, purposely large for better aerodynamic arrangement and purposely spacious for more convenient operating utility. As a companion model to the passenger-carrying craft, a single-place freighter version was also available. The type certificate number for the model C5-301 (C5-300) was issued 12-26-30 and it seems likely that 4 examples of this model were built by the Spartan Aircraft Co. at Tulsa, Okla.

Listed below are specifications and performance data for the "Spartan" model C5-301 as powered with the 300 h.p. "Wasp Jr." (R-985) engine; length overall 32'8"; height overall 8'11"; wing span 50'0"; wing chord 80"; total wing area 299 sq.ft.; airfoil Clark Y; wt. empty 2632 lbs.; useful load 1543 lbs.; payload with 85 gal. fuel 805 lbs.; crew wt. (freighter) 170 lbs.; gross wt. 4175 lbs.; max. speed 145; cruising speed 124; landing speed 58; climb 875 ft. first min. at sea level; ceiling 14,500 ft.; gas cap. 85 gal.; oil cap. 7.7 gal.; cruising range at 16 gal. per hour 610 miles; price at factory $13,350. All figures would also apply to the model C5-300.

The construction details and general arrangement of the C5-301 were typical to that of the models C5-225, C4-300, and C4-301 as described in the chapters for ATC # 310-383-394 in this volume, including the following. The fifth seat in the C5 was located on right side of the cabin to the rear, with a baggage compartment for an allowance of 125 lbs. to the left of

this rear seat; baggage was accessible from the outside only. Cabin appointments included rich upholstery, dual controls, and unusually complete instrument panel, adjustable cabin ventilators, cabin heater, cabin lights, fire extinguisher and all windows were of shatter-proof glass. Windows in the forward section could be rolled down for convenience during ground maneuvering. The wide tread outrigger landing gear used "Aerol" shock absorbing struts and wheels were low pressure type in either 9.50x10 or 9.50x12 sizes; wheel fairings and Bendix brakes were standard equipment. Engine cowling was a modified Townend-ring type and a long tail-pipe directed gases well below the cabin level. The tail-group was built up in a composite structure and both rudder and elevators were aerodynamically balanced; the horizontal stabilizer was adjustable for trim during flight. A metal propeller, navigation lights, electric engine starter, a battery, swivel-type tail wheel with 12x4 tire, and choice of colors were also standard equipment. The next development in the "Spartan" monoplane was the model C4-301 described in the chapter for ATC # 394 in this volume.

Listed below are C5-301 entries as gleaned from registration records:
NC-986N; Model C5-301 (# H-1) Wasp Jr. 300.
NC-11006;   "   (# H-2)   "
Registration number for serial # H-3 and # H-4 unknown; ser. # H-4 also eligible as C5-301-P with crew of 3 and 114 gal. fuel for special-purpose work at 4175 lbs.

Fig. 278. Fairchild model 100 with 575 h.p. "Hornet" B engine; ship shown here in prototype.

Sherman Fairchild clung to the idea that airplanes should be built to carry useful burdens, burdens of all sorts that would prove the unmatched utility of the airplane; it is this idea that prompted the efforts and activities within his airplane manufacturing company more than anything else. For example his first, the FC-1 of 1925, was designed to carry Fairchild's aerial cameras in better comfort and more utility. The FC-2 and FC-2W became notable for activities in the Canadian "bush country", the larger FC-2W2 carried cargo across the high Andes of So. America and helped develop air-mail service in Mexico; the improved model 71 became a popular transport for many of the early air-lines, and became quite popular with progressive men of business. Sherman Fairchild insisted that each of the company's designs could work at varied tasks dependably, without complaint, and upon these demands a long-held tradition was based. Up to this time the 7-place "Fairchild 71" was the most popular airplane produced by this company and its history of achievement was down-right amazing, but progress in aviation has been ever-linked to the needs of more payload at greater speeds so a new transport design was already in the making. With the suggested plan to retain proven characteristics and other useful features, Fairchild engineers designed the "Model 100" as an airplane that came at a rather bad time in this company's history, but was later developed into a very useful machine. Plagued with the un-

certainty caused by the "economic depression" in this country, the Fairchild corporate structure certainly had its ups and downs, finally emerging as the American Airplane & Engine Corp., later in 1931, with a hasty revival of the "Model 100" design which became the "Pilgrim 100-A". Because of this curtailment in the normal course of development, the actual Fairchild "Model 100" was rare and scarce in number, but it did set the stage for a very successful and popular design concept.

As shown here, the Fairchild model 100 was a high-winged cabin monoplane of fairly large proportion that had seating for 9 and was arranged primarily for the transport of passengers on the shorter routes; as an added feature, all passenger seating was easily removable to convert the "One Hundred" into a profitable cargo-hauling freighter. Of slightly larger size than previous "Fairchild" transports, the Model 100 retained most proven features, plus the addition of modifications to structure and arrangement that were more suited to a larger craft. Powered with the 9 cyl. Pratt & Whitney "Hornet B" engine of 575 h.p. the Model 100 carried a substantial amount of payload with ease in a broad range of very good performance. Typical of all "Fairchild" aircraft that preceded this model, the "One Hundred" was kept rather plain and simple for care-free dependable service, a maximum in operating utility, and no need for intricate and time-consuming maintenance. "Fairchild" airplanes had built up an enviable

Fig. 279. Built in one example only, Model 100 became basis for design of famous Pilgrim 100-A.

record for dependable service in scattered corners of the globe and the Model 100 design, in a later version, was to prove itself capable of matching its accomplishments with the best of its forerunner. No lore was left behind by this particular craft because of its scarcity but indications are that it surely possessed all of the characteristics that made the "Fairchild" monoplane such a great favorite, and it possessed a definite personality that later sired the better-known "Pilgrim 100" series. The type certificate number for the Fairchild Model 100 was

issued 1-14-31 and it seems that only one example of this model was built by the Fairchild Airplane Mfg. Corp. at Farmingdale, L.I., N.Y. As of April in 1931, Fairchild became the American Airplane & Engine Crop. as a division of the huge Aviation Corp. F. G. Coburn was president; V. E. Clark and Thurman H. Bane were V.P. and Virginius E. Clark was chief engineer.

Listed below are specifications and performance data for the Fairchild Model 100 as powered with the 575 h.p. "Hornet B" engine; length overall 38'0"; height overall 11'6"; wing span

Fig. 280. Approved version of Model 100 incorporated many changes.

Fig. 281. Model 100 offered comfort for 8 passengers and good vantage point for pilot.

57'0"; wing chord 84"; total wing area 418 sq.ft.; airfoil Goettingen 398; wt. empty 3700 lbs.; useful load 2800 lbs.; payload with 120 gal. fuel 1820 lbs.; crew wt. 170 lbs.; gross wt. 6500 lbs.; max. speed 141; cruising speed 120; landing speed 59; climb 900 ft. first min. at sea level; climb to 10,000 ft. was 15.5 mins.; ceiling 18,000 ft.; gas cap. 120 gal.; oil cap. 12 gal.; cruise range at 32 gal. per hour was 432 miles; price at factory not quoted. The following figures are for a later modification of this same airplane; wt. empty 3850 lbs.; useful load 2650 lbs.; payload with 120 gal. fuel 1670 lbs.; gross wt. 6500 lbs.; all other figures were typical.

The fuselage framework was built up of welded steel tubing in various grades, faired to shape with duralumin formers and fairing strips, then fabric covered. The main cabin was arranged to seat 8 passengers with entry through two large doors in the rear; the pilot's cabin forward was elevated above the main cabin to a point of better visibility, with a separate entry door on the right side. A large compartment for mail and baggage was under the pilot's floor with access door on the right side. The semi-cantilever wing framework was built up of girder-type spar beams of welded steel tubing with wing ribs of riveted dural tubing; the leading edges were covered with dural metal sheet and the completed framework was covered in fabric.

The folding wings were hinged to a center-section built into the upper fuselage and braced by two vees of streamlined duralumin tubing; one fuel tank was mounted in each wing half. The wide-tread landing gear was of the outrigger type using oleo & spring shock absorbing struts; wheels were 32x6 and Bendix brakes were standard equipment. A full-swivel tail wheel was fitted with a hard rubber tire. The fabric covered tail-group was built up of dural spars and ribs in a riveted structure; both rudder and elevators were aerodynamically balanced and the horizontal stabilizer was adjustable in flight. Engine compartment temperatures were regulated by adjustable louvers in the nose-section and the engine was shrouded in a Townend-ring fairing. A metal propeller, navigation lights, and electric inertia type engine starter were also standard equipment. The next development in the Fairchild 100-series was the "Pilgrim" 100A described in the chapter for ATC # 443; the next Fairchild development in the KR-series was the "Model 22" described in the chapter for ATC # 408.

Listed below is the only known "Fairchild 100" entry as gleaned from registration records:
NC-754-Y; Model 100 (# 6000) Hornet B.
Serial # 6000 later in 1931, also listed as the Pilgrim 100.

Fig. 282. Bellanca "Airbus" model P-200 with 575 h.p. "Cyclone" engine.

Following the development of the earlier "Airbus" model P-100 (ATC #360) quite closely, the model P-200 was introduced to the industry also as a versatile single engined air transport; a specially designed transport for hauling mixed loads of mail-cargo and passengers with an economy of upkeep and operation that pointed only to improved service at substantial profit. Gaining several hundred pounds in useful load by the installation of a large air-cooled engine, an installation undoubtedly more compatible to this particular design, the new "Airbus" posed in its majestic splendor as one of the brightest offerings of the new year. Available as a 12 or 15 place carrier by simple adjustment of fuel or cargo load, the "Bus" was also available as a good-performing float-seaplane (model P-200-A). It was used for a time by the New York & Suburban Airlines and service was offered from points in the outlying Long Island area to a dock on the East River, unloading at the foot of Wall St. in downtown New York City. Another example, a P-200 Deluxe landplane, was owned privately and fitted with a handsome custom-club interior, a gay color scheme and scads of special equipment. Due to circumstances guided by our stifled economy at this time, airlines were not buying the "Airbus" as was hopefully expected, but further development and progressive improvement of the series continued. By

1932 the "Airbus" was available with the 650 h.p. Wright "Cyclone" F or the Pratt & Whitney "Hornet" S3D1-G of 650 h.p., gaining a considerable boost in its performance with the ability to now carry 15 and yet a larger load of mail and cargo. Also in 1932, the Army Air Corps bought four of the "Airbus" fitted with 550 h.p. "Hornet" (geared) engines for utility transport work and labeled them the YIC-27; in 1933 ten more of the "Airbus" were delivered and put into service as the C-27-A with 650 h.p. "Hornet" engines and in turn, in 1934, one of these was converted to a C-27-B with 675 h.p. "Cyclone" F engine, 9 were converted to C-27-C with the 750 h.p. "Cyclone" engine and the four original YIC-27 were also converted to the C-27-C. Available in these more powerful versions through 1934, with gross weight allowances increased by nearly a ton, the "Airbus" by then was most often being called the "Aircruiser" but still no commercial examples of these had been built. In 1935, with the basic airframe modified slightly and still more power added, Bellanca finally introduced the new, bigger "Aircruiser" series on ATC #563. Powered also with large "air-cooled radial" engines as in the P-200 and P-300 versions, the Bellanca "Airbus" concept was finally given the chance for better expression in various versatile combinations; in a class definitely by itself, the "Air-

Fig. 283. "Airbus" P-200 Deluxe seated 9 in plush interior.

bus" was a type of craft that probably represent-ed the maximum in practical utility and effi-ciency with the inherent ability to often go way beyond the bounds of its intended capacities.

As the forerunner to the very ultimate in single -engined transports (the "Aircruiser"), the Bellanca "Airbus" model P-200 was a large high-winged monoplane of the sesqui-plane type with a long and spacious cabin interior adaptable to a variety of loads and seating. Normally as a 12-passenger (P-200) or a 15-passenger (P-300) carrier, the "Airbus" was also available in custom interior arrangements for 9 people (P-200 Deluxe), or all seating could be removed for conversion to an all-cargo carrier. Particularly adaptable to installation of twin-float gear be-cause of its odd configuration, the "Airbus" (P-200-A) operated well as a seaplane. Powered at first with the aircooled 9 cyl. Wright "Cy-clone" R-1820-E engine of 575 h.p. there was a slight detriment to overall performance as com-pared to the earlier model P-100 (Curtiss GV-1570 of 600 h.p.); however a gain in utility and reliability offset these losses to some ex-tent and increased power ratings of the "Cy-clone" engine offered progressive performance increases. Developed through four different Army Air Corps models, some of which were used to haul airmail in the 1934 fiasco, the Bellanca "Airbus" finally evolved into the bigger, faster and more powerful versions of the famous "Aircruiser" type. With useable ef-ficiency uppermost in mind, the "Airbus" made no trades with utility or performance to attain that goal, and as a special treat, it has been said the pilot had 'most everything going his way; pilots always spoke highly of the big

"Airbus" and literally jumped at the chance to fly it. It is one of the quirks of aviation his-tory that an airplane with so much promise was so sadly neglected by those it was actually de-signed to help. The type certificate number for the Bellanca "Airbus" model P-200 was issued 1-26-31 and shortly after amended to include the P-200-A seaplane, the P-200 Deluxe club-plane and the 15-place model P-300. Four commercial examples were built in the P-200, P-300 series and the subsequent 14 airplanes were delivered to the Army Air Corps as C-27 utility transports by the Bellanca Aircraft Corp. of New Castle, Dela. Giuseppe Mario Bellanca was president and chief of engineering; Andrew Bellanca was V.P. and secretary; Robert B. C. Noorduyn was V.P. and assistant engineer; veteran Geo. W. Haldeman was chief pilot.

Listed below are specifications and perform-ance data for the Bellanca "Airbus" model P-200 as powered with the 575 h.p. Wright "Cyclone" R-1820-E engine; length overall 42'9"; height overall (tail up) 11'6"; height over-all (tail down) 10'4"; wing span 65'0"; wing chord 94"; total wing area 652 sq. ft.; airfoil "Bellanca"; empty wt. 5155 lbs.; useful load 4435 lbs.; payload with 200 gal. fuel 2945 lbs. (11 pass. at 170 lb. each & 1075 lb. baggage-cargo); payload with 200 gal. fuel (model P-300) was 14 pass. at 170 lb. each & 565 lb. baggage-cargo; gross wt. 9590 lbs.; max. speed 143 (with speed-ring cowling); cruising speed 122; landing speed 55; climb 700 ft. first min. at sea level; ceiling 14,000 ft.; gas cap. max. 200 gal.; oil cap. 12 gal.; cruising range at 32 gal. per hour was 6 hours or 720 miles; price at factory field $32,900; P-200 and P-300 series available

Fig. 284. "Airbus" P-200-A seaplane on Edo floats; configuration was particularly suitable for float installation.

with 600 h.p. "Cyclone" engine by March 1931 for slight increase in all-round performance; weights and specs remained same.

Following figures are for the model P-200-A seaplane as mounted on Edo twin-float gear; wt. empty 5800 lbs.; useful load 3800 lbs.; payload with 150 gal. fuel 2635 lbs. (11 pass. at 170 lb. each & 765 lb. baggage-cargo); gross wt. 9600 lbs.; max. speed 135; cruising speed 115; landing speed 60; climb 680 ft. first min. at sea level; ceiling 13,000 ft.; price upon request; all other specs and data comparable.

Following figures for improved "Airbus" as offered in 1932-33; powerplants either the Wright "Cyclone" R-1820-F or Pratt & Whitney "Hornet" S3D1-G (geared), both of 650 h.p.; wt. empty 5300 [5400] lbs.; useful load 4870 [4770] lbs.; payload with 200 gal. fuel 3405 [3305] lbs.; figures in brackets are for "Hornet" engine; gross wt. 10,170 lbs.; max. speed 154

Fig. 285. Bellanca K of 1928 was basis of "Airbus" design.

Fig. 286. Same "Airbus" arranged to seat 15 as the model P-300.

(with cyl. baffles & speed-ring cowling); cruising speed 133; landing speed 60; climb 750 ft. first min. at sea level; ceiling 16,000 ft.; gas cap. 200 gal.; oil cap. 12-14 gal.; cruising range at 34 gal. per hour was 5.8 hours or 770 miles; price at factory not quoted; following figures are for cargo version of this same series; payload with 150 gal. fuel, 12 gal. oil and pilot only was 4920 lbs.; gross wt. increased to 11,400 lbs.; all other specs and data remain same, performance comparable.

The construction details and general arrangement of the model P-200 and P-300 was typical to that of the model P-100 as described in the chapter for ATC #360 in this volume, including the following, which in most cases applies to all versions. The fuselage framework for all versions was built up of welded chrome-moly steel tubing of rather large cross-section at points of greater stress, amply trussed with tubing in the forward section and braced with heavy steel tie-rods in the aft section; faired to outline and shape with formers and fairing strips, the completed framework was covered in fabric. The cabin was arranged with either 12 or 15 seats with a small lavatory in the rear; entry or exit was provided through a large door on each side and a convenient fold-in step. A large cargo door to permit loading of bulky freight was also available. The lower stub-wings being of a modified "Clark Y" airfoil section, had sufficient depth to provide a metal-lined cargo compartment of 30 cu. ft. capacity on each side; beside the space provided for personal hand-baggage under each seat, additional baggage-cargo space was provided in the top side of the nose-section just ahead of the pilot's station. This metal-lined compartment had capacity of 18 cu. ft. with allowance for up to 450 lbs. The thick lower stub-wings were built up of welded steel tube spar beams and steel tube truss type wing ribs covered in fabric. The air-foiled wing bracing struts were also built up of heavy steel tube spar beams, steel tube truss-

type former ribs and fabric covered; being of a special "Bellanca" airfoil section, these struts contributed heavily to the lateral stability and the overall lifting area. Long span, narrow chord Friese-type ailerons in the upper wing provided good lateral control even at stalling speeds; all control bearings were equipped with "Alemite" grease fittings or oil-less bronze bushings. Two fuel tanks of 100 gal. capacity each were mounted in the upper wing flanking the fuselage; the oil reservoir was mounted in the engine nacelle. The landing gear fork assemblies were each equipped with two oleo-spring shock absorbing struts but the large and squashy Goodyear "airwheels" absorbed most of the shock. Adaptable to various cabin arrangements, the "Airbus" was also available in rich custom interiors or as an all-cargo carrier. A 3-bladed metal propeller, "speed ring" engine cowling, electric engine starter, steerable tail wheel, navigation lights, cabin lights and cockpit lights, a battery and an unusually complete instrument panel were standard equipment. The next "Airbus" development was the model 66-75 "Aircruiser" as described in the chapter for ATC #563. The next immediate "Bellanca" development was the "Pacemaker" model E described in the chapter for ATC #476.

Listed below are Bellanca "Airbus" entries as gleaned from registration records:

NC-684W;	Model P-200 (#701) Cyclone 575	
NC-785W;	" (#702)	"
NC-786W;	" (#703)	"
NC-10796;	" (#704)	"
-10797;	" (#705)	"
-10798;	" (#706)	"

Serial #701 first as model P-100 with 600 h.p. Curtiss "Conqueror" engine; ser. #702 also as P-200-A seaplane and 15-place model P-300; ser. #703 also as 9-place P-200 Deluxe; ser. #704 later as model 66-67 on Group 2 approval numbered 2-519; ser. #705-706 later delivered to Army Air Corps as YIC-27; ser. #707 through #718 delivered to Air Corps as C-27 series.

Fig. 287. Viking "Kitty Hawk" model B-8 with 125 h.p. Kinner B5 engine.

As originally introduced back in 1928 by Allan Bourdon, the "Kitty Hawk" was a compact and very delightful little airplane, a 3-place open cockpit biplane of more or less normal configuration that was very well arranged and done up into a rather pert machine. Dainty and petite in comparison with other craft of this type at this time, the "Kitty Hawk" was nevertheless designed to perform all of the duties generally expected from a general-purpose airplane, one that would be used by private-owners and fixed-base operators but with the thought in mind that it was not really necessary to haul a lot of extra airframe and bulk around in performance of these normal duties. Consequently, the "Kitty Hawk" biplane was a smaller, snappy and efficient airplane that performed very well 'most anywhere in the best economy. Developed earlier to this time in two different models with a production of some 24 examples, the model B-8 now under the "Viking" banner, was the latest "Kitty Hawk" and also the last development in this all-purpose series. Definitely improved but still quite typical, the model B-8 stands as a companion offering to the earlier models B-2 and B-4 as described in the chapters for ATC # 134 and # 166, both chapters in U.S. Civil Aircraft, Vol. 2.

In the progressive development of any series, it was of course quite normal to add some accessories, add operating conveniences and pilot aids or any number of other useful details that added up to a heavier airplane. Also, to keep somewhat in pace with the consistent desires for performance increases it was most often mandatory to boost the horsepower, at least enough to allow for the piling on of added accessory weight, yet have some surplus in power; enough to assure at least a comparable performance to previous lighter-weight models, or perhaps just a slight increase. As a consequence of fhis modification and general improvement, the "Kitty Hawk" in the model B-8 was a slightly heavier airplane, better fitted and better equipped, mounting the extra muscle that was now provided by the 125 h.p. Kinner B5 engine. Casual analysis and comparison seems to prove that Viking Aircraft had a good airplane to sell in the model B-8, but it was belatedly introduced at a time when competition was perhaps the greatest in a rapidly narrowing market. Despite all its honest promise for increased utility and somewhat better performance, it was only in limited production for a year or two more and then quietly slipped off the scene entirely. It is at least heartening to know that of the 30 or so

Fig. 288. Sideview of "Kitty Hawk" B-8 shows sweep-back of wings.

examples built in all 3 models of the "Kitty Hawk", at least 17 were still flying actively by 1939.

The Viking "Kitty Hawk" model B-8 was an open cockpit biplane with seating arranged for 3 and basically typical of earlier models of this design except for various minor improvements, a boost in horsepower and some modifications necessary to this new combination. Somewhat on the small side for an airplane of this type, the B-8 however was well arranged to offer ample room for its payload with practical utility and good operating economy. Normally operated as a landplane for sport or in all-purpose service, the B-8 "Kitty Hawk" was also eligible to operate as a seaplane on "Edo" twin-float gear, with very little detriment to all-round performance in view of the added bulk. Now powered with the 5 cyl. Kinner B5 (R-440) engine of 125 h.p., the model B-8 was an easy-going combination that delivered a nimble and very satisfying per-

formance with seemingly little effort. Although quite plain in outline and simple of mechanical form, the "Kitty Hawk" B-8 was a well mannered machine with a loveable, subtle and girlish charm; general opinion voiced it a nice airplane to own and a very pleasant airplane to fly. The type certificate number for the Viking "Kitty Hawk" model B-8 was issued 1-29-31 and some 6 or 7 examples of this model were built by the Viking Flying Boat Co. at New Haven, Conn. Robert E. Gross was president; Courtlandt S. Gross was V.P., treas. and sales manager; R. Proctor was secretary; Jacob L. Freed was general manager; and Franklin T. Kurt was chief engineer. Suffering from the effects of the general business depression, as were all other aircraft manufacturers for that matter, Viking built very few airplanes during 1931; by early 1932, things were falling off rapidly and operations were getting real tight so Viking was prompted to move its facilities from the plant

Fig. 289. Small and dainty, B-8 realized good performance from 125 h.p. engine.

Fig. 290. "Kitty Hawk" B-8 on Edo floats; shown in setting of summer-time fun.

on Causeway Island to a rented hangar on the New Haven, Conn., municipal airport. Here they could still carry on with token manufacture but were also outfitted to conduct general service repairs and even offered flying instruction as a means to hang on hopefully until things got better, if they ever would.

Listed below are specifications and performance data for the "Kitty Hawk" model B-8 as powered with the 125 h.p. Kinner B5 engine; length overall 22'11"; height overall 8'8"; wing span upper 28'4"; wing span lower 28'0"; wing chord both 54"; wing area upper 123.4 sq.ft.; wing area lower 110 sq.ft.; total wing area 233.4 sq.ft.; airfoil USA-27; wt. empty 1178 lbs.; useful load 772 lbs.; payload with 35 gal. fuel 370 lbs.; gross wt. 1950 lbs.; max. speed 112; cruising speed 95; landing speed 42; climb 730 ft. first min. at sea level; ceiling 14,500 ft.; gas cap. 35 gal.; oil cap. 3 gal.; cruising range at 7 gal. per hour 425 miles; price at factory $4250. The following figures are for the model B-8 as seaplane on Edo 2260 floats; wt. empty 1490 lbs.; useful load 780 lbs.; payload with 35 gal. fuel 373 lbs.; gross wt. 2270 lbs.; max. speed 103; cruising speed 88; landing speed 50; climb 650 ft.; ceiling 13,000 ft.; gas cap. 35 gal.; oil cap. 3 gal.; cruising range 380 miles; all other dimensions and data remain same.

The fuselage framework was built up of welded 1025 steel tubing, lightly faired to shape with wooden fairing strips then fabric covered. A large door was on the left side for entry to the front cockpit and a convenient step was provided for the pilot. The cockpits were deep, roomy and well protected by large windshields; both cockpit interiors were carefully upholstered in leatherette. The rear cockpit was provided with a streamlined head-rest and a baggage compartment with allowance for 30 lbs. was behind the pilot's seat with access from the outside. The robust wing framework was built up of

solid spruce spars that were routed to an I-beam section with Warren truss-type wing ribs of spruce members reinforced with plywood gussets; the leading edges were covered with dural metal sheet and the completed framework was covered in fabric. The upper wing was in two panels that were joined together on center-line atop a cabane of inverted vee-struts; all 4 panels were rigged with 2 deg. of sweepback to improve the weight distribution and aid in directional stability. The 35 gal. fuel tank was mounted high in the fuselage ahead of the front cockpit and a direct-reading fuel gauge projected through the cowling. The split-axle landing gear of 88 in. tread used long telescopic legs with oleo-spring shock absorbers to iron out even the worst landing; wheels were 26x4 without brakes and the tail skid was of the steel spring-leaf type. 6.50x10 semi-airwheels with or without brakes and a 10x3 tail wheel were optional. The fabric covered tail-group was built up of welded 1025 steel tubing; the fin was ground adjustable and the horizontal stabilizer was adjustable in flight. A wooden Hartzell propeller, 26x4 wheels, spring-leaf tail skid, navigation lights, first-aid kit, and a fire extinguisher were standard equipment. A metal propeller, Heywood air-operated engine starter; 6.50x10 semi-airwheels, wheel brakes, a battery, 10x3 tail wheel and dual controls were optional.

Listed below are "Kitty Hawk" model B-8 entries as gleaned from registration records:
NC-868Y; Model B-8 (# 25) Kinner B5.
NC-753Y;   "   (# 26)   "
NC-996M;   "   (# 27)   "
NC-975M;   " ´   (# 28)   "
NC-794N;   "   (#   )
Serial # 26 and # 31 later on Group 2 approval numbered 2-463; serial number for NC-794N unknown; no listing found for ser. # 29-30-31 and up; approval expired 9-30-39.

A.T.C. #393
(1-29-31)
WACO, MODEL MNF

Fig. 291. "Waco" model MNF with 125 h.p. Menasco C4 engine.

One of the neatest, and perhaps most interesting versions in the new "Waco F" series, was the racy looking model MNF; a companion model in this 3-place biplane line that was powered with the Menasco "Pirate" C-4 engine. Though quite rare and scarce in number, the model MNF did have certain qualities and some innovations to its credit. Because of the inverted in-line engine, frontal area of the nose was at a minimum, allowing somewhat better vision around and over the top of the engine cowl and because the cylinders were inverted, most everyday maintenance to the engine could be more easily performed from the ground without resorting to standing on boxes or rickety stepladders. Baffled for efficient cooling, the engine was completely enclosed by removable cowling panels, and this presented a clean and neat nose-section with no drag inducing assemblies protruding into the wind-stream. The new Menasco C-4 "Pirate" was a 4 cyl. inverted in-line engine of the aircooled type with a rating of 125 h.p. at 2175 r.p.m.; with 363 cu.in. displacement and a dry weight of 288 lbs. the C4-125 was a small, compact and well designed power package. Smooth-running with a "sporty" bark from its exhaust pipes, the "Menasco" delivered eager power and proved to be a compatible combination in the "Waco F" biplane; it is sometimes wondered why this version sold

only in such small number. One of the proudest owners of this version (MNF) was Al Menasco who used the MNF in promotion of the "Menasco" engines.

The "Waco" model MNF was an open cockpit biplane seating 3 and was typical of the three other "F" versions described here previously, except for its engine installation and the modifications necessary for this particular combination. Because of its comparative scarcity not much lore has been handed down or written about this MNF version but we can safely assume that it was a good airplane which shared all the qualities that made the "Waco F" such a great favorite. In offering this version as a companion model to the three models already in production, Waco Aircraft proved that it was also one of the several aircraft manufacturers that were forcing the engine manufacturers to promote and share the burden of selling airplanes; prosperity for both hinged on the sale of airplanes so if an engine manufacturer expected to sell engines, he also had to demonstrate and help sell airplanes. In a sagging market where other manufacturers found it increasingly hard to stay in business, "Waco" was selling its bargain biplanes by the hundreds, a testimonial that indicated the F-series was just what the flying public had wanted, and was buying in spite of close-out sales and near-cost prices that

Fig. 292. Neat cowling of engine and wheel pants offered boost in speed.

other builders were featuring. The type certificate number for the "Waco" model MNF was issued 1-29-31 and at least 4 or more examples of this model were manufactured by the Waco Aircraft Co. at Troy, Ohio.

Listed below are specifications and performance data for the "Waco" model MNF as pow-

ered with the Menasco "Pirate" C4-125 engine; length overall 22'0"; height overall 8'4"; wing span upper 29'6"; wing span lower 27'5"; wing chord both 57"; wing area upper 130.5 sq.ft.; wing area lower 111 sq.ft.; total wing area 241.5 sq.ft.; airfoil (NACA) M-18; wt. empty 1166 [1195] lbs.; useful load 734 [705] lbs.; payload

Fig. 293. Inverted Menasco engine offered better vision over the nose.

Fig. 294. "Waco" MNF with Al Menasco flying; reputed to be one of his favorites.

with 32 gal. fuel 354 [325] lbs.; gross wt. 1900 lbs.; wts. in brackets with metal propeller; max. speed 118; cruising speed 99; landing speed 35; climb 800 ft. first min. at sea level; ceiling 16,000 ft.; gas cap. 32 gal.; oil cap. 3 gal.; cruising range at 8 gal. per hour 365 miles; price at factory field first quoted at $4475, and hiked to $4675 in April of 1931. An amendment later allowed gross wt. of 1938 lbs., this boosting the useful load and payload approx. by 38 lbs.; performance was only slightly affected.

The construction details and general arrangement for the "Waco" model MNF was typical to that of the models RNF-KNF-INF, as described in the chapters for ATC # 311-313-345 in this volume, including the following. In the fuselage structure, all steel tubing ends were milled with a curve to assure close-fitting contact with longerons for better welded joints. The wing bracing struts on the Waco F series were now of the N-type and fabricated from chrome-moly steel tubing of a streamlined section with adjustment provided for rigging incidence; the interplane bracing was of heavy gauge stream-lined steel wire. Rigging geometry included 2 deg. positive incidence for upper and lower

wings, 28.5 in. stagger between panels, and 1 deg. 45 min. dihedral in both upper and lower wing panels. There was an aileron on each panel, connected together in pairs by a streamlined push-pull strut; aileron action was differential as an aid in making smoother turns. A Hartzell wooden propeller, Heywood air-operated engine starter, 6.50x10 semi-airwheels, caster action tail wheel, compass, navigation lights, first-aid kit and fire extinguisher were standard equipment. A ground-adjustable metal propeller, dual controls, wheel brakes and wheel streamlines (wheel pants) were optional. Wheel brakes were dual controlled and were operated in conjunction with the rudder; a hand-lever arrangement, combined with the throttle lever, set tension for both brakes and a locking arrangement was provided for parking. The next development in the "Waco" biplane was the cabin-type model QDC described in the chapter for ATC # 412.

Listed below are MNF entries as gleaned from registration records:

NC-11213; Mod. MNF (# 3408) Menasco C4-125
NC-11222;   "   (# 3442)   "
NC-11239;   "   (# 3445)   "
NC-11246;   "   (# 3484)   "

Fig. 295. Spartan model C4-301 with 300 h.p. "Wasp Jr." engine; plump beauty was best of series.

The "Spartan" C4-series monoplane was an extremely attractive airplane in a configuration that was primarily developed as a plush and practical family-type airplane, an airplane that could also double in duty as a comfortable, efficient air-taxi, especially for the business man. As the third and last version in this particular series, the model C4-301 was every inch a "Spartan" and recognizable at first sight. Faired-out lavishly to a plump rounded figure with careful streamlining at nearly every point, the 300 h.p. Pratt & Whitney "Wasp Junior" engine was fitted to this model, then bedecked with an engine fairing that could best be described as being the better features of a "Townend-ring" and the full NACA type cowling. Other features being typical, the model C4-301 bore some signs of a slight progressive development that came with the experience gained since the introduction of the first example in the "Cee-Four" series, nearly a year previous. Broad, tall, and stately, the C4-301 reminds one of a lady that has spent the best part of an afternoon parading in front of a mirror while attaching varied baubles and assorted finery to achieve the best results and now she poses for your enjoyment and of course, your expected approval. Though slightly over-dressed in comparison with other typical craft, the C4-301 nevertheless was a beautiful lady and behaved herself very much like a beautiful lady should. Standing in comparison with other craft of this

type, the model C4-301 could hold its head up in company with the very best.

The "Spartan" model C4-301 was a high-winged cabin monoplane of rather large and buxom proportion, with more than ample seating for 4, in accomodations of good taste and high quality. One can't help but agree that the "Spartan" monoplanes were beautiful airplanes of distinctly feminine character, custom craft that were leveled at the owner who could well afford the best. It seems folly that Spartan Aircraft should undertake to develop such a craft at this particular time when the economy was so shaky and money was so hard to come by, but had things taken a turn for the better it is no doubt that "Spartan" would have gotten a good piece of the market. The C4-301 was made "clean" and of good proportion to insure a well-behaved character and a pleasant easy-going nature. Easy to fly and pleasantly stable, the C4-301 was ideal for relaxed cross-country flying or for flights that had no other purpose than to allow the comfortable enjoyment of the unfolding scenery below. Powered with the 9 cyl. "Wasp Junior" (R-985) engine of 300 h.p. the C4-301 could be judged slightly under-powered, and certainly was no "tiger," but this did not hamper its general performance ability to any great extent and once air-borne it was not left straggling behind by too many others. The type certificate number for the model C4-301 was issued 1-29-31 and only one example of

Fig. 296. Spartan C4-301 being readied for export.

this model was built by the Spartan Aircraft Co. at Tulsa, Okla. Lawrence V. Kerber was president; Rex B. Beisel was V. P. and chief engineer; C. H. Reynolds was secretary and E. W. Hudlow was treasurer. The C4-series were built in 3 different versions and no more than 7 or 8 examples appear in record. It is a pity that these graceful monoplanes didn't inhabit our airways in more numerous amounts; this was a design surely deserving of a better fate. Struggling earnestly to achieve some success with this series, but to no avail, "Spartan" developed a small trainer-type monoplane later in 1931 with which they had a little more success. In May of 1931, 3 four-place monoplanes and one 5-place monoplane were coming through the assembly line, and tooling was being prepared for the production of 25 of the "Spartan" C2-60 light trainer monoplanes.

Listed below are specifications and performance data for the "Spartan" model C4-301 as powered with the 300 h.p. "Wasp Jr." engine; length overall 32'7"; height overall 8'11"; wing span 50'0"; wing chord 80"; total wing area 299 sq. ft.; airfoil Clark Y; wt. empty 2608 lbs.; useful load 1448 lbs.; payload with 85 gal. fuel 710 lbs.; gross wt. 4056 lbs.; max. speed 145; cruising speed 124; landing speed 57; climb 880 ft. first min. at sea level; ceiling 14,500 ft.; gas cap. 85 gal.; oil cap. 7.5 gal.; cruising range at 16 gal. per hour 610 miles; price at factory first announced as $12,350 lowered to $11,900 in May of 1931.

The construction details and general arrangement of the model C4-301 was typical to that of the models C4-225 and C4-300 as described in the chapters for ATC #310 and #383 in this volume, including the following. The 710 lb. payload of the C4-301 was ample to seat the pilot and 3 passengers at 170 lbs. each, and still have an allowance for up to 200 lbs. in baggage. With careful attention to detail, all exposed fittings and strut junctions were generously streamlined by formed metal covers and cuffs. The outer wing strut junction was braced in a truss similar to that used on the Bellanca "Pacemaker" and the Curtiss "Robin," to stiffen the long wing and not allow any flexing during abnormal aileron loads. All moveable control surfaces were aerodynamically balanced to insure minimum operating pressures. This model had the typical forked-type landing gear with 9.50x 10 low pressure semi-airwheels covered with large tear-drop "wheel pants." Side windows in the forward portion of the cabin could be rolled down for extra ventilation and cabin heat was available for cold-weather flying. The engine was fitted with a long tail-pipe to direct exhaust gases well below the cabin level. Dual wheel-type controls, wheel brakes, swivel-type tail wheel, metal propeller, navigation lights, Eclipse electric engine starter and battery were standard equipment. The next "Spartan" development was the model C2-60 trainer described in the chapter for ATC #427.

Listed below is the only known example of the model C4-301 as gleaned from registration records:

NC-988N; Model C4-301 (#F-1) Wasp Jr.

Fig. 297. Keystone-Loening "Air Yacht" model K-85 with 525 h.p. Wright "Cyclone" engine.

Although the configuration and arrangement of the earlier Loening "Amphibian," namely the C2C and C2H "Shoe-horn" type, were quite practical and exceedingly efficient for amphibious dock-to-dock service, they were somewhat unusual looking and fast becoming antiquated in the eyes of the general flying public. Much more in favor now was the larger flying-boat type with an enclosed cabin area in the hull, an arrangement better suited to seat and house the pilot and the passengers together, and in more comfort. If for naught but to induce the feeling of strength, safety and comfort, the flying-boat hull was the better choice for a craft relegated to strenuous short-trip service. Whether the Keystone-Loening model K-85 "Air Yacht" was designed and developed for air-line shuttle or air-ferry service, we cannot definitely say, but it would have been an excellent craft for either service, or perhaps as a luxurious airborne "yacht" for some wealthy sportsman. The model K-85 was a valiant but useless try to perpetuate further the famous "Loening" designs, but outside of some amphibious service machines (Army-Navy-Marines) and a few specials, this proved to be about the last of the breed.

The Keystone K-85 series was actually developed by the Loening Engineering Co. as the model C4C, a new flying-boat version of the amphibian that was actually designed into the basic wing cellule as used on the earlier C2C (ATC #90) or the C2H (ATC #91); in essence, a standard boat-type hull with an enclosed cabin for all the occupants was designed to replace the classic "shoe-horn type" as had been used before. Tested earlier in 1930 as the model C4C with seating for 6 and subsequently approved on a Group 2 approval numbered 2-298, this series designation was changed by Keystone to their model K-85 with seating increased to eight.

The Keystone-Loening "Air Yacht" model K-85 was a large cabin biplane of the flying-boat type with a retractable wheeled undercarriage that permitted operations off land or off water. Patterned somewhat after the smaller "Commuter" (K-84), the model K-85 had a large hull of deep proportion with an enclosed cabin area that seated 8 in ample room and plush comfort. Designed to provide practical utility in all-purpose service, the K-85 interior was offered in several arrangements, including clear floor area for all-cargo service. Powered with the big 9 cyl. Wright "Cyclone" R-1750-E engine of 525 h.p., the performance of the new "Air Yacht" was noticeably improved over the C2C type because of aerodynamic advantages in the flying-boat hull. The engine was mounted to a streamlined nacelle structure faired into the

Fig. 298. Amphibious K-85 operated from land or water.

center-section panel of the upper wing; driving a 3-bladed propeller in a "tractor" fashion, this installation was placed in this manner to minimize danger to occupants from the whirling blades. Because of its relative scarcity, operational history of the K-85 is practically nonexistent, but it is safe to assume by its form and performance figures that it was a practical all-purpose carrier with many inherent qualities of the basic "Loening" design. The type certificate number for the "Air Yacht" model K-85 was issued 1-30-31 and perhaps no more than 2 examples of this model were built by the Keystone Aircraft Corp. at Bristol, Pa., a div. of the Curtiss-Wright Corp. Edgar N. Gott was president; C. T. Porter was V. P. and chief of engineering; Geo. H. Prudden, formerly with Stout Airplane Co. and the Atlanta Aircraft Co. was chief of design and development in all-metal construction as used in various Keystone aircraft.

Listed below are specifications and performance data for the Keystone-Loening "Air Yacht" model K-85 as powered with the 525 h.p. Wright "Cyclone" engine; length overall 37'2"; height wheels up 13'2"; height wheels down 15'9"; wing span upper 46'8"; wing span lower 46'8"; wing chord both 72"; total wing area 517 sq. ft.; airfoil "Loening" 10-1 (or as 10-A); wt. empty 4209 lbs.; useful load 2091 lbs.; payload with 140 gal. fuel 1001 lbs.; gross wt. 6300 lbs.; max. speed 138; cruising speed 107; landing speed 60; climb 800 ft. first min. at sea level; climb in 10 min. was 6200 ft.; ceiling 13,800 ft.; gas cap. 140 gal.; oil cap. 10 gal.; cruising range at 28 gal. per hour 500 miles; price at factory first quoted as $30,050 lowered to $28,700 in June of 1931.

The hull framework was built up of spruce members joined together by heavy duralumin gussets into Warren-truss girders; bulkheads divided the structure into several water-tight compartments and outer covering was of "Alclad" metal sheet with all seams water-proofed

by strips of glue-impregnated fabric. The pilot's cabin forward seated two with entry through a forward hatchway but this compartment was actually directly connected to the main cabin area; dual "Dep" wheel controls were provided. The main cabin area normally seated 6 but could be fitted with several custom arrangements; cabin entry doors on either side were to the rear with baggage stowed forward and space for a small lavatory was in back. Mooring gear for docking or anchoring was stowed in the nose-section of the hull. The double-bay wing framework was built up of solid spruce spar beams with stamped-out "Alclad" metal wing ribs; the leading edges were covered with dural metal sheet, the aileron frames were of wood and the completed wing framework was covered in fabric. The streamlined engine nacelle was built into the center-section panel of the upper wing and braced to the hull by N-type steel tube struts; the oil reservoir tank was mounted in the engine nacelle. The folding landing gear of 96 in. tread used "Aerol" (air-oil) shock absorbing struts and was extended or retracted mechanically by hand; wheels were 36x8 and Bendix brakes were standard equipment. The fabric covered tail-group was a composite structure of spruce with dural metal spars and ribs; the large fin was built as integral part of the hull and the horizontal stabilizer was adjustable in flight. A 3-bladed metal propeller, Eclipse electric engine starter, anchor and mooring gear, navigation lights and fire extinguisher were standard equipment. Special interior arrangements and custom color combinations were optional.

Listed below are the only known entries of the Keystone K-85 as gleaned from registration records:

NC-10588; Model K-85 (#299) Cyclone 525
X-769W;   "   (#300)   "
X-62K was the Loening model C4C (ser. #600) as approved on Group 2 approval numbered 2-298, issued 11-5-30, the forerunner to the K-85 series.

Fig. 299. Aeronca C-3 with 36 h.p. Aeronca E-113 engine.

The enthusiastic acceptance of the "Aeronca" model C-2 during the latter half of 1930, more or less assured the low-powered light airplane a bright, promising future and paved the way for further development. The avid interest that centered around the single-seated C-2 brought occasional comment that pointed to a dire need for a similar airplane with ample seating for two. Of course the folks at "Aeronca" had this in mind all along and several of the earlier models C-2 had been converted to 2-seaters in the initial development of extra seating. By the dawning of 1931, two prototypes of the new model C-3 had been through exhaustive testing and by March of 1931, production lines were beginning to roll. A factory demonstrator was sent off on a 13,000 mile tour through 17 states and the cordial reception enjoyed by the new C-3 was very gratifying. Being a more practical type of airplane for the average private flyer, and an economical boon to the many small flying schools that found their operations at a near stand-still, orders for the new C-3 two-seater were piling in like fan-mail. By the end of 1931 nearly 100 examples of the "Duplex" and the "Collegian" had already been built.

Despite the added capacity for another passenger, with only the addition of 10 more horsepower to handle the larger load, the model C-3 was every bit as good as the earlier C-2 and many

liked it even better. Private-owner fliers now found that they could actually afford their own airplane with which to continue regular flying and the flying schools, due to the reduced charges now possible, were finding their enrollments swelling by leaps and bounds. Soon many of the flying schools were offering "learn to fly for $65.00" and solo-flying time was about $5.00 to $7.00 an hour. Participation in this new way of flying brought some of the enthusiasm back to flying in general and some of this enthusiastic energy was naturally converted to record attempts of all sorts; like the C-2, the "Aeronca" C-3 was also the holder of many records that included distance, load, duration, height and the like. To show off the new C-3 country-wide in a gala affair and to prove to the multitudes that it was not just a putt-putt plaything restricted to the visible boundaries of the home 'port, "Aeronca" entered the new "Collegian" in the 1931 National Air Tour. With George Dickson flying, the hustling C-3 averaged 64 m.p.h. for the 4858 mile grind and became the first light airplane ever to complete this grueling test of performance and stamina. Whether authentic or not, the following story has been told regarding the "Aeronca" in the 1931 "tour". In the formula for performance, additional points were based on "Stick" time (time for landing roll to a stop) so the eager pilot, not having any wheel brakes,

Fig. 300. Aeronca C-3 for 1934.

would reach out and grab his spinning tires with both hands (gloved) to give him brake effect and lessen his ground roll!

Happily, the "Aeronca" C-3 enjoyed much success and lasting popularity so it was able to continue in production for many years. The company always kept careful note on repeated criticism and kept pace with each demand by introducing new and improved versions of the C-3 quite frequently. For 1932 the open-cockpit "Aeronca" was fitted with a detachable cabin enclosure that was snug enough for foul weather flying and could be easily removed for balmy summer flying; numerous other detailed improvements were added too. For 1933 the cockpit of the C-3 was made somewhat wider, more leg-room was provided and the seat-back was higher for extra comfort; an improved motor mount and the new E-113A series engine cut down on vibration. For 1934 the cabin enclos-

Fig. 301. Aeronca "Collegian" showing optional cabin enclosure.

Fig. 302. Aeronca C-3 for 1935.

ure was somewhat improved, upholstery was now of leather and a new cantilever single-strut landing gear was introduced; the new E-113B series engine also incorporated several improvements. In 1935 the fuselage lines of the "Aeronca" were almost drastically changed to eliminate the long-familiar "razor back" shape and a full cabin layout was now offered for year-round flying; added too was a cabin heater and wheel brakes. During this time "Aeronca" was also developing the sporty LA-LB-LC series of low-winged cabin monoplanes. For 1936 the "Aeronca" C-3 was improved only in some details and now mounted the improved E-113C series engine, an engine that was by now recognized as one of the finest available for the light airplane. Production of the C-3 still held up well into 1937 but by then the new "Aeronca" K had

been groomed for the market and the faithful model C-3 was finally discontinued. For more than 6 years the "Aeronca" C-3 was a favorite all over the land and even in many other parts of the world; it was not often that an airplane enjoyed such a long span of popularity.

The "Aeronca" model C-3 was a high wing wire-braced light monoplane arranged in somewhat comical lines that could never be mistaken for any other. With side-by-side seating quite ample for two, the cockpit was first arranged in a semi-cabin layout with panels offered later that could be added to enclose the sides for cold-weather flying. Blessed with a popularity that sponsored a production run of some 6 years, improvements were continually added to make it a better airplane and by 1935 the C-3 was sporting a full cabin enclosure that

Fig. 303. Cabin C-3 had now lost its razor-back fuselage, providing all-weather comfort and better performance.

Fig. 304. Aeronca PC-3 flying high over Pacific coast.

was as comfortable as the best. First powered with the 2 cyl. "Aeronca" E-113 engine of 36 h.p., this engine was also continually improved through the models E-113A, the E-113B and the E-113C; the E-113C of 40 h.p. was then also used to power the 1937 "Aeronca" model K. Performance of the "Aeronca" C-3 in any version was a veritable delight and anybody that could learn to put up with it really "had a ball"; not all learned to love it but never a harsh word has ever been said about its character or its behavior. It was an airplane in which you sit back relaxed and fly entirely by feel and sound; it was sort of a vibrating platform from which to view the slowly unfolding scenery below. True, the C-3 was limited in many ways but the enjoyment possible on a minimum budget was reason enough to hail it as one of the best airplanes built during this particular period. The type certificate number for the "Aeronca" model C-3 as powered with the 36 h.p. "Aeronca" E-113 series engine was issued 1-31-31 and amended from time to time to allow additional equipment and increases in allowable gross weight; the gross weight varied from 875 lbs. for 1931 model to 1006 lbs. for the last models built. The "Aeronca" was manufactured by the Aeronautical Corp. of America at Lunken Airport in Cincinnati, Ohio. Taylor Stanley was president; Conrad G. Dietz was V.P. and general manager; Robert Taft was sec.; Walter Draper was treas.; and Robert B. Galloway was chief engineer. By 1933 H. H. Fetick was V.P. and Roger E. Schlemmer was chief engineer. In 1936 this same group was intact except for addition of C. S. McKenzie as general manager and J. C. Welsch as sales manager.

Listed below are specifications and performance data for the "Aeronca" model C-3 as powered with the 36 h.p. Aeronca engine; length

overall 20'0"; height overall 7'6"; wing span 36'0"; wing chord 50"; total wing area 142 sq.ft.; airfoil Clark Y; wt. empty 466 lbs.; useful load 409 lbs.; payload with 8 gal.; fuel 190 lbs.; gross wt. 875 lbs.; baggage 25 lbs.; max. speed 80; cruising speed 65; stall speed 40; landing speed 35; climb 500 ft. first min. at sea level; climb in 10 min. 4000 ft.; ceiling 14,000 ft.; gas cap. 8 gal.; oil cap. 3 qt.; cruising range at 2.5 gal. per hour 200 miles; price at factory in 1931 was $1895 for the "Duplex" and $1695 for the "Collegian" trainer. For 1932 model, specs and performance were typical except as follows; wt. empty 461 lbs.; useful load 414 lbs.; payload with 8 gal. fuel 195 lbs.; gross wt. 875 lbs.; max. speed (with enclosure) 85; cruising speed 70; price at factory for open model $1730 and $1790 with winter enclosure. Gross wt. for landplane models 1931-34 amended to 900 lbs.; seaplane allowed 972 lbs. on Edo D-990 floats and 953 lbs. on Warner A-1900 floats. The next significant change in "Aeronca" specs was in 1935; all figures typical except as follows; wt. empty 569 lbs.; useful load 437 lbs.; payload with 8 gal. fuel 213 lbs. (25 lbs. baggage), gross wt. 1006 lbs.; max. speed 90; cruising speed 75; landing speed 38; climb 450 ft. first min. at sea level; ceiling 12,000 ft.; cruising range at 3 gal. per hour 190 miles. 1935 seaplane allowed 1069 lbs. gross wt. with Edo D-1070 floats (20 lbs. was for mooring gear). Float installations averaged 135 lbs. (minus 46 lbs. for tripod landing gear and tail skid). There were no significant changes for the rest of the series to 1937. The "Aeronca" C-3 was also available on the time payment plan with down payment of $450 and 12 equal monthly payments.

The fuselage framework was built up of welded 1025 chrome-moly steel tubing lightly faired with steel tube and wooden fairing strips, then

Fig. 305A. 16-year-old son of Peter Bowers on his second solo flight in restored C-3 almost twice his age.

Fig. 305B. Chummy interior of "Collegian"; enclosure removable for summertime flying.

fabric covered. The triangular portion of the fuselage aft of the wing had its apex at the top and this earned the earlier C-3 the nick-name of "razor-back"; in 1934-35 additional fairing was fitted to eliminate this triangular portion and the fuselage now had near-normal lines. The cockpit had side-by-side seating with a large windshield in front that formed a semi-cabin with open sides; in 1932 and add-on cabin enclosure was available to close in the drafty sides and models of 1935 and up offered a full cabin with open sides; in 1932 an add-on cabin ment with allowance for 25 lbs. was behind the seat-back. The wing framework, its attachment and bracing was typical as described in the chapter for ATC # 351. The landing gear on 1931 models of the C-3 was of the stiff-legged tripod type fitted with soft-cushion airwheels; due to various complaints of "porpoising" during landings, the tripod gear was fitted with "Aeronca" oleo-struts; 1934 models introduced the cantilever single-strut landing gear with the shock absorbing mechanism housed in the lower fuselage. The fabric covered tail group of varying shapes throughout the series, was built up of welded steel tubing; the fin was an integral part of the fuselage and the horizontal stabilizer was ground adjustable for load trim. 1935 models introduced the cabin heater and wheel brakes; other standard equipment included 16x7 Goodyear airwheels, first-aid kit and fire extinguisher. All models of the C-3 were available as seaplanes on APC or Edo twin-float gear; the installation amounting to an increase of 86 lbs. The next "Aeronca" development was the single-seated "Cadet" described in the chapter for ATC # 447.

Listed below is partial tally of "Aeronca" C-3 entries as gleaned from registration records:

X-657Y;	Aeronca C-3 (# A-99) Aeronca E-113.		
X-658Y;	"	(# A-102)	"
NC-11277;	"	(# A-107)	"
NC-11284;	"	(# A-116)	"
NC-11285;	"	(# A-117)	"
NC-11286;	"	(# A-118)	"
NC-11287;	"	(# A-119)	"
NC-11288;	"	(# A-120)	"
NC-11289;	"	(# A-121)	"
NC-11291;	"	(# A-123)	"
NC-11292;	"	(# A-124)	"
NC-11293;	"	(# A-125)	"
NC-11294;	"	(# A-126)	"
NC-11295;	"	(# A-127)	"
NC-11296;	"	(# A-128)	"
NC-11297;	"	(# A-129)	"
NC-11298;	"	(# A-130)	"
NC-11400;	"	(# A-131)	"
NC-11401;	"	(# A-132)	"
NC-11402;	"	(# A-135)	"
NC-11403;	"	(# A-136)	"
NC-11404;	"	(# A-137)	"
NC-11405;	"	(# A-138)	"

Serial # A-122 was prototype C-1 "Cadet"; ser. # A-128, A-131 unverified; no listing for ser. # A-133, A-134, A-139; NC-11406 was ser. # A-140 and numbers ran consecutively to NC-11416 which was ser. # A-150; ser. # A-151 was a C-1 "Cadet"; NC-11418 was ser. # A-152 and numbers ran consecutively to NC-11424 which was ser. # A-158; no listing for ser. # A-159; NC-11425 was ser. # A-160; NC-11491 was ser. # A-161 and numbers ran consecutively to NC-11495 which was ser. # A-165; NC-12400 was ser. # A-166 and numbers ran consecutively to NC-12421 which was ser. # A-187; ser. # A-185 was a PC-3 on floats; no other listings checked beyond 1931 aircraft register. Serial numbers roughly coincide with A-200 for 1932, A-300 for 1933, A-400 for 1934, and A-500 for 1935 and so on. There was no accurate record or tally easily available on the number of C-3 examples built in the period from 1931 to 1937 but serial numbers above A-500 had been recorded so we can assume that at least 400 or more had been built. Four of the C-3 were operating in Brazil, one as a landplane and 3 were on floats.

A.T.C. #397
(1-31-31)
CURTISS-WRIGHT "JUNIOR", CW-1

Fig. 306. Curtiss-Wright "Junior" model CW-1 with 45 h.p. Szekely engine.

The Curtiss-Wright "Junior", because of its rather unusual arrangement, was one of the most intriguing airplanes to come out of this particular period. In the words of Curtiss-Wright, the "Junior" was not only a new airplane but a new kind of airplane. It was designed especially for the amateur pilot, the low-budget sportsman pilot, and the fixed-base operator who wished to offer flying time at a rate more within reach of the flying populace. Easy to buy, easy to fly, and cheap to operate, the "Junior" soon became just about the most popular flivver-type airplane of the next few years. Endowed with particularly good performance in spite of its measley 45 horsepower, the "Junior" was soon literally pulled by its boot-straps into various feats of astonishment. Test pilot H. Lloyd Child, who nurtured the new design from its very inception, flew one of the early examples to over 11,000 ft. with a husky passenger aboard; lady-pilot Edna Rudolph flew solo to 14,600 ft. for a junior-woman's record and an 18-year-old boy-pilot nursed a "Junior" to near 20,000 ft. for another record. Because of its peculiar hull-type fuselage and high-mounted "pusher" engine, which placed the occupants in a front-row seat, the "Junior" was soon widely used in hunting, for sport and profit, from the air. The "gunner" was in the front cockpit of course and had a clear undisturbed shot from many different angles. This feature providing a clear, practically un-obstructed platform was ideal for air-to-air photography also. Fixed-base operators who had been looking for means of decreasing their mounting expenses to in turn lower their charges for flying time, had found the "Junior" ideal for

their purpose and a boon to poor aspiring pilots; they now charged from $5.00 to $8.00 for solo time and still made a fair profit. Most of the pilots that have had experience this far back will surely agree that the Curtiss-Wright "Junior" was just about the most provocative airplane of this decade. For a scant few it was a source of unkind laughter and ridicule, no doubt prompted by prejudice, but for literally thousands it was a craft that waddled and sput-tered its merry way into the very depths of their hearts. Flying in this craft is rather hard to describe because it was a continuous panoramic picture of warm pleasures and small annoyances, but it was thoroughly enjoyable with all sorts of little experiences that built memories to last, even to this day. Though seemingly far-fetched at this late date, many opinions relate that the "Junior", the "Aeronca" C-3 and the "Eaglet" of this early 1930-31 period provided more flying pleasure and priceless experience for the private-owner flyer than any other type of airplane ever built.

The Curtiss-Wright "Junior" had an interest-ing and scarcely known development period that started out from a design called the "Buzzard". A "Bud" Snyder designed the little "Buzzard" sometime in 1927-28 and it has been described as a wooden structured parasol mono-plane with a high-mounted "pusher type" engine. A single seater, this little flivver mono-plane was powered at one time with a 4 cyl. Continental A-40 engine of 37 h.p. and among other oddities, it used an all-movable horizontal stabilizer. Feeling the urge and the necessity to offer competition against the recently released

Fig. 307. Test-hop of early "Junior" over St. Louis.

"Aeronca" C-2 and the forthcoming American Eagle "Eaglet", Curtiss-Wright became actively interested in the flivver-plane movement, so in 1930 Walter Beech and Ralph Damon arranged for purchase of the Snyder "Buzzard" design rights. Flown in test by H. Lloyd Child on 7-19-30 and again on 7-24-30, the "Buzzard" proved somewhat lacking in stability, and control tended to be a little erratic, so discussed opinion decided a redesign was certainly in order. Karl H. White, project engineer on the "Moth" biplane, which had been transferred by now to the St. Louis Div. for manufacture, was called in to work on the design and engineering of a new concept. Although Karl White has often been credited as designer of the new "Junior" concept, both Walter Beech and H. Lloyd Child played a large role in its final development. The new Curtiss-Wright design finally flew its maiden flight in Oct. of 1930 and was

Fig. 308. Prototype for the "Junior" series was called "Skeeter".

Fig. 309. Sales agency displays "Junior" at Holland air-show.

first called the "Skeeter"; the 2 cyl. ABC "Scorpion" (British), the 4 cyl. Continental A-40 and various other small engines were alternately mounted for test before it was decided to use the 3 cyl. Szekely SR-3-0 of 45 h.p. as the standard powerplant. The "Skeeter" was finally groomed for mass production and now its only resemblance to the "Buzzard" was the fact that they were both pusher-type monoplanes. Approval tests were in full swing by Dec. of 1930 and by this time the name of the new craft had been changed more appropriately to "Junior"; by March of 1931 production was tentatively set for 3 airplanes per day and "Juniors" began leaving the production lines like hot-cakes. By June of 1931 at least 125 examples had been built and sold. Priced initially at $1490. the "Junior" was a terrific value and orders poured in from all over the country-side; railroad box-cars were filled to capacity with 8 disassembled "Juniors" to the car and were shuttled to all points of the land. In the space of less than two years time, "Juniors" were gayly flying in all corners of the U.S. and thousands of pilots, both amateur and seasoned veteran, were flying it and enjoying it.

The Curtiss-Wright "Junior" CW-1 as shown here in various views, was a 2-place open cockpit monoplane of the "parasol" type, but this hardly describes it properly because the engine was mounted on top of the wing in a "pusher" fashion and the fuselage was a hull-type affair with cockpits in its forward end. Looking for all the world like a "flying boat", the forward occupant from his vantage point had almost unlimited visibility whether on the ground or in the air. Waddling off in short order, the "Junior" was usually airborne in a few hundred feet and from then on there was no doubt that you were "flying", flying into an experience that was pleasant, noticeably different but sheer enjoyment. Repeated flights in a span of time usually promoted a sort of togetherness with this cheerful-natured craft, a feeling that is usually reserved between man and his dog. Most people got acquainted quickly and grew very fond of its sporty atmosphere. Though top speed was near to 75 m.p.h. and cruising speed was hardly more than 60, it didn't matter too much because "Junior" pilots hardly ever went very far from the home 'port and they were getting their kicks by the hour anyhow. If you had time to spare the "Junior" could be carefully nudged to 14,000 ft. and the trip back down was akin to a ride in a sailplane. With a stalling speed of about 35 m.p.h. you could set "Junior" down at less than 30 m.p.h. and it hardly required more than 150 ft. to come to a stop. Roly-poly tires and a good stout landing gear allowed operation from all sorts of so-called unimproved surfaces which were usually a farmer's pasture, a sandy beach, the desert floor, golf course, an open highway, and of course even some airports. They at Curtiss-Wright (St. Louis Div.) surely had the intent to come up with an airplane the likes of which

Fig. 310. Unusual hull-type fuselage seated occupants in excellent vantage point.

had rarely been seen, also a craft that was sure to endear itself to the average flyer, at least in part, by its homely charm. One cannot argue that this hadn't been done.

Though available figures could not be easily verified, it shows that at least 270 C-W "Juniors" were built through 1931 and 261 of these were issued registration numbers. Active production was finally discontinued sometime in 1932 because of the crippling depression when sales had fallen off to drastic levels, but "Juniors" flew on to pile up many more years of faithful service; at least 85 were still active in 1939 and 15 were still registered in 1954. A few have been rebuilt to fly again in the sixties and many people are still searching barns, haylofts, and dusty hangars, hoping to find a decrepit "Junior" somewhere that could be made to fly once more. The type certificate number for the Curtiss-Wright "Junior" model CW-1 was issued 1-31-31 and were manufactured by the Curtiss-Wright Airplane Co. on Lambert Field in St. Louis (Robertson), Mo. Walter Beech was now

the president; Ralph S. Damon was V.P. and general manager; H. Lloyd Child was test pilot and chief engineer; Karl H. White was project engineer on "Junior" development.

Listed below are specifications and performance data for the Curtiss-Wright "Junior" model CW-1 as powered with the 45 h.p. Szekely engine; length overall 21'3"; height overall 7'4"; wing span 39'6"; wing chord 58"; total wing area 176 sq.ft.; airfoil "CW-1" (first listed as CR-1); wt. empty 555-570 lbs.; useful load 420-405 lbs.; payload with 8.5 gal. fuel 184-173 lbs.; gross wt. 975 lbs.; later versions were 15 lbs. heavier empty therefore there was a loss in useful load and payload; max. speed 80; cruising speed 70; landing speed 32; climb 580 ft. first min. at sea level; climb in 10 min. 4200 ft.; ceiling (normal) 12,000 ft.; gas cap. 8.5 hal.; oil cap. 2 or 1.5 gal.; cruising range at 2.75 gal. per hour was about 25 miles per gallon or 200 miles; price at factory field $1490., raised to $1595. in June 1931.

Fig. 311. "Junior" being tested on Edo 990 floats.

Fig. 312. This "Junior" looks like a space-eater; every flight was a frolic.

The hull-type fuselage framework was built up of welded chrome-moly steel tubing, heavily faired to a shape with steel tube and wooden fairing strips, then fabric covered. Tandem open cockpits were forward in the fuselage frame, cockpits were roomy, deeply cowled and well protected; due to center of gravity location, solo-flying was permitted from the front cockpit only. Empty weight varied slightly throughout the series so baggage allowance varied from 14 to 3 lbs. The wing frame in two halves, was built up of solid spruce spar beams with spruce and plywood truss-type wing ribs; the leading edges were covered with aluminum alloy sheet and the completed framework was covered in fabric. The wing halves were fastened to a center-section panel mounted high above the cockpits on a steel tube cabane that was an integral part of the fuselage; the wings were then braced to the lower longerons by parallel steel tube struts of streamlined section. The Szekely engine was mounted backwards to the rear edge of the center-section panel with fuel and oil tanks immediately in front; the tanks were so shaped as to form a streamlined nacelle for the engine. The direct-reading fuel gauge was mounted on front of fuel tank and visible from front cockpit only. The squat landing gear was of two short cantilever stubs fitted with low-pressure airwheels; wheel brakes and mud-guards (fenders) were optional. The fabric covered tail-group was built up of welded steel tubing; the vertical fin was integral part of the fuselage and the horizontal stabilizer was adjusted on the ground only. The rudder and the elevator halves were interchangeable, the ailerons were of the off-set hinge type and all controls were cable operated. A steel spring-leaf tail skid was provided but a tail wheel was optional. Cockpit covers, an engine cover, a propeller cover and dual controls were optional equipment. The propeller

was a wooden Flottorp of reverse-pitch "pusher" design. The next Curtiss-Wright development was the Travel Air "Sport" model 12-Q described in the chapter for ATC # 401.

Listed below is partial tally of the Curtiss-Wright "Junior" entries as gleaned from registration records:

NC-623V;	Junior CW-1	(# 1012)	Szekely 45.
NC-630V;	"	(# 1013)	"
NC-631V;	"	(# 1014)	"
NC-632V;	"	(# 1015)	"
NC-633V;	"	(# 1016)	"
NC-634V;	"	(# 1017)	"
NC-635V;	"	(# 1018)	"
NC-636V;	"	(# 1019)	"

NC-638V thru NC-660V are consecutive as ser. # 1020 thru # 1042; NC-661V thru NC-688V are consecutive as ser. # 1043 thru # 1070; NC-689V thru NC-699V are consecutive as ser. # 1071 thru # 1081; NC-10900 thru NC-10925 are consecutive as ser. # 1082 thru # 1107; NC-10929 thru NC-10982 are consecutive except NC-10941 as ser. # 1108 thru # 1160; NC-10983 thru NC-10999 are consecutive as ser. # 1161 thru # 1177; NC-11800 thru NC-11804 are consecutive as ser. # 1178 thru # 1182; NC-11808 thru NC-11826 are consecutive except NC-11816 as ser. # 1183 thru # 1200; NC-11827 thru NC-11860 are consecutive as ser. # 1201 thru # 1234; NC-11865 thru NC-11898 are consecutive except NC-11895 is in doubt, others as ser. # 1235 thru # 1268; ser. # 1269 may have been NC-11899 or NC-12299; NC-12300 was ser. # 1270; NC-12301 was ser. # 1271; NC-12305 was ser. # 1272, the last registered "Junior" as of 1932 registry. This type certificate for ser. # 1012 and up; type certificate expired as of 4-26-37; ser. # 1164 and # 1224 later modified with 40 h.p. Salmson engine as model CW-1S on Group 2 approval numbered 2-525.

A.T.C. #398
(2-6-31)
GEE BEE "SPORTSTER", E

Fig. 313. The Gee Bee "Sportster" model E with 110 h.p. Warner engine.

Various good airplanes have been especially designed for the sportsman-pilot but so far none had the singular purpose of the little Gee Bee "Sportster". Nearly all of the craft touted as the airplane ideal for the average run of flying sportsmen had seats for 2 or 3, or more. They could also be used as advanced training airplanes, they often substituted in the chores of business, or even hauled a party of fun-seekers to some distant lake for a little fishing. As such, these were airplanes designed for the sportsman and his friends or they doubled in some other useful duty, but the single-seated Gee Bee "Sportster" was designed for the sportsman-pilot and just he alone. It came to life and it flew for him only. Together they scooted merrily across the skyways as one, sometimes on errand or sometimes just on whim, but they shared their pleasures together and no one else was there to interrupt the appreciation of each other. One might think it selfish perhaps, but the intimate association of "just you and me" became an unspoken bond of friendship and a deeper understanding between man and his machine. Saucy, squat and stubby, in a rather pleasing way, the Gee Bee "Sportster" basically had no other purpose than to answer to the varied whims of the owner-pilot, especially a type of pilot that flew hard and would demand the maximums of an airplane

from the start to finish. Strong of frame, strong of heart and with playful nature, the little "Sportster" was best described as a good and even match for the pilot with the same qualifications. There isn't any doubt that the single-seated "Sportster" was the very kind of airplane that hundreds of good pilots occasionally dreamed of owning because it always attracted a covey of impressed onlookers. Even as it sat, it fairly radiated fun and high adventure but, being practical for a moment, it was just not the type of airplane that most pilots could afford to own. During these lean times an airplane almost had to help earn its keep, in part at least, or not be considered at all. Though comparatively rare because of the small number built, the Gee Bee "Sportster" nevertheless spred its wings across a big chunk of sky, to leave a trail of records in the record-books and many-faceted memories in the hearts and minds of airmen.

The history and background development of the "Gee Bee" sport-type monoplane actually stems from the Model X "Cirrus Derby Racer" that was flown to second place by Lowell Bayles in the All-America Derby of 1930. Better known to most as the "Cirrus Derby" this race was sponsored by American Cirrus Engines, Inc. and open to all entrants that were powered with either of the A.C.E. engines, whether upright

Fig. 314. Spunk and dash of Gee Bee E leveled at the sportsman.

or inverted and standard or supercharged; it was quite an interesting line-up that faced the starter's flag that day in July. The "Gee Bee" X, flying easy and not pushed too hard, followed the Command-Aire "Bullet" to the finish tape after 5500 grueling miles; Lowell Bayles grinned happily and said it was all good sport. So then it takes very little imagination to see the "Model X" with slight modification here and there to emerge as the "Sportster" line, a series of high performance single-seaters for the use of sportsman-pilots. Coming up with a versatile basic airframe that could be fitted with any number of in-line or small radial engines, Gee Bee had a nice selection of models to choose from. The first of these came out late in 1930 as the Model B with an inverted Cirrus "Ensign" engine of 110 h.p.; a companion model followed shortly, powered with the new 4 cyl. inverted in-line Menasco "Pirate" B-4 engine of 95 h.p. as the

Model C, and the swifter Model D was powered with the Menasco "Pirate" C-4 engine of 125 h.p. The spunky 7 cyl. Warner "Scarab" radial engine of 110 h.p. was installed in the popular Model E. The new 6 cyl. inverted in-line Fairchild 6-390 engine of 125-135 h.p. was reserved for the proposed Model F. A fast, dependable sport-flying airplane in any of the versions, the 'Sportster' inherited natural abilities applicable to closed-course air racing and although Granville Brothers had no intention of specializing in the design of racing airplanes, a forceful destiny was shaping things to go in that direction. The success of the Model X spawned two new designs called the models Y and Z; it was the barrel-shaped Model Z, flown by Lowell Bayles, that won the 1931 Thompson Trophy Race at the National Air Races held in Cleveland. The chain of events that followed in the history of Granville Brothers Aircraft, Inc. saw

Fig. 315. Gee Bee model X, second place winner in "Cirrus Derby" was basis for "Sportster" design.

Fig. 316. First model E had spatted landing gear, later models had "pants" over wheels.

development of some of the most daring designs the air-racing fraternity had ever seen.

The Gee Bee "Sportster" model E was a single-seated, low winged, wire-braced monoplane arranged specifically for the sporting pilot. It was capable of fast cross-country trips to points at least 500 miles apart or acrobatic nip-ups within sight of the home-port for all to see. Because of its relatively high speed on nominal power, the "Sportster" was a natural for closed-course air racing around the pylons and was just about the fastest stock airplane in this horse-power class. Flying the Model E with throttle to the firewall, Lowell Bayles was in 4th place at the finish of the National Air Tour for 1931, averaging over 140 m.p.h. for the distance. At the 1931 National Air Races, Robert Hall, Granville's design-engineer, who was also an accomplished pilot, flew a Model E to first place in a trophy dash at over 128 m.p.h. Of course there were also the many minor air-meets around the country where the "Sportster" was bound to show up and usually show its tail to all the others. Though often used for air-racing, more for the fun than the profit, the "Sportster" E was not basically a racing airplane; its stock in trade was its relatively high all-round performance. Climbing out like a home-sick angel, it could be at 11,000 ft. in less than 10 mins., its excellent maneuverability could seduce the heart of a pilot in less than 5 mins. and even at normal cruising speed it could fly over 2 miles of scenery in less than a minute. Powered with the 7 cyl. Warner "Scarab" engine of 110 h.p., the Model E differed from the other "Sportster" models because the models B-C-D and the proposed F were all powered with 4 or 6 cyl. inverted in-line engines; standing out by itself the Model E was the most popular of any in the series. We can safely venture to say that under more favorable circumstances at another time,

the playful "Sportster" surely would have graced the skies all over the country instead of the pitifully small number that were actually built. The type certificate number for the Gee Bee "Sportster" model E was issued 2-6-31 and some 5 examples of this model were built by Granville Brothers Aircraft, Inc. on Springfield Airport in Springfield, Mass. Zantford D. Granville was pres.; James Tait, one of the 4 Tait brothers who were principal financers of the company, was treas.; Robert L. Hall was the chief design-engineer and Lowell Bayles was the chief pilot who handled most of the promotional flights. With Zantford "Granny" Granville as the spark-plug of the operation, various chores in the shop such as fabrication and assembly, were split up amongst the 4 other Granville brothers so it actually approached the status of a family affair where all worked together to achieve a desired goal.

Listed below are specifications and performance data for the Gee Bee "Sportster" model E as powered with the 110 h.p. Warner "Scarab" engine; length overall 16'9"; height overall 6'0"; wing span 25'0"; wing chord at root 56"; total wing area 95 sq. ft.; airfoil (NACA) M-6; wt. empty 912 lbs.; useful load 488 lbs.; payload with 37 gal. fuel 243 lbs. (170 lbs. for pilot & 73 lbs. for parachute, equipment and/or baggage); gross wt. 1400 lbs.; max. speed 148; cruising speed 127; landing speed 52; climb 1500 ft. first min. at sea level; climb in 10 min. 11,500 ft.; ceiling 19,000 ft.; gas cap. 37 gal.; oil cap. 3 gal.; cruising range at 7 gal. per hour 570 miles; price at factory field $5230 with standard equipment.

The stubby fuselage framework was built up of welded chrome-moly steel tubing faired to an approximate oval section with spruce formers and fairing strips, then fabric covered; a detachable steel tube mount allowed fitting of various

Fig. 317. Spirited nature of Gee Bee E at its best when rounding pylons, seen frequently at air-races.

engines. The cockpit was deep and well protect-ed with a drop-down cowl panel for ease of entry; the cockpit was detailed with rich, durable upholstery and several handy pockets for log books, gloves, maps and the like. A metal-lined baggage locker of 2 cu. ft. capacity with allow-ance for 40 lbs. was high in the fuselage just behind the metal firewall. The wing consisting of a center-section and two outer panels, was built up of solid spruce spar beams with spruce and plywood truss-type wing ribs; the ailerons were of sheet steel ribs fastened to a tubular spar, the leading edges were covered with dural metal sheet and the completed framework was covered in fabric. The wing's center-section was braced to the top longerons by 2 parallel steel tube struts on each side and the outer wing panels of elliptical form were fastened to the faired-in wing stubs; the wire-braced wing used streamlined steel tie-rods that fastened to the landing gear structure on the under-side and a steel tube pylon structure high inside the fuse-lage on the upper side. A fuel tank was mounted in each stub end of the center-section panel and fuel was fed to a small gravity tank high in the fuselage and ahead of the cockpit by a hand-operated wobble-pump. This fuel system was designed to give proper fuel flow for acrobatics and extended upside-down flying. The landing gear fastened to outer ends of the wing stubs and incorporated attach points for landing gear and wing bracing wires; oleo shock absorbers of 6 in. travel worked in conjunction with rubber compression rings to absorb landing and taxiing loads. The landing gear structure was fully

faired on the later examples and 20x9 low-pres-sure airwheels were encased in large stream-lined wheel pants; wheel tread was 62 in. The tail skid was of the steel spring-leaf type with a removable hardened shoe. The fabric covered tail-group was built up of welded chrome-moly steel tubing and rigidly braced to the fuselage by streamlined steel tie-rods; the rudder was actuated by braided steel cables, the elevators were actuated by a metal push-pull tube and the ailerons were torque-tube operated. A Curtiss Reed metal propeller, navigation lights, wheel brakes and a Townend-type "speed ring" engine cowling were standard equipment. A Heywood air-operated engine starter was optional. The "Sportster" was available in several two-tone color combinations done up in the familiar scal-loped design; some of the popular color combinations were red and white, blue and white, green and cream, black and white and two-tone brown. The next development in the Gee Bee "Sportster" series was the Menasco-powered Model D described in the chapter for ATC #404.

Listed below are "Sportster" model E entries as gleaned from registration records; it seems that various confusing serial numbers were used so the aircraft are listed in their approximate order of manufacture:

NC-856Y;   Model E (#   ) Warner 110.
  NC-46V;   "   (#   )   "
  NC-72V;   "   (#   )   "
NC-11041;   "   (#   )   "
NC-11044;   "   (#   )   "

A.T.C. # 399
(2-10-31)
SWALLOW "SPORT", HC

*Fig. 318. Swallow "Sport" model HC with 165 h.p. Continental A-70 engine;
was last of the "Swallow" biplanes.*

With the intent to offer the flashy "Sport" biplane in a variety of models, powered by several different powerplants, Swallow Aircraft prepared the Model HC with the new and increasingly popular Continental A-70 engine, as a running-mate to the models HA and HW. Comparatively speaking, the new model HC was typical of the two previous offerings; but because it could stand toe-to-toe with its sister-ships under any comparison, was equipped with a fine engine that compared favorably with the very best, and because of its sharply reduced price, the new Model HC posed as quite a bargain and might easily have been the star of the series. It is of some interest perhaps that the Model HC was not a version initially developed as such, but a modification of an earlier example that had first been built as a Model HA; the terminated supply of Axelson engines no doubt had a detrimental effect on the future life of the HA. Because of policies dictated by the crippling business depression, the Axelson Aircraft Engine Co. suspended normal operations as of June 1930 and no more B-7-R engines were being produced; in effect this move just about cancelled the feasibility of producing a "Sport" model powered with the Axelson engine, even though a fair supply of these engines were still in stock. Of the two Axelson-powered HA that were originally built, one was converted to a Model HW (refer to ATC # 379) and one was converted to the Model HC as described here; the "Comet" 7-E engine was to have been an installation for yet another version but this proposed model must not have gone any further than the planning stage. Typical of many other aircraft manufacturers of this particular time, all of whom were sharing the same problems, Swallow was progressively hard-hit by the sagging market and was finding it increasingly difficult to stay in business. The new "Sport" (Special) created a scattered interest but was left begging for customers. The TP trainer biplane was still selling occasionally but by 1932, Swallow Aircraft was in dire straits, so they were offering the remaining TP fitted with OX-5 engines, at a sell-out price of $995. to clear the inventory. To all intents this was the end of the road for Swallow Aircraft and the "Sport" model HC had also the distinction of being the last of the "Swallow" biplanes; some 5 years later "Swallow" manufacture was revived in the form of a sleek monoplane arranged for 2 but then, that is another story.

The Swallow "Sport" model HC was an open cockpit biplane with seating for 3 and although quite suitable for the varied chores of all-purpose work, it was primarily leveled at the sporting type of flyer. Typical of the two previous models (HA and HW) the HC differed only in its engine installation and some minor modifications necessary to this combination. All three models being basically alike, it is interesting to note the careful grouping of the useful load in close proximity about the C.G. (center of gravity) to require less "trimming" for balance in flight. Because of the large amount of interplane stagger and the placement of cockpits in relation to the wings, visibility was quite good in all directions. Powered with the 7 cyl. Continental A-70 engine of 165 h.p., the model HC had power reserve for an exceptional performance. Take-offs were short, quick and clean, landing runs averaged about 300 feet and the steep climb-out must have been a joy to pilots operating out of small fields surrounded with obstacles. Landing speed was an easy 42 m.p.h. but it must have required a little technique to fend off the jars that would normally come through a "stiff-legged gear", in spite of the big roly-poly "airwheels" that were supposed to absorb the shocks. However, oleo-spring shock absorbers were available and wheel brakes were also optional. We can only guess as to the flight characteristics and general behavior but anyone familiar with airplanes will agree that everything on or about the "Sport" looked "right" for effortless maneuverability and a spirited nature. We'd venture to guess that this was the best airplane that "Swallow" ever built. The type certificate number for the "Sport" model HC was issued 2-10-31 and only one example of this model was built by the Swallow Airplane Co. at Wichita, Kan. W.M. Moore was president; Earl Hutton was V.P.; Geo. R. Bassett was general manager and Dan Lake was chief engr.

Listed below are specifications and performance data for the Swallow "Sport" model HC as powered with the 165 h.p. Continental A-70 engine; length overall 22'3"; height overall 8'4"; wing span upper 31'0"; wing span lower 23'0"; wing chord upper 60"; wing chord lower 48"; wing area upper 150 sq. ft.; wing area lower 90 sq.ft.; total wing area 240 sq.ft.; airfoil (NACA) M-12; wt. empty 1375 lbs.; useful load 825 lbs.; payload with 40 gal. fuel 377 lbs. (2 pass. at 165 lb. each & 47 lb. baggage); gross wt. 2200 lbs.; max. speed 135; cruising speed 115; landing speed 42; climb 1200 ft. first min. at sea level; service ceiling 16,000 ft; gas cap. max 40 gal.; oil cap. 5 gal.; cruising range at 8 gal. per hour 500 miles; price at factory field $3995.

The construction details and general arrangement of the "Sport" model HC was typical to that of both models HA and HW, as described in the chapters for ATC # 341 and # 379 of this volume, including the following. The Continental A-70 engine was normally equipped with a front-mounted exhaust collector ring with ample volume for expansion to quiet the exhaust noises, but the "Sport" was equipped with curved "short stacks" on option for those that wouldn't mind the racket or judged the staccato noises necessary to impart a more sporty feeling. Because of a slightly less empty weight, the model HC was allowed 47 lbs. of baggage that was stowed in a bin under the front seat. The stiff-legged landing gear was fitted with low-pressure "airwheels" and not normally equipped with shock absorbing struts; however, oleo-spring shock struts were available and wheel brakes were optional. The placing of the pilot's cockpit, in relation to the wings, afforded good visibility in all directions and a large cut-out in the trailing edge of the upper wing afforded good visibility overhead. All wing bracing struts were of chrome-moly steel tubing in a streamlined section and interplane bracing was of heavy gauge streamlined steel wire, all this arranged in a pattern quite different from any previous series of "Swallow" biplanes. Angle of wing incidence upper and lower 4.5 deg; stagger was 28 in. and dihedral angle of upper and lower wings was 1.5 deg. Wiring for navigation lights, Goodyear airwheels and a metal propeller were standard equipment. Oleo-spring shock absorbers, wheel brakes, navigation lights and engine starter were optional. Various two-tone color schemes were available on prior order.

Listed below is the only known example of the "Sport" model HC as gleaned from registration records:

NC-110V; Model HC (# 2003) Continental A-70.

Serial #2003 was first as Model HA with serial # 103.

Fig. 319. Emsco model B-3-A with 420 h.p. "Wasp" engine; its buxom, flowing lines were handsome as well as efficient.

As one of the handsomest large cabin monoplanes of this period, the buxom but shapely "Emsco" model B-3-A was designed for shuttle service on the smaller airlines, or for the business executive requiring room enough to transact company chores while in flight to an urgent appointment. Bearing the loving touch of several outstanding airplane designers during its development, the B-3-A was a proud lady of majestic air and very good manners, bearing the mark of someone's unmistakable flair for good proportion with soft rounded lines of graceful flow. Those who can remember will agree that some of the most beautiful airplanes of this time came out of the Emsco factory in Downey. Gleaming in their bright colors, with careful attention to streamlining and the flowing form, the "Emsco" monoplanes were an eye-catcher and a crowd-former in any of their several different versions. Backed by an organization that seemed to have ample funds and resources, promotion for the varied line was adequate but a late start launched "Emsco" right into the face of the on-coming depression. Vying with other established companies for what little business there was to be had, was a losing battle, so there was naught to do but turn to building custom-made ships for specialized service or turn to foreign markets; precious little was to be had in either case, just then. A large "Emsco" transport, believed to be

the model B-3, was reported delivered to Mexico for the launching of a new airline there; one B-3 was later delivered to the Prince of Roumania, another B-3, after finding its way into the hands of a used-airplane dealer, was to be used in 1933 by Francisco Sarabia on his T.A.C. line in Mexico for hauling coffee beans.

The first airplane built by Emsco Aircraft was a tri-motored airplane that was provided with integral fittings to convert it to either a twin-engined version of some 500-600 h.p. or a single-engined version of some 300-450 h.p. Of those that were built, at least one tri-motor was converted to a "twin" (model B-5) and at least one was converted to a single-engined model B-3. A racy looking open sport plane for two, called the model B-4, was also one of the early developments; as a wire-braced mid-wing monoplane, the B-4 was powered with the 4 cyl. upright "Cirrus" engine and its sporty appearance alone would have normally been a guarantee of its success. Modified into a similar but larger version called the model B-7, these were powered with aircooled radial engines in the 165 h.p. range. As time went on, a few special-purpose airplanes were built, plans were laid for a huge 4-motored transport and also a twin-engined amphibian; a large "flying wing" type of airplane was also built for an assault on the non-refueled endurance record and later a proposed

Fig. 320. The "Albatross" B-1 was inspiration for Emsco design.

flight across the Pacific to Japan. Busy as a bee-hive, the plant at Downey was always buzzing with something daring and something new and no one can say there was a lack of ambition, but the absence of cash-customers for their varied products finally shriveled that ambition to sober thoughts of caution.

The purchase of the Albatross Aircraft Corp. of Long Beach, Calif. by E. M. Smith and Associates, was the nucleus for the beginning of the Emsco Aircraft Corp. early in 1929. While its new 45,000 sq. ft. million-dollar factory was under construction on a 73 acre site in Downey, "Emsco" began construction of two of its 8-place tri-motors in the "Albatross" plant on the Long Beach airport. Charles F. Rocheville, accomplished pilot and airplane designer, was more or less inherited with the Albatross deal

Fig. 321. The first Emsco-built airplane was tri-motored B-2 with 3 Curtiss "Challenger" engines.

Fig. 322. B-3-A as personal airplane for Prince of Roumania.

and it was his untiring energies that were responsible for a complete redesign of the "Albatross" concept as originally created by Albin K. Peterson, into the various early models offered by Emsco (E. M. Smith Co.). The first tri-motored "Emsco", powered with 3 Curtiss "Challenger" engines, was delivered to the Emsco Derrick & Equipment Co. (a subsidiary of the E. M. Smith Co., the parent-firm of all "Emsco" organizations that was founded in 1911) and embarked on a 4-month tour of promotion, piling up over 25,000 miles while visiting scattered Emsco factories. Meanwhile, by Oct. of 1929, the Downey, Calif. plant and factory airport was dedicated and Walter L. Seiler was taken on as chief pilot to relieve Rocheville of some of his many duties; Ted Lundgren became the sales manager. By Feb. of 1930, Emsco was building a Wasp-powered model B-3, one of the tri-motors was being converted into a "twin" with 2 Wright J6-9-300 engines and production got slowly underway on a few of the Cirrus-powered model B-4 sport mid-wing; a large 32 passenger 4-motored transport was under construction, a twin-motored "amphibian" was in design and a "flying wing" type was started for a proposed flight to Japan. With activities geared to many diversified developments by late 1930, Roger Q. Williams, famous trans-Atlantic flyer who was formerly connected with Bellanca, Uppercu-Burnelli and the General Airplanes Corp., was taken on as chief pilot and consulting engineer. Charles Rocheville of energetic personality and a dreamer of dreams, chafed

fretfully at the undetermined future at Emsco and finally left to seek his fame and fortune elsewhere. Gerard F. Vultee, formerly chief engineer for Lockheed, served a short term at the Curtiss-Wright Institute in Los Angeles and then came on as Emsco's chief engineer. With not enough to keep him busy and somewhat enamored with plans of his own, Vultee left by Sept. of 1931 to be replaced by T. V. Van Stone. By then, development had narrowed to the new line of sport mid-wing monoplanes that Vultee had refashioned and there was naught else to do but hang on and hope for better times.

The "Emsco" model B-3-A was a rather large high-winged cabin monoplane with seating normally arranged for 8; as a coach-style transport, seating was increased to 8 passengers and a pilot, and special custom interiors with variable seating were also offered. Spacious and quite plush, the B-3-A was a limousine type of airplane but was also adaptable to carrying large loads of mixed cargo. The many faceted technique of using a great number of fairing strips, closely spaced, over the internal structure to achieve a streamlined flow, was one of the interesting points about the B-3-A in particular and all "Emsco" aircraft in general; this method to produce form was even carried out in the novel wheel streamlines which were a faired wooded structure covered in fabric. Perhaps not overly practical but they were handsome and quite efficient. Of good aerodynamic proportion with long and easy moment arms, the B-3-A was docile in character and its aerodynamic efficiency promoted good load-carrying ability suitable to operate out of even the smallest field. Powered with the 9 cyl. Pratt & Whitney "Wasp" R-1340-C engine of 420 h.p., the big B-3-A delivered surprising performance for a ship of this size and its rather clean configuration promoted an exceptional speed range. Weighing the facts against similar craft, it is plain to see that the B-3-A was one of the finest airplanes of this type. The type certificate number for the "Emsco" model B-3-A was issued 2-10-31 and possibly 4 or 5 examples of this model were manufactured by the Emsco Aircraft Corp. at Downey, Calif. E. M. Smith was president; I. W. Fuqua was executive V.P.; and Chas. F. Rocheville was V.P., chief of design, gen. mgr., and chief pilot. Occasional changes in personnel from early 1929 through 1931 were noted in a previous paragraph.

Listed below are specifications and performance data for the "Emsco" model B-3-A as powered with the 420 h.p. "Wasp" C engine; length overall 40'9"; height overall 9'5"; wing span 56'0"; wing chord 108"; total wing area 486 sq.ft.; airfoil Goettingen 398; wt. empty 4199 lbs.; useful load 2401 lbs.; payload with 140 gal. fuel 1311 lbs. (7 passengers at 165 lb. each & 150 lb. baggage); gross wt. 6600 lbs.; max. speed 150; cruising speed 122; landing speed 50; climb 800 ft. first min. at sea level; climb in 10 min. 6800 ft.; ceiling 14,000 ft.; gas cap. 140 gal.; oil cap. 10 gal.; cruising range at 22 gal. per hour 720 miles; price at factory field was $21,500.

The fuselage framework was built up of welded chrome-moly steel tubing, liberally faired with wooden formers and fairing strips to an oval section, then fabric covered. Individual seats were arranged down each side with a center-aisle leading to the pilot's cabin; soundproofed and insulated, the interior was upholstered in rich, durable fabrics. Cabin entry door was to the rear and large picture-windows in the main cabin offered good visibility. Two baggage compartments, one forward and one aft, had capacity for more than 150 lbs. of mixed luggage. The semi-cantilever wing framework, in two halves, was built up of heavy-sectioned spruce spar beams with spruce and plywood truss-type wing ribs; the leading edges were covered with dural metal sheet and the completed framework was covered in fabric. Wing bracing struts were steel tubes of large diameter, encased in fairings of an airfoiled section; the landing gear was built into the wing bracing truss with fittings provided for the mounting of Brewster twin-float seaplane gear. Two gravity-feed fuel tanks were mounted in the wing halves, one flanking each side of the fueselage. The forward section of the fuselage was covered in removable metal panels and the engine was shrouded in a large, deep-chord NACA type low-drag fairing. The outrigger landing gear of unusually wide tread was fitted with Aerol or Gruss shock absorbing struts; wheels were 36x8 and Bendix brakes were standard equipment. The novel wheel fairings were a wooden framed structure covered in a heavy fabric. The pilot's cockpit, arranged for two, was fitted with numerous flight and operating aids and dual wheel controls were provided. The fabric covered tail-group was built up of welded steel tubing into a thick symmetrical cross-section; the fin was ground adjustable and the horizontal stabilizer was adjustable in flight. An adjustable metal propeller, navigation lights, electric inertia-type engine starter, battery, tail wheel, wheel brakes, first-aid kit and 2 fire extinguishers were standard equipment. The next "Emsco" development was the model B-7 sport mid-wing monoplane described in the chapter for ATC # 403.

Listed below are "Emsco" model B-3-A entries as gleaned from registration records:

NR-153W;	Model B-3-A	(# 4)	Wasp 420.
NC-166W;	"	(# 5)	"
NC-823N;	"	(# 6)	"
CV-GOI;	"	(#)	"

Serial # 4 also listed as 3 place; serial # 6 converted from model B-2 tri-motor; serial number for CV-GOI unknown; one B-3 went to Mexico, serial number unknown.

APPENDICES

PHOTO CREDITS

F. C. McVickar — Figs. 67, 69, 80
Chas. W. Meyers — Figs. 182, 183A
Ken M. Molson — Figs. 15, 98, 103
National Aviation Museum, Ottawa, Canada — Figs. 89, 90, 91, 231
Ralph Nortell — Fig. 274
Northrop Corp. — Fig. 250
Roy O. Berg — Fig. 129
Will D. Parker — Fig. 85
Parks Air College — Fig. 26
Gene Poiron — Fig. 313
Pratt & Whitney Aircraft — 34, 79, 112, 114, 116, 118, 184, 185, 208, 249, 281, 295, 322
Wilton R. Probert — Fig. 158
RCAF Photo — Fig. 16
Earl C. Reed — Figs. 23, 63, 268
J. R. Schmidt Photo — Fig. 173
Sikorsky Aircraft Div. — Figs. 146, 147, 148, 149
Edgar B. Smith Photo — Fig. 150
Smithsonian Institution — Figs. 1, 2, 4, 33, 120, 130, 137, 138, 139, 152, 159B, 162, 171,
    179, 180, 183B, 188, 200, 236, 237, 247, 248, 256, 263, 269, 270
Spartan Aircraft Co. — Fig. 296
USAAF Photo — Fig. 243
J. N. Underwood — Figs. 68, 154, 245 (B. Ralph Hall), 260
Alfred V. Verville — Figs. 64, 65, 93, 95, 96
Truman C. Weaver — Figs. 27, 28
Western Airlines — Fig. 20
Gordon S. Williams — Figs. 11, 13, 25, 30, 38A, 41, 51, 102, 168, 169, 172, 174, 177,
    187, 197, 198, 212, 213, 214, 217, 225, 226, 228, 234, 254, 257A, 259, 283, 286,
    287, 299, 304, 306, 312

BIBLIOGRAPHY

BOOKS:
Revolution In The Sky; Richard Sanders Allen.
The Ford Story; Wm. T. Larkins.
Aircraft Year Book (1930-31-32); Aero. Chamber of Commerce of America.
The Gee Bee Story; Chas. G. Mandrake.
A Chronology of Michigan Aviation; Robert S. Ball.
Jane's All The World's Aircraft; C. G. Grey.
PERIODICALS:
Flying   Western Flying
The Pilot   Popular Aviation
Aviation   Air Transportation
Aero Digest   Journal of A.A.H.S.
Air Progress   American Modeller
American Airman   Lightplane Review
Antique Airplane News   Model Airplane News
SPECIAL MATERIAL:

Licensed Aircraft Register by Aero. Chamber of Commerce of America, Inc.
Characteristics Sheets by Curtiss Aeroplane & Motor Co.
Factory brochures and promotional literature.

CORRESPONDENCE WITH FOLLOWING INDIVIDUALS:
H. Lloyd Child   Earl C. Reed
Chas. W. Meyers   James W. Bott
Melba Beard   John W. Underwood
Peter M. Bowers   John H. Livingston
Alfred V. Verville   Tom Towle
Joe Christy

INDEX